THE MASTER PLAN

ALSO BY HEATHER PRINGLE

The Mummy Congress:
Science, Obsession and the Everlasting Dead

THE MASTER PLAN

HIMMLER'S SCHOLARS AND THE HOLOCAUST

HEATHER PRINGLE

FOURTH ESTATE • *London*

First published in Great Britain in 2006 by
Fourth Estate
An imprint of HarperCollins*Publishers*
77–85 Fulham Palace Road
London W6 8JB
www.4thestate.co.uk

1

A catalogue record for this book is
available from the British Library

ISBN-13 978-0-00-714812-7
ISBN-10 0-00-714812-7

Interior Maps drawn by Signy Fridriksson-Fick

Set in Dante MT and Spectrum

Printed in Great Britain by Clays Ltd, St Ives plc

TO GEOFF AND MY BROTHER ALEX

The sleep of reason brings forth monsters.

—FRANCISCO GOYA

He marveled at how the past could be
refigured to suit the present, at how fragile reality
truly was when you started to twist it.

—ALAN FURST

CONTENTS

CONTENTS

COMPARATIVE RANKS

SS RANK	US EQUIVALENT
SS-Anwärter	Recruit
SS-Mann	Private
SS-Sturmmann	Private First Class
SS-Rottenführer	Corporal
SS-Unterscharführer	Sergeant
SS-Scharführer	Staff Sergeant
SS-Oberscharführer	Technical Sergeant
SS-Hauptscharführer	First Sergeant
SS-Sturmscharführer	Master Sergeant
SS-Untersturmführer	Second Lieutenant
SS-Obersturmführer	First Lieutenant
SS-Hauptsturmführer	Captain
SS-Sturmbannführer	Major
SS-Obersturmbannführer	Lieutenant Colonel
SS-Standartenführer	Colonel
SS-Oberführer	(none)
SS-Brigadeführer	Brigadier General

SS-Gruppenführer	Major General
SS-Obergruppenführer	Lieutenant General
SS-Oberstgruppenführer	General
Reichsführer-SS und Chef der deutschen Polizei	General of the Army

Source: George Leaman. *Holdings of the Berlin Document Center: A Guide to the Collections.* Berlin: BDC, 1994, pp. 276–277.

THE MASTER PLAN

1. FOREIGN AFFAIRS

IN THE FALL OF 1938, in the small industrial town of Offenbach am Main just outside Frankfurt, the renowned firm of Gebrüder Klingspor received an important commission from one of the most prominent men of the Third Reich. The owner of the business, Karl Klingspor, was an influential typographer and aesthete, a master of ink and paper who transformed the creative fantasies of others into some of the most beautiful books of his age.[1] To seduce the eye, Klingspor retained artists and painters to design sleek new typefaces so words would scroll stylishly across the page. To woo the sense of touch, he selected handmade papers of unusual size and heft—rich and thick and textured. For Klingspor and his colleagues, making a book was rather like making love, and bibliophiles from Berlin to Boston sighed with pleasure as they thumbed through his exquisite productions.

The commission in question had arrived from Heinrich Himmler, the head of the Gestapo, the Security Service, and the Security Squad or SS, a paramilitary organization that ran Germany's concentration camps, controlled a profitable network of business enterprises, and provided Adolf Hitler's personal bodyguard. The Reichsführer-SS was a busy man, but he remained, in his personal life, something of a bookworm. He read avidly, owned a substantial private library, and carried his favorite volumes with him wherever he traveled. He often recommended books to his subordinates and presented copies as gifts to family members and close associates.[2]

It was in this frame of mind that he resolved to produce a special gift for Hitler on the occasion of his fiftieth birthday.

For months, prominent Nazis had been drawing up plans for a gala celebration for Hitler, searching feverishly for presents. The leaders of the Confederation of German Industry had quietly purchased the complete manuscript scores of Richard Wagner's early operas, as well as fair copies of parts of *Der Ring des Nibelungen*, the composer's masterpiece.[3] Rudolf Hess had acquired a rare collection of letters written by one of Hitler's heroes, Frederick the Great, the eighteenth-century monarch who transformed Prussia into a major European power.[4] And the National Socialist German Workers' Party, or Nazi party, had thrown caution completely to the wind, building the Eagle's Nest, a teahouse and conference center situated atop a mountain near the Bavarian town of Berchtesgaden.[5]

Himmler had no intention of being left out of this slavish display. He planned to present a fine equestrian portrait of Frederick the Great by the German artist Adolf von Menzel, a painting that would fit nicely into Hitler's private study.[6] But he also wanted to give Hitler something more personal, a set of leather-bound books that would artfully present the SS chief's lesser-known contributions to Hitler's Nazi state.[7] The most important of these books, he decided, would be a large portfolio produced by the creative staff of Gebrüder Klingspor. It would be entitled:

The Research and Educational Society,
The Ahnenerbe:

Evolution,
Essence,
Effect.[8]

The Ahnenerbe was an elite Nazi research institute that Himmler had founded in 1935 with a small group of associates. Its name derived from a rather obscure German word, *Ahnenerbe* (pronounced AH-nen-AIR-buh), meaning "something inherited from the forefathers."[9] The official mission of the Ahnenerbe was twofold. First, the institute was to unearth new evidence of the accomplishments and deeds of Germany's ancestors, as far back as the Paleolithic or Old Stone Age if possible, "using exact scientific methods."[10] Second, it was to convey these findings to the German public by means of magazine articles, books, museum shows, and scientific conferences.

In reality, however, the elite organization was in the business of mythmaking. Its prominent researchers devoted themselves to distorting the truth and churning out carefully tailored evidence to support the racial ideas of Adolf Hitler. Some scholars twisted their findings consciously; others warped them without thought, unaware that their political views drastically shaped their research. But all proved adept at this manipulation, and for this reason, Himmler prized the institute. He made it an integral part of the SS and housed it in a grand villa in one of Berlin's wealthiest neighborhoods. He equipped it with laboratories, libraries, museum workshops, and ample funds for foreign research, and personally befriended several of its senior scientists. By 1939, the Ahnenerbe would count 137 German scholars and scientists on its payroll and employ another 82 support workers—filmmakers, photographers, artists, sculptors, librarians, laboratory technicians, accountants, and secretaries.[11]

THE NAZI LEADERSHIP went to enormous lengths to maintain a public facade of rationality, fairness, and middle-class decency. Hitler and the senior members of his government craved admiration and respect from the world, and to obtain it they continually attempted to display their political ideas—particularly on the subject of race—in the best possible light. To assist in this difficult work, they actively recruited German scholars and scientists who could command respect both at home and abroad and make Nazi ideas sound plausible and reasonable to others.

Those who studied the ancient past figured prominently in these recruitment efforts, for Adolf Hitler held strong views on prehistory and history. He believed that all humankind in its astonishing richness and complexity, all human societies in the past, from the Sumerians on their ziggurats to the Incas in their mountain citadels, could be parsed into just three groups. These he described as "the founders of culture, the bearers of culture, the destroyers of culture."[12] Hitler was convinced, based on his own highly selective reading of history, that only one racial group fell into the first category. These were the Aryans, a fictional race of tall, willowy, flaxen-haired men and women from northern Europe. According to Hitler, only the Aryans had possessed the spark of genius needed to create civilization; invent music, literature, the visual arts, agriculture, and architecture; and advance humanity by putting their shoulders to the heavy wheel of progress. Most modern Germans, Hitler claimed, descended from the ancient Aryans, and as such they had inherited their forefathers' brilliance.

This was the most positive side of the human ledger. On the negative side, Hitler placed the Jews. These, he claimed, were the destroyers of culture. He insisted upon categorizing all the world's Jews together as a single race, although scholars of the day agreed they were a diverse collection of peoples united by their religious faith.[13] Hitler declared that the Jews posed a serious threat to humankind, insisting that they possessed a singular talent for undermining and corrupting the cultures of other races. He was especially fond of likening "the Jew" to a type of germ—"a noxious bacillus [that] keeps spreading as soon as a favorable medium invites him. And the effect of his existence is also like that of spongers; wherever he appears, the host people dies out after a shorter or longer period."[14]

Serious scientists and scholars outside Germany in the 1930s dismissed these ideas as nonsense.[15] There was simply no disputing, for example, that Jews had contributed brilliantly to human civilization. Between 1901 and 1939 alone, twenty-one Jewish scientists and scholars won Nobel Prizes, from Albert Einstein for his contributions to theoretical physics to Otto Loewi for his pioneering work on the chemical transmission of nerve impulses.[16] Indeed, nearly 30 percent of all Nobel laureates from Germany during this period were Jewish, although Jews accounted for just 1 percent of the German population.

Hitler's notions about the Aryans were equally far-fetched. Scholars had failed to uncover any proof of a tall, blond-haired race of ur-Germans who first lit the torch of civilization and gave birth to all the refinements of human culture. The first cities, the first system of writing, the first successes in agriculture had all arisen along the green river valleys and hills of the Near East and Asia, many thousands of miles away from the dark, cold forests of northern Europe. And this knowledge posed a problem for ardent Nazis, such as Himmler, who took an interest in scholarship and intellectual discourse. How could they persuasively portray ancient Germans and their modern descendants as a master race if indeed they had played little, if any, part in the great early advances of human civilization?

The answer to this problem, in Himmler's mind, lay in more German scholarship—scholarship of the right political stripe. So he created the Ahnenerbe. He conceived of this research organization as an elite think tank, a place brimming with brilliant mavericks and brainy young upstarts—up-and-comers who would give traditional science a thorough cleansing. Men of this ilk would not balk at sweeping away centuries of careful scholarship like so much dust and useless debris. With much fanfare, they would publicly unveil a new portrait of the ancient world, one in which a tall, blond race of ur-Germans would be seen coining civilization and bringing light to inferior races, just as Hitler claimed.

This was the primary work of the Ahnenerbe. Privately, however, Himmler nurtured another hope for his creation. He believed, like many other prominent Nazis, that an almost magical elixir—pure Aryan blood—once flowed through the veins of the ancient Germanic tribes. Undiluted and undefiled by later racial mixing, this superior hemoglobin supplied Germany's ancestors with heightened powers of creativity and intelligence, or so Himmler supposed.[17] If Ahnenerbe researchers could recover this primeval Germanic knowledge through archaeology and other sciences, then they might find superior ways of growing grain, breeding livestock, healing the ill, designing weapons, or regulating society.[18] All this would greatly benefit the Reich.

For this important work, Ahnenerbe officials recruited a diverse assortment of scholars and scientists—archaeologists, anthropologists, ethnologists, classicists, Orientalists, runologists, biologists, musicologists, philologists, geologists, zoologists, botanists, linguists, folklorists, geneticists, astronomers, doctors, and historians. Himmler intended them to work together, sharing their findings. Each researcher, he suggested, would contribute a part of what he sometimes liked to call the "hundreds of thousands of little mosaic stones, which portray the true picture of the origins of the world."[19]

Despite his many pressing responsibilities, Himmler clucked contentedly over the scholars' reports and papers. He pondered their many theories, paid careful heed to their conclusions, and often discussed their ideas with his SS subordinates over dinner. And when the opportunity presented itself, he employed their research to fuel and justify the Holocaust.

I FIRST LEARNED of Himmler's calculating use of the past while researching a book on mummies and mummy scientists. I was holed up at the time in a tourist hotel in the small city of Assen in the eastern Netherlands. I had journeyed there to see some of the world's most famous mummies, the ancient dead of Europe's northern peat bogs. Preserved naturally by the bogs' dark, clear broth, dozens of Roman-era men and women had resisted oblivion for nearly two thousand years. But what struck me most about these resilient survivors was not so much their immortality, impressive as it was, but the violent way in which many had met their end—garroted, stabbed, slashed, decapitated, and hung.

At the museum in Assen, I bought a book by a Dutch archaeologist, Wijnand van der Sanden, describing the history and science of these strange cadavers. That evening in my hotel, I pulled the book from my bag, curious to see how van der Sanden interpreted the gruesome deaths.[20] I put my feet up on the bed and began casually thumbing through the pages. One section of the book in particular caught my attention. It concerned a speech that Himmler had given behind closed doors in February 1937 to a group of senior officers at the SS officer-training school at Bad Tölz.[21] In this address, Himmler aired his personal views on homosexuality.

The Reichsführer-SS regarded gay men as a great blight upon society. They contributed little more than red ink, in his opinion, to "the sexual balance sheet," rarely fathering children.[22] This was a serious failing in the Third Reich, where fatherhood was deemed one of the prime patriotic duties of all German men. Worse still, Himmler was convinced that homosexuality was a communicable disease. He believed it could infect straight men and he worried that it might reach epidemic proportions in Germany, particularly in such hotbeds of male bonding as the SS. If this happened, he warned, homosexuality could weaken the essential fiber of the SS, destroying one of the most powerful arms of the Nazi state.[23]

During his speech at Bad Tölz, Himmler mulled over methods of eradicating this imaginary peril from the Reich. And this was where the bog bodies came in. Researchers had long pondered the violent deaths of these ancient Europeans, developing competing hypotheses to explain them. Some archaeologists thought these people were murdered prisoners of war. Others argued that they were honored members of society selected as precious sacrifices to the gods. But a few German researchers, including one of Himmler's favorite young archaeologists, Herbert Jankuhn, advocated a much harsher hypothesis. Jankuhn believed these individuals were social pariahs, specifically deserters and homosexuals put to death for their transgressions against ancient Germanic laws.[24]

All the various hypotheses, however, were interpretations of scarce data. None could be proven. But the mere speculation that Iron Age Germans had once executed tribesmen accused of homosexuality pleased Himmler, providing exactly the kind of justification he was looking for. "Homosexuals were drowned in swamps," he stated categorically to the SS audience at Bad Tölz in 1937. "The worthy professors who find these bodies in peat, do not realize that in ninety out of a hundred cases they are looking at the remains of a homosexual who was drowned in a swamp along with his clothes and everything else." Then he drew his own conclusion. "That was not a punishment, but simply the termination of an abnormal life."[25]

With other leading Nazis firmly behind him, Himmler acted on these ideas. His police officers rousted suspected homosexuals from their beds and bars and bathhouses, beating them and seizing their address books to make further arrests. They sent some to regular prisons and dispatched as many as fifteen thousand to concentration camps.[26] There they issued them uniforms with distinctive pink triangle badges, setting them apart from other prisoners. Tragedy awaited them. Medical experimenters castrated some and subjected others to procedures designed to transform them into heterosexuals. SS guards added to the misery, starving and working many to death. Today no one knows how many perished, but one leading historian, Rüdiger Lautmann, estimates that as many as 60 percent of the pink-triangle prisoners perished in Third Reich camps.[27]

Sitting in my chair in the hotel room in Assen, I felt a deep chill, and when I later pieced together the story more fully, I understood its terrible significance. At Bad Tölz, Himmler had transformed a simple piece of archaeological speculation into a hard, murderous fact. He had cloaked his own hatred of others under the respectable mantle of science. He had disguised the Nazis' brutal agenda of mass murder as a venerable tradition of the German people, worthy of modern emulation. In Himmler's hands, the distant past had become a lethal weapon against the living.

IN THE SUMMER of 2001, I began delving into the history of the Ahnenerbe. I expected to find a rich body of literature on the subject, but this did not prove to be the case. Only a handful of articles on the organization had been written in English, most in scholarly journals, and surprisingly little had appeared in German. Indeed, just one major study of the brain trust had made its way into print at the time. Entitled *Das "Ahnenerbe" der SS 1935–1945*, the book was published in 1974 by a Canadian historian, Michael Kater, who had studied in Germany. By all rights, Kater's superb scholarly study should have sparked a major investigation. But few researchers chose to look into the matter further. Many German scholars tended to dismiss Nazi-era prehistorians as a group of "harmless fellow travelers."[28] Others feared digging into the Ahnenerbe. Several of the organization's former members held prominent academic positions in West Germany after the war, and they deeply resented young historians and archaeologists probing into their pasts.[29] So during the 1980s most German scholars let sleeping dogs lie.

The fall of the Berlin Wall in 1989 brought this era of complacency to an abrupt end. Achim Leube, a prominent East German archaeology professor, had long harbored an interest in archaeology during the Third Reich. Living in East Berlin, however, he could not easily examine the Ahnenerbe's records, most of which lay in West German archives. When the two Germanys finally reunited, Leube, a generous man with a wry sense of humor, embarked on a study of Ahnenerbe archaeologists. He urged others to do the same. In November 1998, he organized an international conference in Berlin on National Socialism and prehistory.[30] Nearly 150 scholars from twelve countries attended. The conference sparked a major scholarly reevaluation of the Ahnenerbe.

Few European historians, however, focused on what I strongly suspected would be the most fascinating part of the Ahnenerbe story. This was the foreign research—specifically the peacetime expeditions and the wartime missions of the Ahnenerbe and the ways in which Himmler put them to use. During the early 1970s, Kater had uncovered a wealth of letters and reports describing in detail one such journey—a large expedition to Tibet in 1938 and 1939 led by zoologist Ernst Schäfer. In addition, Kater had also chanced upon numerous hints and suggestions concerning several other research trips and expeditions—to northern Africa and South America, the Middle East and Scandinavia.[31]

Based on the fragmentary evidence at hand, however, the Canadian historian concluded that few of these foreign ventures had ever taken place. Indeed, he inferred from the available sources that most references to these trips were simply wishful thinking, mere projections of the overweening ambitions of the Ahnenerbe's leaders.[32] For nearly thirty years, historians had accepted this view. But Kater's passing mention of these foreign trips fascinated me. What possible interest, I wondered, could the Ahnenerbe have had in such exotic locations—the Canary Islands, Iraq, Finland, or Bolivia? What conceivable political use could Himmler have made of prehistoric research in these countries?

I set out to find the answers, assisted by a small team of translators and researchers. Working together, we began combing the vast microfilm collection of captured German documents at the National Archives and Records Administration in Maryland, and poring over the original Ahnenerbe files at the Bundesarchiv in Berlin. As we slowly exhausted our leads in these places, we expanded the search to other German archives, twenty-three in all, from the offices of the German Archaeological Institute to the collections of the former East German intelligence agency, the Stasi. From there, we stretched farther afield, to archives in Norway, Finland, Sweden, Poland, and Britain and library collections in Iceland and Russia.

For nearly two years, we waded through a vast, seemingly bottomless sea of letters, memoranda, minutes, reports, evaluations, accounting records, personnel files, equipment lists, expense accounts, unpublished articles, and published books. In all, nearly 961 voluminous Ahnenerbe files—occupying 180 linear feet of shelf space at the Bundesarchiv—survived the war. This, however, represented only a fraction of the original total. Much had been lost. At the end of the war, Ahnenerbe scientists incinerated stacks of incriminating documents, fearing that their letters and reports would be used against them in war-crime trials.

As the archival research proceeded, I began tracking down surviving members of the Ahnenerbe, as well as friends, relatives, and close colleagues of those who are now dead. I was constantly aware of the ticking clock. Some sources were far too frail or advanced in years to grant interviews; others, small in number, fell seriously ill or died between the first contact call and a second to set an appointment. In the end, I had the feeling that if I had waited even another few months to begin work on the project, I would have found scarcely a soul who had any firsthand knowledge of the Ahnenerbe's foreign research.

Adding to these difficulties was the inevitable reluctance of many sources to talk. Some former employees of the Ahnenerbe refused to relive the past, while children and spouses of those who had died often feared blackening the memories of fathers or husbands. In the end, however, a surprising number of people agreed to talk, perhaps because I was so obviously an outsider from Canada—and therefore burdened with less emotional baggage than most German authors—and because I was known as a science writer, rather than as a historian of the Nazi era. So as the research proceeded, I talked to dozens of sources, crisscrossing Germany from Sylt in the north to Lake Constance in the south, and then further expanding this research, from Austria, France, Norway, and Sweden to Italy, Finland, and the Netherlands.

What emerged from all this research shed a disturbing new light on the Holocaust. For years, most historians dismissed Himmler's intense interest in the ancient past as pure quackery and half-baked mysticism, foolishness that had no serious part in Himmler's plans for the Nazi state. But our new information on the Ahnenerbe's foreign travels painted a very different picture. Soon after the founding of the institute, Himmler began to sponsor expeditions and research trips across Europe and Asia. Through this foreign research, the Reichsführer-SS intended not only to control Germany's ancient past, but to master its future. Baldly stated, Himmler, the architect of the Final Solution, planned to use tall, blond-haired SS men and selected women to scientifically rebreed a pure Aryan stock. With knowledge gleaned by Ahnenerbe scientists, he intended to tutor SS men in ancient Germanic lore, religion, and farming practices, teaching them to think as their ancestors had. When the time was ripe, he proposed to plant SS agricultural colonies in Germany, as well as in specific parts of the East—places where he believed Germany's ancient ancestors had particularly flourished. There he hoped they would reverse the decline of Western civilization and rescue humanity from its mire.

To bring this about, Himmler was prepared to sacrifice and destroy the lives of millions of people. Most of the modern inhabitants of the supposed Aryan territories, he believed, would have to be dispossessed of their homes, forcibly deported, and either slaughtered or enslaved. All of the world's Jews would have to be exterminated, down to the very last man, woman, and child. In this way, no Jewish "germs" would survive to infect and destroy the fledgling Aryan colonies. The pandemic of "Jewishness" would at last be eradicated.

To lay part of the groundwork for this monstrous scheme, Himmler dispatched Ahnenerbe scholars on eight foreign expeditions or research trips before the war. With the assistance of Wolfram Sievers, the Ahnenerbe's managing director, and Walther Wüst, its soft-spoken superintendent, researchers journeyed across Europe and Asia—to remote Bronze Age rock carvings in Sweden and the rural homes of shamans in Finland, to inscription-covered palace walls in Croatia and the toppled temples of Parthian kings in Iraq, to mysterious Paleolithic caves in France and the enigmatic ruins of ancient settlements in Greece, to sprawling monasteries in Tibet and the sweeping coastal dunes of Libya.[33] In addition, Ahnenerbe officials drew up plans for at least four other expeditions—to Iran, the Canary Islands, the South American Andes, and Iceland—during the initial phase of the Nazi regime. Only the outbreak of the Second World War forced Himmler to postpone them indefinitely.

The research journeys, both planned and executed, were remarkably varied affairs. But most, if not all, were intended to prove Aryan supremacy or retrieve some form of ancient Aryan knowledge. Herman Wirth, an eccentric Dutch spendthrift with immense reserves of personal charm, set out to Sweden to decipher what he believed to be the world's oldest writing system: a lost Aryan script. Yrjö von Grönhagen, a handsome young Finnish nobleman who once auditioned for the cinema, roamed remote eastern Finland to record and film ancient magical rites; Grönhagen believed they were religious rituals handed down through the centuries from the Aryans. The urbane classical historian Franz Altheim and his photographer-mistress Erika Trautmann journeyed first to Croatia and Serbia, then later to Iraq, to study the role of blond-haired Aryans in the Roman Empire. (En route, they gathered intelligence on Iraqi pipelines and tribal leaders for the SS Security Service.) The renegade Dutch prehistorian Assien Bohmers scoured for clues to the origins of Aryan ritual and art in the painted caves of southern France. Ernst Schäfer, a mercurial man with a hair-trigger temper, and Bruno Beger, an SS expert in racial studies, trekked to Tibet to uncover proof of the ancient Aryan conquest of the Himalayas. While he was at it, Schäfer drew important maps of Himalayan passes and gathered valuable military information on local leaders and their loyalty to Britain.

Himmler received detailed reports and personal briefings on these trips. He was fascinated by the way in which Ahnenerbe scholars were piecing together the "hundreds of thousands of mosaic stones." But the Allies' declaration of war in 1939 convinced him that something more was required. He harnessed the Ahnenerbe's work directly to the war effort. After the German blitzkrieg through Poland, he sent archaeologist Peter Paulsen and a detachment of scholars to Warsaw to loot the city's most important prehistory museums, stealing all valuables deemed to be of German origin. After Operation Barbarossa in 1941, he dispatched Herbert Jankuhn, the researcher who had championed the homosexual bog-body theory, and a small team to hunt for proof of an ancient Germanic empire in the Crimea. The findings, Himmler hoped, would bolster Germany's claim to the region and justify his plans—approved in principle by Hitler— to execute or deport most of its inhabitants and plant colonies of SS men and their wives.

Himmler also enlisted the help of Ahnenerbe scholars in solving "the Jewish problem." In 1942, mobile SS killing squads at work in the Crimea and the Caucasus had encountered an unexpected difficulty. Jews and Muslims had lived side by side in these regions for millennia, intermarrying and trading age-old customs, religious traditions, and languages. The result was an ethnic pastiche that dumbfounded SS killing squads. Who was Jewish? Who wasn't? And what physical traits set the Jewish "race" clearly apart from all others? In search of answers, Himmler ordered two of his most trusted Ahnenerbe scientists, Ernst Schäfer and Bruno Beger, to mount a scientific expedition to the Caucasus. A year later, Beger's secret quest for data on the Jews culminated in one of the more notorious war crimes of the Second World War: the Jewish Skeleton Collection.

TO DISGUISE ALL these activities as science, however, the Ahnenerbe cultivated an air of solid professional integrity. Its staff went about their lives quietly and performed all the tasks that scientists and scholars are accustomed to doing. They published scholarly articles and books. They gave scientific lectures. They mounted museum shows and hosted scientific conferences. And in 1939, when the firm of Gebrüder Klingspor produced its lavish portfolio on Himmler's orders, the Ahnenerbe presented itself as a typical research institute, gravely declaring its founding principles: "Never narrow-mindedly bound to dogmas and doctrines / Truthful and strict in research and science."[34]

The Ahnenerbe was one of Himmler's great masterpieces of deception. To understand it thoroughly, one must begin at the beginning, by understanding something of its founder, Heinrich Himmler.

2. THE READER

TRAVELERS BOARDING THE TRAIN to Passau in the old Bavarian city of Landshut on September 24, 1924, may scarcely have noticed a pale, anemic-looking man with a slender book tucked beneath his arm. As a newly minted organizer for the National Socialist Freedom Movement, an offshoot of the recently banned Nazi party, Heinrich Himmler settled into his seat and waited impatiently for the train to depart, taking note of the exact time so he could record this event in his diary later. He was a precise, meticulous man, with a thin, spindly frame that had never known much exercise and a head that was one or two sizes too small for his body. He wore thick horn-rimmed glasses and a neatly trimmed mustache, and his straight brown hair was cut extremely short. His hands were small and doll-like. He was, all in all, a rather far cry from the men he sought to attract to the Nazi cause—brawny farmers and burly street fighters with a knack for breaking heads. Himmler, as one associate later joked, resembled "a half-starved shrew."[1]

If the studious party official felt cowed in any way by the new company he was keeping, he did not let on. At twenty-three, he possessed a haughty, condescending manner with a thin smattering of charm. He was bright, well educated, connected to the best social circles, and he possessed a romantic sense of his destiny. He believed that fate held something in store for him, a moment of greatness when he would someday be called upon to perform a momentous duty for his country.[2] He intended to be

15

ready. To prepare for that day, he read extensively. For the train trip from Landshut, where he was living, he had selected a book of classical history borrowed from friends.[3]

Germania was the work of Cornelius Tacitus, a Roman historian who had lived as a boy through the scandalous excesses of Emperor Nero. Fearing that Rome was rushing headlong down a road to ruin in the first century A.D., Tacitus decided to pen a kind of wake-up call to his fellow Romans. He searched about for a more moral society to hold up as an example, finally settling on those of the diverse barbarian tribes living east of the Rhine and north of the Danube, a region known to Romans as Germania. *Germania* read a little like modern anthropology, but Tacitus was clearly lacking in intimate knowledge of his subjects. According to modern scholars, he gathered his information from older texts and from stories he heard from tribesmen living in the Roman capital.[4] The resulting description of the Germanic tribes—their valor in war, their abhorrence of adultery, their boisterous hospitality—was terribly flawed. But Himmler was captivated. He raced through *Germania* in its German translation, finishing it that very day on the train. He then added it to his booklist, summing up, as he always did, his impressions in few sentences. *Germania*, he observed, was a "wonderful portrait of how high, pure and capable our ancestors were. This is how we will become again, or at least part of us."[5]

It was an immensely revealing remark. Already, at the age of twenty-three, Himmler had begun to reach toward and formulate a deceptively simple idea that would guide both the SS and the Ahnenerbe in later years and influence the course of European history. Already, he had begun to think of Germany's ancestors and their primeval traditions as a kind of template for the future, a blueprint for a better Reich.

By 1924, the die had clearly been cast. But how, one wonders, did Himmler's deep, abiding interest in the past come about?

THE SINGLE MOST important influence on Himmler as a child was his father, a prominent Bavarian educator. Gebhard Himmler was the son of a police sergeant turned district official, but as a young man Gebhard had aspired to greater things. In 1884, he passed the entrance examinations for the University of Munich, joining the privileged ranks of its students. Over the next ten years, he studied philosophy, polished his Latin and Greek grammar, and majored in the arcane field of philology. The latter, according to the *Athenaeum* in 1892, was "a master science, whose duty is to present to us the whole of ancient life, and to give archaeology its just place by the side of literature."[6]

On graduation, the elder Himmler found work as an assistant teacher at a Munich school and, through his uncle's connections as a court chaplain, landed a post as tutor to Prince Heinrich, a member of the famous Wittelsbach family.[7] The Wittelsbachs had supplied Bavaria with a succession of kings, scholars, art connoisseurs, generals, eccentrics, and madmen for nearly seven hundred years. Their medieval-looking palace in the center of Munich bustled with interesting people. Prince Heinrich's aunt Therese, for example, was a distinguished ethnographer, zoologist, botanist, and anthropologist who had mounted expeditions to such far-away parts of the world as the Brazilian rain forest; her apartment in the palace brimmed with strange and wonderful collections.[8]

In 1897, Gebhard Himmler married Anna Maria Heyder, the daughter of a prosperous merchant, and the young couple took up residence in the old medieval heart of Munich. By then, the young Wittelsbach prince had outgrown his need for a tutor, so Gebhard took a teaching position at a prominent Munich high school offering a classical education in Latin and Greek. There he grew a "Velásquesian beard," dressed stylishly in fashionable suits, wore an air of "pampered elegance," and frequently alluded to his royal and aristocratic connections.[9] He was, as one former student recalled, an insufferable social climber who was "laughably pushing and fawning towards the upper classes."[10]

His young charges were too frightened, however, to poke fun at him in his presence. Gebhard Himmler possessed a talent for ferreting out the weaknesses of others, and he combined this with a broad streak of cruelty: young pupils who failed to live up to his exacting standards were verbally assaulted and eviscerated in front of their peers in the classroom. His victims apparently never forgot these sessions. So traumatized was one famous student, Alfred Andersch, that he later published a ninety-five-page-long account describing the schoolmaster's pitiless verbal attack during a visit to Andersch's class.[11]

Munich society saw little of this side of Gebhard's character, however. He enjoyed a solid reputation in the Bavarian capital, and in the summer of

1900 the young Prince Heinrich willingly agreed to stand as godfather for Gebhard's second son. The baby, born on October 7, 1900, was named Heinrich in the prince's honor, and in a spacious flat on Hildegardstrasse—a short walk away from the Wittelsbach palace—the Himmlers set about raising their young family in the cultured comfort of the upper middle class.

Gebhard Himmler naturally took a strong hand in the education of his sons. As a teacher, he closely supervised their schoolwork, even reading and correcting Heinrich's personal diary for spelling and grammatical errors. As a devout Catholic, he took his children every Sunday to the splendor of St. Michael's Church, where many of the royal Wittelsbachs were buried, to attend Mass among the wealthy and powerful of Munich society. And as a strong German nationalist, he made certain that his sons received a thorough grounding in the German classics.[12]

Often in the evenings, he and his wife read aloud to the boys from books on German history or from the sagas of the old European bards.[13] In this way, they introduced Heinrich to the *Nibelungenlied*, a famous medieval tale of Siegfried and the fall of the Nibelung dynasty, and an important source of inspiration for Richard Wagner's grand operas, *Der Ring des Nibelungen*.[14] The old saga riveted the young Himmler.[15] And he was equally enthralled by the *Edda*, a collection of Old Norse sagas and poems that invoked, among other things, the magical world of Thor, Freya, Loki, and other Norse divinities. Himmler never forgot them.[16]

Gebhard also saw to it that his children developed a suitable appreciation for the antiquity of their family. The schoolteacher had somehow learned of a Himmler house in Basel that dated back to the thirteenth century. It was older, he sometimes boasted, than the castles of some Bavarian nobles.[17] To better display the family lineage, he set aside one room in the family apartment as an *Ahnenzimmer* or "ancestor room."[18] German kings and princes and nobles often built *Ahnensäle* in their castles—large gilded halls crowded with portraits of their ancestors, all the better to show off their illustrious bloodlines.[19] But the German middle class could ill afford such luxury. Ancestor rooms were unknown in their more modest apartments. Gebhard, however, did not let this stop him. His ancestor room seems to have displayed an assortment of family heirlooms, pictures, and records, as well as a collection of Roman coins and other antiquities that he collected as a hobby.[20]

At the time, archaeology was all the rage among the upper crust in Germany. Kaiser Wilhelm himself had enthused over excavations of an old Roman site at Saalburg and had even gotten his hands dirty while taking part in a dig in Corfu in 1911.[21] Gebhard followed suit enthusiastically. In his spare hours, he visited local archaeological sites and hunted for antiquities.[22] When his sons were old enough to accompany him, they took excursions into the countryside, visiting places mentioned in histories and old Bavarian tales and searching ruins for rune stones to read and coins, weapons, and potsherds to collect.

Heinrich thoroughly absorbed his father's enthusiasms. He passed many evenings watching his father classify the finds and catalog them in a filing system he had set up in the ancestor room.[23] Archaeology during that era was largely a science of classification. Its disciples sought to identify and sort artifacts into precisely defined categories as a first step in making sense of the chaos of objects recovered from the ground. The pleasure the boy derived from watching his father classify dusty treasures took a more ominous form later in life. Under his direction, concentration-camp officials issued prisoners color-coded badges so that individuals could be classified at a glance into one of eighteen categories—from political prisoners to Gypsies.[24]

BY THE TIME Heinrich Himmler was ten, he had memorized details of Germany's most famous historic battles.[25] A few years later, when he reached high school, his knowledge of ancient German weaponry and warfare equaled that of some of his teachers. He had begun to consider himself something of an authority on the ancient German past, but his father saw to it that he did not neglect his other studies. "He appears to be a very keen pupil," wrote one of his classroom teachers in a 1914 report, "whose tireless application, burning ambition and active participation in class have produced excellent results."[26] Consistently, young Himmler finished near the top of his class.

For all his academic aptitude, however, he demonstrated little leadership ability. He was sallow, physically weak, and clumsy, plagued by lung infections and a stomach ailment that produced severe cramps which could incapacitate him. His shrill voice squeaked on the high notes and his laughter sounded like a "perfunctory cackle," unpleasant to the ear.[27] He was unable to take part in most sports, and he was so physically ungainly as a teen that he struggled to master a bicycle, repeatedly falling and skinning his knees and hands.

In 1913, the family moved to Landshut so that Gebhard Himmler could take a position as a deputy headmaster. There the local students discovered that Heinrich regularly reported their pranks to his father, and they began to shun him.[28] They turned down his friendly overtures, going silent at his approach and only resuming their conversations when he was safely out of earshot. To avenge this silent treatment, Heinrich began supervising the after-class punishments that his father liberally doled out. This further incensed his classmates. On one occasion, they stuffed him headfirst and flailing into a garbage can in a locked room.[29]

Away from his hostile classmates, he dreamed of a military career. When World War I broke out in 1914, he followed news of the German campaigns avidly, sopping up details of the latest carnage from newspapers and dispatches at the local telegraph office. When his older brother Gebhard departed for officer-training school in 1916, he was desperate to follow. Isolated from all the terrible bloodshed at the front, he dreamed of glory. And neither the mortal wound that his own royal patron, Prince Heinrich, sustained at the front in 1916, nor the mounting toll of German casualties could dampen Heinrich's schoolboy ardor.

He hounded his father to let him join up, and in the end the elder Himmler, fearing perhaps that his son might be conscripted into the rank and file, wrote to several influential friends to see if a spot might be found in officer training for the boy. One of his letters hit the mark, and in December 1917, Heinrich joined the Eleventh Bavarian Infantry Regiment as an officer trainee. To the boy's intense disappointment, however, he spent the rest of the war in training, never setting so much as a foot in the trenches.

LIKE MOST OF his friends and acquaintances, the young Himmler was stunned by the surrender of the Imperial German Army in late 1918 and the abdication of Kaiser Wilhelm. The bedrock foundation of German society seemed to be crumbling, and when the Allies announced the Treaty of Versailles in June 1919, Himmler felt a sense of outrage. Not only did the newly inked treaty call for Germany to relinquish all of its colonies and part of its homeland, it also forced the country's military commanders to strip down its once mighty army, leaving just one hundred thousand officers and men, all volunteers.

The news shattered Himmler's dream of a career in the German army, leaving him at loose ends in Landshut. Throughout Germany, a groundswell of discontent washed over the ranks of the newly demobilized, and in nearby Munich, a bloody state of anarchy broke out. Taking advantage of the growing power vacuum, German communists attempted to seize control of the city, but they were soon beaten back and massacred by armed paramilitary groups of right-wing extremists, the *Freikorps*.

Himmler's sympathies lay firmly with the *Freikorps*, but he had a future to think about. He was eager to escape from under his father's large thumb and talked of studying agriculture, a career he likely saw as a ticket to someplace far away. He even considered becoming a colonist in Eastern Europe, a destination that might just as well have been at the end of the earth for a boy who had never traveled more than fifty miles from his birthplace.[30] His father, who must have hoped for a grander career for his son, something more suited to the family's lofty royal connections, eventually consented to the plan. He had begun to worry apparently about Heinrich's unhealthy attraction to right-wing politics; agricultural studies, he must have hoped, would take his son's mind off the *Freikorps*.[31]

So Himmler enrolled at what is now called the Technical University of Munich, attending classes in chemistry, physics, and botany. In the evenings, he cultivated new friends. He took dance lessons to master the Boston waltz, went skating, attended a costume ball as an Arab sultan, and mooned for a while over a young woman who was in love with someone else. He joined a fraternity, learned how to fence, and took part in a friendly duel that left him with a few small scars—a long-standing fraternity tradition among German university students.[32] In his diary, he regularly chided himself for talking too much at social gatherings. He was becoming garrulous.

He argued heatedly with his father about money.[33] There was suddenly very little to go around anywhere, as Germany struggled to meet the Allies' heavy demand for war reparations. In a doomed attempt to pay the mammoth bill, the new Weimar Republic churned out more and more paper money. The result was something that economists clinically call hyperinflation, but that ordinary people recognized at once as utter madness. In just one year, between June 1922 and June 1923, prices in Germany soared 10,000 percent. Over the next five months, they exploded, shooting up more than 10,000,000 percent. Factory managers began paying their employees once a day, and providing a short break immediately after handing over the cash. The workers then tore off frantically to the nearest shops to spend the entire packet before prices rose further. In the streets, people stole anything of value, even brass fittings, and upper-middle-class families such as the Himmlers watched in despair. Their savings had evaporated into thin air, and they were forced to scrimp and cut corners like everyone else.

In Munich, right-wing extremists found the atmosphere of panic very conducive to recruiting. They papered the city with posters, announcing evening speeches in the *Festsäle* of the city's huge beer halls. There, demobilized soldiers and newly impoverished tradesmen, teachers, students, and white-collar workers rubbed shoulders, draining great earthenware steins of foamy beer as they listened to, heckled, and applauded the political pitchmen. Munich was a cauldron of ultraright sentiment, and those who drank from this poisonous brew increasingly turned their ire on the struggling Weimar government and on an imaginary conspiracy of Jewish bankers and businessmen said to be profiting merrily from the misery of the German people.

One of the most strident of these pitchmen was Adolf Hitler, a young Austrian artist, drifter, and former soldier. Often dressed in heavy boots, a plain dark suit, white shirt, and a leather waistcoat, Hitler did not strike an imposing figure. He resembled, as one close associate of the day later remarked, "a waiter in a railway station restaurant."[34] But the young political leader possessed a talent for powerful oratory. Speaking in rented halls, he attracted a small but intensely loyal following for what was soon to be known as the National Socialist German Workers' Party. His thuggish supporters regularly took to the Munich streets, armed with heavy rubber truncheons and looking for trouble—which they often found. "The Jew with flat feet and hooked nose and crinkly hair," they chanted, "he dare not breathe our German air? Throw him out!"[35]

On one particularly memorable evening in February 1920, at Munich's most famous beer hall, the Hofbräuhaus, Hitler took the stage in front of a large audience of two thousand men standing shoulder to shoulder. As a throng of communists and socialists shouted insults from the rear—only to be assaulted and beaten and silenced by Hitler's ham-fisted supporters—he spoke for nearly four hours, delivering his party's platform. Hitler promised to unite all Germans—no matter where they resided, whether in Austria, Czechoslovakia, or Poland—in a Greater Germany and tear up the hated Treaty of Versailles. He planned to forge a strong new army and execute citizens he deemed traitors. He proposed to censor the press and take control of Germany's entire cultural system, beginning with the schools. Most ominously of all perhaps, he intended to extend the privilege of German citizenship only to those possessing German blood: no Jews would be allowed.

Like many other young right-wing supporters in Munich, Himmler was greatly drawn to these proposals, redolent as they were of rewriting history and wiping out the shame of the recent war. Moreover, he agreed that some action had to be taken against the Jews. His parents had not particularly raised him as a virulent anti-Semite, but as early as 1919, he had begun to explore what he and many others called "the Jewish question."[36] He wanted to appear coldly rational in his views—a character trait that became his trademark in later years—so he initially dismissed writers who approached the subject of Jews emotionally, from a position of blind hatred. Himmler needed the pretense of reason, and in early February 1922, he came across a booklet entitled *Rasse und Nation* ("Race and Nation"). The British-born author, Houston Stewart Chamberlain, was the son-in-law of Richard Wagner and had sopped up the family's intense anti-Semitism. Himmler read Chamberlain's elaborate rationales with great interest. They supplied, as he later noted in his booklist, "a truth that convinces you that it is objective and not filled with hatred and anti-Semitism. This is why it is so much more successful. This awful Jewry."[37]

Himmler took out a membership in the Nazi party in the summer of 1923, a year after graduating from the Technical University. His father greatly disapproved—he supported the Bavarian People's Party, a more moderate Catholic group—but Himmler remained adamant.[38] That November, Hitler and a group of fellow conspirators attempted to seize power in Bavaria with a haphazard putsch. Himmler eagerly offered his services as a combatant, manning a barricade with his brother outside Munich's War Ministry building, but the plot ended in failure. Bavarian authorities charged Hitler with high treason and outlawed the Nazi party.

The failed coup did little to dampen Himmler's newfound ardor, however. He went to work in the summer of 1924 as an organizer for a local Nazi party boss from Landshut, a chemist named Gregor Strasser. It was Himmler's job to help build up one of the ultraright groups, the National Socialist Freedom Movement, that had suddenly sprouted up to replace the Nazi party. Weedy as Himmler looked, he seemed a born organizer. In his private life, he was accustomed to jotting down the exact time of day, sometimes down to the precise minute, when he received letters and birthday greetings from friends and family members, and this mania for record keeping set him apart from other party recruits.

In 1925, Strasser made Himmler his deputy in Bavaria, Swabia, and the Palatinate. It was an important step up the political ladder for a young Nazi official, particularly after the Bavarian prime minister released Hitler from prison in 1925 and lifted the ban on the Nazi party, observing with reckless optimism that the "wild beast is checked; we can afford to loosen the chain."[39]

Himmler took his new responsibilities very seriously, and when the first volume of Hitler's political autobiography *Mein Kampf* came out in 1925, he quickly picked up a copy.[40] Many other party members, entranced by Hitler's oratory, eagerly bought the book as well. They flipped through the first few pages and swiftly lost interest in their leader's disjointed ramblings—on the evils of Communism, the dangers of syphilis, the foppishness of modern fashions, the menace of an international Zionist conspiracy, the magic of the Wagnerian opera *Parsifal*, the merits of boxing, the perils of racial mixing, the degeneration of modern art, and the necessity of correctly educating the young. But Himmler read both volumes avidly. "There are a tremendous number of truths in it," he later noted approvingly in his booklist.[41]

One section of *Mein Kampf* delved into a subject particularly near to Himmler's heart: the greatness of Germany's ancestors. Privately, Himmler liked nothing better than to while away the hours reading historical novels or historical nonfiction. One-third of the books noted on his booklist before February 1927, when he finished the second volume of *Mein Kampf*, explored the past.[42] Imaginatively and intellectually, Himmler lived in a world inhabited by the dead—feudal lords and kings, soldiers and peasants, Teutonic knights and Roman emperors—and he must have felt a shock of pleasure when he realized just how much Hitler shared this interest.

In the pages of *Mein Kampf*, Germany's future Führer described for his readers what he saw as the noble antecedents to modern Germans. These progenitors, he explained, were the ancient Aryans, and all humanity lay forever in their debt:

> All human culture, all the results of art, science, and technology that we see before us today, are almost exclusively the creative product of the Aryan. This very fact admits of the not unfounded inference that he alone was the founder of all higher humanity, therefore representing the prototype of all that we understand by the word 'man.' He is the Prometheus of mankind from whose bright forehead the divine spark of genius has sprung at all times, forever kindling anew that fire of knowledge which illumined the night of silent mysteries and thus caused man to climb the path to mastery over the other beings of this earth. Exclude him—and perhaps after a few thousand years darkness will again descend on the earth, human culture will pass, and the world turn to a desert.[43]

Himmler was enthralled by this vision of the past. The fact that it was purely fictional—a work of imagination—did not seem to cross his mind.

3. ARYANS

THE IDEA OF RACE is a rather modern invention. Before the first Europeans landed on the shores of the New World in the late fifteenth century, the notion of racial difference does not seem to have existed.[1] Earlier travelers remarked upon the appearance of those in foreign lands—the hue of their skin, the texture of their hair, the color of their eyes—but they took no interest in classifying strangers according to physical traits. They were far more concerned with religious differences, describing foreigners as "idolators" or even "infidels." But after Europeans conquered the Americas and became colonists on a grand scale, forcing others into servitude and slavery, the language of difference began to change. The word "race" appeared in the English language in the sixteenth century, borrowed either from French or Italian.[2] Its counterpart *Rasse* appeared in German nearly two centuries later, in 1791.[3]

It was a maddeningly vague word, however, and it gave rise to considerable confusion. Some writers used it to describe all of humanity, as in the phrase "human race." Some politicians brandished the word to mean nationality, speaking of the French race, while scientists tried to define it more specifically, as one of the great divisions of humankind that could be distinguished scientifically by certain unvarying physical traits. All this confusion, however, made race a very valuable concept for Hitler and the early Nazis. They could use a word like *Rasse* to mean whatever was most expedient for them politically. And they added considerably to this confusion by referring to a mythical race, the Aryans.

The notion of the Aryans, however, did not begin with the Nazis as they schemed and plotted in the smoky beer halls of Bavaria in the 1920s. Nor did it first emerge from the overheated rhetoric of middle-class German nationalists in late nineteenth-century Germany. Its roots were much older and they were far more deeply entangled in European scholarship and science. The concept of the Aryans—a tall, slim, muscular master race with golden hair and cornflower blue eyes—traces its origins back to a line of legitimate scientific inquiry in eighteenth-century Great Britain. This research blossomed in India and took off in a strange new direction in Germany during the Romantic era. Only in the late nineteenth century did the concept of the Aryans undergo a final, virulently racist twist in the hands of German nationalists, becoming something truly malignant.

IT WAS A British naturalist, James Parsons, who planted the first unwitting seed of Aryanism. A contemporary of Voltaire and Dr. Samuel Johnson, Parsons had fallen under the influence of the European Enlightenment, an age when gentlemen in powdered wigs asked profound questions about the way the world worked and had a pronounced inclination to seek out answers. Parsons was a doctor by training, taking his medical degree in Rheims in 1736, but like many gentlemen scholars of the age, he possessed a daunting range of interests. He penned papers on the anatomy of crabs and corals, the preternatural conjunction of two female children, the fossil fruits of the Island of Sheppey, and the double horns of a rhinoceros.[4] In his spare hours, he read widely in ancient history and joined the Antiquarian Society.

Parsons was passionately curious about the murky origins of the Irish and the Welsh, and he turned to one of his favorite reference books for possible clues: the Bible. In the book of Genesis, he noted, God instructed the three sons of Noah to go forth into the world and multiply. One of these sons was said to have fathered the Semitic peoples—the Arabs, Jews, Bedouins, and other related groups in the Middle East. The second had given rise to the Egyptians, Ethiopians, and other African peoples. And the third son, Japhet, had founded most of the remaining lineages of humankind, including the Europeans.[5]

But the Bible offered devout Christians few details about the peopling of the earth. Parsons hungered to know more about his ancestors. Were the Irish and the Welsh and for that matter all other Europeans truly the descendents of Japhet? And, if so, how could one prove this? Parsons took a novel approach to the problem. He decided to trace human origins by searching for possible affinities between major European and Asian languages. If the Europeans as well as the Persians and the Indians all descended from Japhet, as the Bible suggested, then their respective tongues should contain at least a few old shared words handed down from their common ancestor. Parsons decided to investigate.

He focused his research on words for numerals, reasoning that since numbers were essential to the culture of "every nation, their names were most likely to continue nearly the same, even though other parts of the languages might be liable to change and alteration."[6] He consulted foreign-language dictionaries, drawing up lists of the relevant words. Then he patiently compared the terms from one language to another. Striking similarities existed between many European and some Asian tongues. The word for three, for example, was identical in Irish, Welsh, Russian, and Bengali—*tri*. Moreover, the word phonetically resembled *tre* in Danish and Swedish, *treis* in Greek, *tres* in Latin, and *drei* in German.[7] From such studies, Parsons concluded that Bengali, Persian, English, Irish, Latin, Italian, Spanish, French, Danish, and German all sprang from a common ancestral language, one that could have been spoken by Japhet as he lit out into the world after the deluge.

Parsons had devised an enormously powerful new tool for tracing the origins of peoples—comparative linguistics—and he astutely recognized its importance, publishing *The Remains of Japhet, being historical enquiries into the affinity and origins of the European languages* in 1767. The scientific world, however, ignored this book. Readers familiar with Parsons's scattered papers on corals, fossil fruits, and rhinoceros horns wondered what an anatomist was doing meddling in fields he knew nothing about. Others dismissed Parsons's ideas after stumbling upon his linguistic slipups, convoluted biblical theories, and lines of dubious evidence. However, in 1786, a prominent Orientalist, Sir William Jones, discerned these same linguistic affinities independently, and in the process of this research, he helped coin the term "Aryan."

Jones was the founder of the Asiatic Society in Calcutta. He was a lawyer by profession and had earned a solid reputation as an Orientalist, mastering nearly two dozen languages, including Hebrew and one of the Chinese tongues. In 1783, he sailed for India as a new justice in the Supreme Court. There, surrounded by the splendor of the British Empire, he began applying himself to acquiring Sanskrit. He intended to write a digest of Hindu and Muslim law, and it was while in the throes of this study that he began to discern a strong resemblance between the words and grammar of Sanskrit, Greek, and Latin.

Intrigued, Jones mentioned these linguistic correspondences in a now famous paper he gave at the Asiatic Society on February 2, 1786. As an attentive audience of colonial officials and merchants, physicians and engineers gazed on, Jones observed that "the Sanskrit language, whatever may be its antiquity, is of wonderful structure; more perfect than the Greek, more copious than the Latin, and more exquisitely refined than either; yet bearing to both of them a stronger affinity, both in the roots of the verbs and in the forms of the grammar, than could have been produced by accident; so strong that no philologer could examine all the three without believing them to have sprung from some common source, which, perhaps, no longer exists."[8]

Jones's remarks were extremely well received. Over the next eight years of his life, he diligently expanded this new field of research, and it was in the thick of this work that he introduced a new word to European scholars—*Arya*.[9] The word came from Sanskrit and it meant "noble." In India, it was often employed to describe those who worshipped the Hindu gods, as opposed to the divinities of other Indian religions.[10] Jones rather fancied the term and he borrowed it to describe those who spoke a particular group of Indian languages.

Inspired by Jones, other scholars began systematically comparing words and grammar from parlances across Europe and Asia, looking for possible affinities. They discovered that more than forty of the world's major languages shared strong similarities to Sanskrit—from English, French, German, Danish, Swedish, and Irish Gaelic to Serbo-Croatian, Yiddish, and Romany.[11] But no one could quite agree on what to call this newly identified group of languages. British linguists tended to favor the term "Indo-European," while German scholars rather liked the sound of "Indo-German." Others, however, happily settled on a word that Jones had first drawn to their attention: Aryan.

All these linguistic similarities raised many perplexing questions. How did Sanskrit, the language of ancient Hindu scholars, come to resemble German? Could the Brahmans of Calcutta be related in some way to the burghers of Bavaria or Hesse? In the old cathedral city of Cologne, a prominent nineteenth-century German scholar wrestled with these questions.

A SURVIVING PORTRAIT of Friedrich Schlegel shows a rather plain-looking man with dark circles beneath a pair of intense eyes and a rumpled collar framing his face. During his youth in Leipzig, Schlegel apprenticed with a banker, but it was the world of literature and ideas—not account books and balance sheets—that beckoned to him. For a time he studied Greek philosophy and culture and tried his hand unsuccessfully at writing a semiautobiographical novel, but eventually he realized that his real métier was literary criticism and history.

Schlegel was a man of restless enthusiasm, forever on the alert for new ideas to weave into his own writings. While in Napoleonic Paris in 1802 and 1803, he took up studies of Sanskrit and comparative linguistics, which convinced him of the antiquity and importance of the Indian culture. "Everything, absolutely everything, is of Indian origin," he declared in a letter to a friend in 1803.[12] This deep-seated conviction soon led him to a daring suggestion: perhaps Germans and other modern European peoples stemmed not from the Holy Land at all, as the Bible claimed, but from the remote valleys of the Himalayas.

In 1808, Schlegel explored this idea in a famous book entitled *Essay on the Language and Wisdom of the Indians*, a strange, poetic concoction of dry scholarship and historical fantasy. During the distant past, he proposed, a brilliant nation of priests and warriors lived quietly in the secret, hidden valleys of the Himalayas. At some point, a terrible crime, "some inconceivable desolation of the human conscience," had transformed them.[13] Abandoning their peaceful, vegetarian ways, they had become war-mongering carnivores, bursting out of their mysterious mountain stronghold.[14] Some swept to the south, where they soon conquered the entire Indian subcontinent. Others surged westward, founding a string of great empires, until at last they arrived in the cold, drizzly forests of Germany and Scandinavia.

Schlegel recognized that some of his fellow scholars might find this a rather tall tale to swallow. Why, after all, would a nation of warriors from the "most contented and fertile region of Asia" want to settle in Europe's chill northern extremities?[15] But Schlegel had an explanation ready at hand. He had discerned references in Sanskrit legends to a sacred mountain in the North much revered by the ancient Hindus. Perhaps the Himalayan holy men had been drawn—as filings were to a magnet—to the northlands of Germany and Scandinavia.

In one fell swoop, Schlegel had added a whole new dimension to the scientific study of Indo-European languages, transforming a well-grounded linguistic concept into an anthropological fable. Eleven years later, he added a final important detail. The ancient Himalayan invaders of his imagination possessed no name. So the German literary critic gave them one—the Aryans. By such means, the Aryans entered into popular culture, a mythical group of Himalayan warriors and holy men with a bizarre fictional history and a borrowed name.[16]

Schlegel, of course, had no idea what he had helped to create or how his ideas could be distorted and used to such catastrophic effect by future German nationalists. If he had known this, he would have been horrified. He had personally campaigned to give German Jews the right to vote.[17] His wife, Dorothea, was Jewish.[18]

FOR ALL THE many liberties that Schlegel had taken in his writings, he had clearly put his finger on two important truths. Some ancient human society had indeed spoken the ancestral tongue that eventually evolved into more than forty major languages across Europe and Asia. And this society had clearly expanded its sphere of influence from a rather confined area to many parts of Eurasia. But scholars needed proof before they would accept Schlegel's vision of ancient Himalayan invaders. And where was evidence to be found? Theodor Benfey, a renowned nineteenth-century Sanskrit scholar, mulled over these problems carefully in the old university city of Göttingen.

Benfey was the son of a Jewish merchant, and as such he had struggled for much of his life against the strong currents of anti-Semitism in German society.[19] Earning a doctorate degree at the tender age of nineteen, he had taken a position at the University of Göttingen as a private lecturer, only to watch in frustration as younger and less accomplished men rapidly climbed above him. Indeed, it was only after twenty-eight years of waiting, and after finally converting to Christianity in his despair, that he finally attained the rank of professor.

Benfey decided to trace the Aryan wanderers back to their homeland by means of language. He could apply precise, well-tested linguistic rules to modern words for plants and animals in the Indo-European languages and determine which, if any, belonged to the lost ancestral vocabulary.[20] Those that did, he surmised, would shed light on the ecology of the old Aryan homeland. So the scholar set about patiently drawing up a list of the ancestral words and comparing them to the flora and fauna in possible Aryan homelands. He published his findings in 1869. Northern Europe, he concluded, with its expansive forests of beech and pine, bore the closest resemblance to the ecosystem preserved in the ancestral vocabulary.[21]

Benfey's work suggested that the Aryans had first emerged from northern Europe, not the Himalayas. It was a slender line of evidence, one later contradicted by other similar studies, but in the simmering cauldron of German politics in the late nineteenth century, nationalists pounced on it gleefully.[22] Germany's diverse royal families had only recently joined forces to create a united German state, and nationalists were keen to stitch the new patchwork together. Benfey's study seemed a particularly handy tool. It implied that it was Germany's forefathers who had swept across the plains of Europe all the way east to India, carrying their distinctive language with them. And in the minds of the nationalists, it seemed to cast their ancestors in the role of energetic conquerors, restless, unstoppable masters of the ancient world.

So enthralled were nationalists with this vision that they immediately began to embroider upon it, filling in the large gaping holes that science could not. They speculated wildly, for example, on the physical appearance of the Aryans, relying heavily on a description of the Germanic tribes penned by the Roman historian Tacitus. In *Germania*, Tacitus had blithely remarked upon the large frames, blue eyes, and red-blond hair of the German tribespeople. Classical scholars had long doubted the veracity of many of his observations, for Tacitus had seemingly done his research from the comfort of his Roman home.[23] But German nationalists were oblivious to such scholarly concerns.[24] They fully embraced the idea of tall, blond-haired and blue-eyed Aryan ancestors.[25] Some even went so far as to suggest that Schlegel's manufactured name, "Aryan," came from an ancestral root word meaning *light,* a reference to the color of Aryan hair and skin.

It was all perfectly absurd, but the idea of the conquering, flaxen-haired Aryans from the north seemed to touch a deep chord among nationalists in Germany. It appealed powerfully to their vanity. And in the decades preceding the rise of the Third Reich, this vision began to merge dangerously with a concept that was rapidly taking hold in German imaginations—the Nordic Race.

SINCE THE EARLY 1800s, researchers had tried to divide all of humankind into races that could be distinguished scientifically by certain unvarying physical traits.[26] To identify these traits, researchers had traveled the world, measuring human bodies, studying the shapes of human heads, puzzling over the texture of their hair, and charting the color of their skin. It had been long, arduous work, with few rewards. Despite many claims to the contrary, anthropologists in the early twentieth century had failed to arrive at any sound scientific criteria for classifying humankind into races.[27] "At present," noted the *Encyclopaedia Britannica* in 1939 in its entry for anthropology, "all the methods in use to determine race are precarious, and their provisional findings must be accepted with the utmost caution."[28]

But that, of course, did not deter politically motivated amateurs from drawing up their own criteria, and in 1920, an obscure German philologist, Hans F. K. Günther, turned his attention to the subject of race. Günther was the son of a violinist in the Freiburg city orchestra. At university, he had studied Hungarian, Finnish, and French, but in the aftermath of the First World War he developed an interest in politics. Soon after Germany signed the Treaty of Versailles, Günther published a book lambasting modern liberalism and urging the Weimar government to pay greater heed to matters of race and racial improvement. The book greatly interested a prominent German ultranationalist, publisher Julius Lehmann, who was searching for someone to write a popular book on the races of Germany. Lehmann decided to test Günther's ability to discern race. He took him on a two-day hike through the Alps, asking him to identify the racial origins of the people they met on their journey. So impressed was Lehmann with Günther's eye for "racial differences" that he handed the scholar a contract for the book.[29]

Günther set to work with a passion. To parse the population of Germany, and later of Europe itself, into what he considered to be its original bloodlines, he proudly employed his own intuition, his personal observations, and odd bits of data he gleaned from a host of fields. He studied facial features carved on ancient Persian statuary and scrutinized the death mask of Frederick the Great. He collected photographs of ordinary people from across Europe and examined their head shapes and proportions. He read psychological and ethnological studies of Europeans from different regions. From all this unscientific hodgepodge, he drew up his own taxonomy of race—a masterpiece of racial stereotyping and racism.

Günther eventually classified five races in Europe. He employed not only visible physical traits for his criteria, but also far more nebulous characteristics, such as intellectual abilities and emotional states. He paid no attention whatsoever to the role of environment in forming human beings, focusing instead on the idea that humans were solely products of their biological nature—their race. Among the five races that he identified, none could hold a candle to the Nordic race, which Günther described with an almost erotic intensity:

> The Nordic race is tall, slender. The long legs contribute towards the stately height, which for the man averages about 1.74 meters. The form both of the whole body and of each of the limbs, as also that of the neck, hands and feet, is one of strength combined with slenderness. The Nordic race is long-headed and narrow-faced. . . . The skin of the Nordic race is rosy and fair: it allows the blood to glimmer through, and so it looks alive, often lustrous, and always rather cool or fresh, 'like milk and blood.' The veins shine through (at least in youth) and show 'the blue blood.' The hair is smooth and sleek or wavy in texture, in childhood it may be curly. Each hair is thin and soft and often 'like silk.' In color it is fair, and, whether light or dark blond, always shows a touch of gold, or a reddish undertone. . . . The Nordic eye, that is, its iris, is blue, blue-grey, or grey. . . . Nordic eyes often have something shining, something radiant about them.[30]

This striking physical appearance was matched, according to Günther, by a long list of mental and emotional traits found in the Nordic race.

These included boldness, a natural aptitude for great undertakings, self-reliance, sound judgment, a love of justice, a deep well of energy and creativity, a broad range of intellectual interests, a gift for storytelling and musicianship, and a talent for warfare.[31] Virtually all were characteristics that German nationalists ascribed to the ancient Aryans.

Günther's racial books became best sellers in Germany, flying off bookstore shelves.[32] Under their powerful spell, German nationalists began to graft the physical characteristics of the imagined Nordic race onto the increasingly popular idea of the ancient Aryans. They began to use words like Aryan, Nordic, Indo-Germanic, and Germanic interchangeably.[33] What had started out in the eighteenth century in Britain as a harmless, purely linguistic investigation into the distant origins of the Europeans had become a deadly racial powderkeg by the early twentieth century in Germany.

4. DEATH'S-HEAD

In JANUARY 1929, ADOLF Hitler appointed one of his most energetic and zealous young followers, Heinrich Himmler, as the head of the Schutzstaffel or Security Squad, better known as the SS. Hitler had created the organization nearly four years earlier as an elite force of personal bodyguards to ward off attacks and foil assassination attempts, but it had never really lived up to his expectations. Party recruits with an appetite for violence tended to ignore the SS, gravitating instead to the brown-shirted Sturmabteilung, the Storm Detachment or SA, the political combat troops whose work it was to break bones in the streets. By the beginning of 1929, the SS was a pale shadow of an organization. It counted fewer than three hundred members, and its most visible public role was as a sales force, hawking advertising space in the Nazi party newspaper, the *Völkischer Beobachter*.[1]

Heinrich Himmler was determined to transform the motley organization into the pride of the Nazi party. During a sixteen-month stint as the deputy leader of the SS, he had already begun quietly imposing his own sense of discipline and order. Gone were the chaotic group meetings, where members merely lounged about, smoking and telling war stories and boasting about the communist heads they had smashed. Now, thanks to Himmler's new rules, members paraded in a brisk military drill before each meeting. They sang SS songs. They listened attentively to the political speeches that consumed most of the meetings. And on Himmler's instructions, they gathered and reported intelligence on prominent Jews, Freemasons, and leaders of rival political movements.[2]

The newfound sense of purpose in the SS did not pass unnoticed by Hitler. That was why he decided to put Himmler in charge. Perhaps to others in the party it seemed a modest appointment for a minor party official, but the newly minted Reichsführer-SS did not see things that way. He had important plans. He intended to transform the SS into a racial showplace of lanky, golden-haired, blue-eyed men. With this select gene pool, he intended to breed pure Aryans for the new nobility of the Third Reich. And with all the powerful resources of the SS at his command, he intended to ransack the German past, searching for the ancient lore of the Reich's ancestors—lore that could be used to educate the new Aryan lords to assume their place in history.

FIRST, HOWEVER, HIMMLER had to build up the SS and demonstrate its worth to the party. So in 1929, he embarked on a major recruitment campaign. He drew up advertising posters and crisscrossed Germany on recruitment drives, giving speeches in village halls and town centers. He chose a striking new design for the SS logo and brought order to the SS records. He put in long hours at his desk in the Nazi party headquarters, located behind a photographic studio on Schellingstrasse in central Munich, just a few blocks from his childhood home.[3] It was exactly the kind of work—meticulous, grinding organization—he excelled at. The new Reichsführer-SS was devoted to the force he was creating, and at home in Waldtrudering, just outside Munich, his new wife fretted with loneliness.

Himmler had met his future wife while walking into a hotel lobby in 1926. He nearly crashed into her, then compounded his awkwardness by drenching her with melting snow from his hat as he hastened to introduce himself. His eye settled approvingly on her glossy blond hair. Margarete Boden was the daughter of a West Prussian landowner. Eight years older than Himmler, Marga was a divorcée who owned a small clinic in Berlin specializing in homeopathic medicine and other forms of natural healing.[4] A large, rather humorless woman with a noticeable facial tic, she lacked both charm and beauty and did not make friends easily. But she knew how to wrap Himmler around her finger. She was apparently the first woman ever to take a serious interest in him.

After the initial sparks flew, Himmler seems to have been unclear about what to do next. At age twenty-six, he had very little, if any, sexual experience, and he was very prudish.[5] According to one party associate who knew Himmler at the time, Marga, a domineering, take-charge kind of woman, was forced to do all the seducing.[6] Swept off his feet, the young party official proposed marriage in 1927, but he was deathly afraid to take his fiancée home to meet his father, who was sure to disapprove of his marriage to an older Protestant divorcée. Himmler dreaded the forthcoming battle. "I would rather clear a hall of a thousand communists single-handed," he confided to his brother Gebhard.[7] But the family storm soon passed, and the young couple married on July 3, 1928.

They immediately took up country life. As a trained agriculturalist, Himmler had high hopes of starting a poultry farm, and with proceeds from the sale of Marga's clinic, the couple bought a small parcel of land in Waldtrudering, on the western outskirts of Munich. Like many other young Nazis of the day, Himmler reveled in the idea of returning to the countryside and taking up the simple life of Germany's forefathers. "The yeoman on his own acre," he once observed rather piously, "is the backbone of the German people's strength and character."[8]

It seemed an innocent enough idea, but in Germany of the 1920s, the back-to-the-land movement was steeped in ultranationalist politics. In 1924, for example, right-wing activists had founded a society known as Artamanen, which was ostensibly devoted to sending groups of city kids to work on rural estates where they could hoe beets, dig potatoes, and bale hay while living in what resembled miniature military training camps. The founders of Artamanen enthused a great deal about the "joy in healthy work" and were fond of portraying their activities as an agricultural education for Germany's deprived urban youth. But their real goal was far more subversive: to school a future generation of ultranationalists.[9]

Artamanen's leaders and supporters dreamed of expelling all Polish migrant workers from the German countryside. Moreover, they envisioned their young members as a future fighting force of *Wehrbauern*, or "defense farmers," that could be mobilized to guard Germany's eastern border from Slavic attacks.[10] To prepare the students for this future work, they gave classes on the importance of racial purity and the Nordic race. They also harped at length on the corrupting influence of cosmopolitan urbanism and city-dwelling Jews, thereby incubating a new generation of virulent anti-Semites. They were very persuasive. By 1927, nearly 80 percent of the society's members had joined the Nazi party.[11]

Himmler had followed the society's activities closely since his early days as a Nazi party organizer. He was particularly impressed by one of Artamanen's notable supporters, Richard Walther Darré. Darré was an agriculturalist by training, a specialist in the genetics of animal breeding, and in 1929, he had published a book entitled *Das Bauerntum als Lebensquell der Nordischen Rasse,* or "Farming as a Source of Life for the Nordic Race," which received glowing reviews in Nazi circles.[12] Darré was convinced that the road to a stronger Germany lay in turning back the clock and returning to the country's ancient agricultural roots. The old German farming traditions, he observed, had refined and biologically honed the Nordic race. In times past, each farmer had picked just one son—the strongest, toughest, and most courageous—to inherit his land.[13] As a result, only the very fittest had farmed the fields over generations, creating a superior human bloodline, or so Darré liked to think.

Germany had eventually abandoned this tradition of land ownership, however, swayed by the fervor of the French Revolution and other foreign movements. Its rulers had written new inheritance laws that permitted estate owners to divide their land equitably among all their heirs.[14] This, Darré claimed, imperiled the Nordic race, undoing all the good of generations of selective breeding. If Germany wished to grow strong once again, he claimed, it had to return to its agricultural traditions. And it had to take serious measures to restore racial purity to the Nordic race. Darré, the animal breeder, did not shy away from spelling out these measures. Germany, he explained, had to exterminate the sick and impure in human society.[15] And it had to encourage members of the Nordic race to use scientific knowledge to select their mates and produce the best human stock.[16]

Himmler found these ideas immensely appealing. His own plan to return back to the land, however, was a dismal failure. The hens on his poultry farm produced few eggs, creating financial problems for the young couple. As Marga confided in one letter to her husband in 1929, "I worry so much about what we're going to live on and how we're going to save for Whitsun [the seventh Sunday after Easter]. Something's always going wrong."[17] The truth was that Himmler no longer cared much for mucking about on the farm himself. Instead he dreamed of applying Darré's ideas to others, particularly in the breeding of SS men.

IN THE EARLY 1920s, the racial writer Hans F. K. Günther had attempted
to estimate just how many Germans belonged to the Nordic race, as he
had defined it. He pored over anthropological studies on height and head
shape, hair color and skin hue of his fellow Germans, and published the results of this study in 1925. In Günther's opinion, just 50 to 55 percent of
Germans possessed any trace of Nordic blood, a rather bleak situation in
his view.[18] In later years, he refined this statistic further. Only 6 to 8 percent
of Germans, he claimed, were Nordic purebreds, while another 45 to 50
percent traced part of their lineage to Nordic ancestors. The remaining
Germans were members of inferior races.[19]

By the end of 1931, the SS had swollen in size, with ten thousand
members and stacks of new applications flooding in daily. To Himmler, the
time seemed ripe to begin transforming the force into a racial elite by ac-
cepting only young males of Nordic blood. To select these individuals in a
way that appeared scientific, Himmler created a new bureaucracy in the
SS. This came to be known as the Rasse- und Siedlungshauptamt der SS,
the Race and Settlement Office of the SS, or simply RuSHA. The head of
the new department was none other than Richard Walther Darré.

RuSHA advisors developed a racial grading system and approached
their work, as Himmler later noted, "like a nursery gardener trying to reproduce a good old strain which has been adulterated and debased; we
started from the principles of plant selection and then proceeded quite
unashamedly to weed out the men whom we did not think we could use
for the buildup of the SS."[20] The examiners required applicants to take a
medical examination and submit both a detailed genealogy chart and a set
of photographs of themselves. In the SS offices in Munich, the examiners
pored over the photographs, searching for supposed Nordic traits—long
head, narrow face, flat forehead, narrow nose, angular chin, thin lips, tall
slender body, blue eyes, fair hair. They rated the bodies of the applicants on
a scale of one to nine, then graded them on a five-point scale from "pure
Nordic" to "suspected non-European blood components."[21] They also
scanned through the applicants' family medical histories, searching for
congenital illness. Finally they arrived at a decision. A green card meant
"SS suited"; red marked rejection.[22]

The examiners were well aware that the Reichsführer-SS was not exactly green-card material, with his small, pigeon-chested body, round face,

sallow skin, dark hair, and receding chin. Indeed, one racial advisor noted with detachment after the war that Himmler was "an unassimilated half-breed and unfit for the SS."[23] But very few people dared say such things to their leader's face. At a social event one evening, however, the wife of a high-ranking SS officer, Dr. Werner Best, broached the problem with Himmler. She observed that the Nazi party would instantly lose its entire leadership—"The Führer, you Herr Himmler, Dr. Goebbels . . ."—if the principles of racial selection were strictly applied.[24] Himmler was not the least disturbed by this remark, however. He brushed off the criticism, replying that while he did not appear Nordic, he certainly possessed a Nordic brain.

Those accepted into the SS were encouraged to think of themselves as a new genetic aristocracy. While most Germans of the day traveled on the country's system of urban trains, for example, a fleet of drivers in private cars chauffeured SS officers around to their appointments. And Himmler made certain that SS men looked elegant. The German firm Hugo Boss supplied SS uniforms.[25] And in contrast to the scruffy brown uniforms of the SA, Himmler's men were decked out impressively in black with silver collar flashes. On their hats, they wore a silver death's-head, an ominous touch that supposedly symbolized "duty until death."[26] Such sartorial splendor clearly served a dual purpose. It intimidated victims and it added to the men's sexual appeal, boosting the chances of "success with the girls," as Himmler once remarked to a potential recruit.[27]

Himmler, after all, was particularly keen on making his men as attractive as possible to women, but like any careful breeder, he did not want his prize stock to mate with just any partner. Potential wives had to undergo racial screening themselves after December 21, 1931, submitting medical reports, genealogy charts, and photographs to RuSHA examiners.[28] If the examiners found fault with the racial quality of a woman, Himmler would deny his permission for the couple to marry. Only in this way, Himmler believed, could the SS breed a pure Nordic nobility: Germany's future depended on this. "Should we succeed in establishing this Nordic race again from and around Germany," he later observed in a speech to SS leaders, "and inducing them to become farmers and from this seedbed produce a race of 200 million, then the world will belong to us."[29]

IT WAS ONE thing to breed a handsome-looking new master race. However, it was quite another to teach this valuable Aryan stock to think like the masters of old and to wean them away from the softness and depravity of modern German cities. Himmler dreamed of turning back the clock, transforming Germany from a heavily industrialized nation into a largely agricultural one. In particular, he yearned to settle the SS back on the land in farm colonies. With each passing generation, however, Germans had forgotten more and more of the old peasant ways that had once made them strong, and few historical records remained of the ancient principles, ideas, and beliefs that once illuminated the lives of Germany's ancient forefathers.

Still, Himmler was not without hope. Ever since he was a boy combing musty ruins and grassy battlefields for old coins and broken pots with his father, he had seen how archaeology and other related sciences could shed light on vanished time. If SS researchers could recapture the minute details of primeval Aryan life, if they could peer into shrouded time to recover the lost thoughts and ideals and deeds of the ancient Nordic conquerors, then Himmler would have the exact blueprint he needed to indoctrinate the young men of the SS. Such men, he truly believed, would be unstoppable.

In the summer of 1932, Nazi candidates won 230 seats in the parliamentary election, the largest number held by any of the parties. Six months later, the aging German president Paul von Hindenburg appointed Hitler as the new chancellor, ushering the Nazi party into power, and on February 28, 1933, the Nazi leader managed to manipulate von Hindenburg into declaring a state of emergency in Germany, suspending all guarantees of civil liberties. It was a moment Himmler had been preparing for. The SS Sicherheitsdienst, or Security Service, widely known as the SD, had created a card index of the party's political opponents, and as the newly appointed police chief of Munich, Himmler began ordering the arrest of those he deemed the greatest threats: journalists, labor organizers, Jewish leaders, communists, socialists. He then saw to it that many were incarcerated in a new prison facility: Dachau concentration camp.[30]

Hitler greatly approved of these strongarm tactics and the plans for a network of concentration camps. The German leader needed someone ruthless, capable, and trustworthy to carry out the party's dirty work, and

Himmler seemed the most logical candidate. The young Bavarian was bright, well organized, and most important of all, he worshipped Hitler. On occasion, he had described the German leader to others as "the greatest brain of all time."[31] Hitler, as it happened, reveled in such adulation. So over the next thirteen months, he permitted the young SS chief to seize control of the political police of all German states. These Himmler ruled from Berlin.

Tucked away in his new headquarters, surrounded by tall blond men in black and silver and doted upon by a pool of secretaries with plaited hair and soft Bavarian accents, Himmler spent his days poring over reports, fretting over security arrangements for Hitler, and presiding over the minute details of the rapidly burgeoning system of concentration camps. But despite all the endless paperwork and phone calls and meetings, Himmler took time to draw up plans for educating and indoctrinating the new SS elite in the Nazi version of history. He called these plans the "education offensive" and put RuSHA, which was responsible for all ideological questions in the SS, in charge.[32]

The resulting offensive incorporated Himmler's pet theories on the German past, which closely mirrored those of Hitler, except in one key area. In speeches and private remarks, Hitler tended to avoid all mention of the early Germanic tribes. The ancient Romans had considered these tribes barbarians, and Hitler himself regarded them as an embarrassment. Indeed, he had developed a rather bizarre theory to account for their simple style of life in the northern European forests. He blamed the climate. The Aryan spirit, he claimed, required a sunnier land in which to flourish. Northern Europe was simply too foul and damp for the early germination of the Aryan genius.[33]

In Hitler's view, the Aryans began to achieve their full potential only when they reached the Mediterranean. "It was in Greece and Italy," he explained loftily to his dinner guests one night, "that the Germanic spirit found the first terrain favorable to its blossoming."[34] And this bizarre conviction deeply colored Hitler's perceptions of archaeology. He was fascinated by classical Rome and Greece and bored to tears by digs at home. "People make a tremendous fuss about the excavations carried out in districts inhabited by our forebears of the pre-Christian era," he declared. "I am afraid that I cannot share their enthusiasm, for I cannot help remembering that, while our ancestors were making these vessels of stone and clay, over which our archaeologists rave, the Greeks had already built an Acropolis."[35]

Himmler was well aware of these views, but he did not share them. He was besotted by old accounts of the fierceness and valor of the Germanic tribes. And he defined these tribes in a loose way to include any who had spoken one of the Germanic languages—Danish, Swedish, Norwegian, Icelandic, Faeroese, Dutch, Flemish, Frisian, Gothic, and, of course, English.[36] He wanted SS educators to concentrate their energies on these tribes. He could not ignore the "Nordic" civilizations of Greece and Rome that Hitler so loved, but it was his own interpretation of the past that became the nucleus of his SS "education offensive."

The frontline troops in the battle were the *Schulungsleiter,* or school leaders—young, racially acceptable men between the ages of twenty-four and forty-five. To ensure that these educators received a thorough grounding themselves in the party line, RuSHA sent them to a political boot camp. Graduates were then assigned to an SS company, where they taught weekly classes and continually impressed upon the men the Nazi point of view.[37] The school leaders posted photographs of Nordic-looking men and women in their offices, and on direct orders from Darré, they gave particular emphasis to "German early history" in their work.[38]

The classes and conversations were only the first salvo, however. The Nazi version of the past had to be underscored by something tangible and real. So RuSHA officials dispatched junior SS officers to Berlin's famous Museum for Pre and Early History, which lay next door to Himmler's headquarters. The museum possessed a rich collection of ancient German weapons, ceramics, and jewels, as well as the famous gold that Heinrich Schliemann unearthed at Troy, and museum officials took very good care of the SS officers. On at least one occasion, director Wilhelm Unverzagt gave a personal lecture to the officers, then led a guided tour of the exhibits.[39] RuSHA officials expected the school leaders to arrange similar outings for the enlisted men. One SS teacher in Munich, for example, planned a series of such events for February 1937—"a celebration hour (theme: Germany), a museum visit and a hike to an archaeological excavation with a comradely evening."[40]

To assist the school leaders, RuSHA developed elaborate slide shows on the origins of the Nordic race and the ingenuity of primeval Germanic farmers and warriors, complete with illustrated booklets apparently edited by Himmler himself.[41] Each month, the department sent school leaders *SS-Leitheft*, a training guide whose articles were to be read aloud at company meetings on Himmler's express orders.[42] *SS-Leitheft* was one of the more bizarre creations of the SS. It blended dime-store pulp, hard-core political doctrine, photos of pretty young blond Fräulein in revealing swimsuits and low-cut dirndls, quotations from *Mein Kampf,* and a kind of Dear Abby column offering tips to SS men for picking suitably Aryan brides.[43] Sprinkled heavily among these features were historical articles on a vast range of subjects—the migrations of the early Germanic tribes, the ancient sayings of the Aryans, the sagas of the Norse bards, the wooden ships of the Vikings, the ancient sun worship of the Aryans.

The articles were short and pithy, and there was no missing the authors' intent. "The more we know our ancestors," noted one writer brightly, "the more proudly we can call on them. . . ."[44]

FOR HIS SENIOR officers, Himmler planned a more exclusive education. In 1933, soon after the Nazi seizure of power, Himmler began drawing up plans for a new "Nordic academy," a cross between a monastic retreat and a finishing school for the upper echelons of the SS.[45] He intended, as one newspaper reporter pointed out, to create a place "of quiet contemplation and intellectual stimulation for the leading accountable men of the Fatherland."[46] Himmler's academy would not only be physically impressive, but it would also be a rather magical place—an old castle intimately linked with a potent German legend or myth. But where to find such a spot?

To narrow down the range of possibilities, he called upon the assistance of a new confidant. This was sixty-seven-year-old Karl-Maria Wiligut, a grizzled former colonel in the Austrian Imperial Army. Wiligut was something of a mystic in ultranationalist circles. He traced his pedigree back to the Norse god Thor and claimed as an ancestor Arminius, a German tribal chief who slaughtered three Roman legions near Teutoberg Forest in A.D. 9. His family, he sometimes explained, had guarded the sacred knowledge of the old German tribes for millennia. "My grandfather K. Wiligut taught me the ancient rune studies of our family," he noted on one official SS record. "My father entrusted our family history to me when I turned twenty-four."[47]

Wiligut's real biography was somewhat less colorful. Born in Vienna in 1866, the son of a mentally deranged army officer, he had seen active service on both the Russian and the Italian fronts during the First World War.[48] When peace was negotiated, he joined one of the right-wing paramilitary organizations in Austria. By then, he had become a violent, unpredictable alcoholic.[49] He carried a loaded revolver about with him, and frequently beat and threatened to kill his wife.[50] His behavior toward their two young daughters was equally worrying. His sister-in-law confided to a local government official that "he kisses the children so often to the point of senselessness that one must consider sexual undertones. As a result his wife has locked the children's doors."[51]

Eventually, Wiligut's wife mustered her courage and committed her husband to a mental hospital in Salzburg. Physicians there described his condition as a "psychosis" and declared that he was "not in the position to care for himself."[52] Over the next two years at the hospital, he lived up to their assessment. He boasted that he alone had prevented a communist takeover in Germany after the war, and informed his fellow inmates that the Ku Klux Klan was preparing to rescue him from the asylum. Most notably, however, he obsessively pocketed pebbles and stones from a nearby gravel pit. He washed and polished them, arranging nearly one thousand on virtually every surface in his hospital room. "He sees round, flat pebbles as amulets and parts of grave sites," observed one physician in his medical history. "A longish stone is a phallus, another stone is part of a throne seat, and another still is an eagle with a snake, symbolic of the fight between light and darkness, and, in a figurative sense, between life and death."[53] Wiligut boasted to his fellow inmates that he could sell the collection for 60,000 marks.

The hospital released him in 1927 under the care of a guardian. By then, Wiligut was eager to share his revelations with the rest of the world. Over the next six years, he became a savant to the more gullible ultranationalists in Germany. With a new name, Weisthor, meaning Wise Thor, in reference to the Norse god, he claimed to channel the ancient knowledge of his forebears, sometimes by entering seizurelike trances, on other occasions by reciting a collection of primeval sayings he claimed to have received from his ancestors. His devoted followers treasured every word. They regarded him as a living encyclopedia of the sacred traditions of the Germanic tribes as far back as creation.[54]

Himmler met Wiligut for the first time in 1933 at a conference of the Nordic Society. He was greatly impressed by the old soldier and at some point he asked for Wiligut's help in locating a suitable castle for the new Nordic academy. Himmler had read a popular romantic poem entitled "At the Birch Tree," which was based on an old German prophecy of a mighty battle that would one day pit the West against East, turning the Rhine red with blood.[55] The legendary battleground sounded to Himmler like an ideal spot for an SS stronghold. But the prophecy supplied few clues to its location. Wiligut promised his assistance.

He arrived one day in Himmler's office with a stack of old books and large manuscripts bound in pig leather, their pages yellowed and grease-stained.[56] By combing such sources, or perhaps by reciting the sayings taught by his father, he had discovered, he announced, that the prophesied battle would take place in the twentieth century. Moreover he had pinpointed its location. Himmler was delighted by this news, proudly announcing to his chief of staff, "We know, we have learned now where the battle will be—in Westphalia and, more precisely still, near the old road of the German heroes that leads from Paderborn via Soest in a westerly direction."[57] Conveniently, this lay close to the old battlefield where the Germanic chieftain Arminius and his troops had slaughtered invading Roman legions—sacred ground for many German ultranationalists.[58] So Himmler set out to find to find an old fortress in the region suitable for conversion into an SS academy.

In early November 1933, Himmler paid a visit to Wewelsburg, a seventeenth-century stone keep situated not far from Paderborn. The castle's heavy, brooding walls loomed over the rolling green Westphalian countryside, a picturesque land of little ravines and forests, small dairy farms and old stone houses. Himmler liked what he saw of the local farmers and their families. He believed them to be blessed with much "original German" blood.[59] He also admired the architecture of the old castle: he considered himself something of a connoisseur in such matters, and he noticed that Wewelsburg possessed an unusual north-south orientation and a triangular-shaped footprint that he deemed a rarity.[60] He immediately decided to lease the castle for the SS. Before long, ominous notices appeared on the sides of buildings in the sleepy little village that lay next to the fortress. They read: "Jew, you have been recognized."[61]

Impressed by his new advisor, Himmler brought Wiligut into the SS.

He promoted him first to the rank of SS-Standartenführer and later to SS-Brigadeführer. He also gave him an office in RuSHA and a private villa in Berlin. Wiligut traveled back and forth to Wewelsburg to offer his unique brand of channeling and ancestral wisdom to Hermann Bartels, the architect assigned to transform the stately old castle into a sprawling new SS complex.[62] Bartels had his work cut out for him. Himmler's plans for the complex kept evolving into something ever grander and more magnificent. So they required a succession of designs—each more complex than the last. To see them through, Himmler eventually ordered the construction of a special concentration camp nearby to supply sufficient quantities of slave labor.[63]

As the construction proceeded, runic designs and strange bits of old lore began to pop up all over the castle. One of the chambers, for example, was renovated and renamed the Grail Room. The castle's staff outfitted it with a large rock crystal displayed on a wooden pedestal and illuminated from below with electrical lights.[64] This strange display was intended to represent the Holy Grail. One of Wiligut's associates, Otto Rahn, had studied medieval legends of the grail. He believed these stories to be of Aryan origin and concluded that the grail was a magical "stone of light" that had fallen from the diadem of an ancient sun god. All those who looked upon the stone would live forever.[65]

Among the other renovations, Himmler wanted a private museum at Wewelsburg—a kind of ancestor room for his senior officers. He found an eager young SS archaeologist and artist, Wilhelm Jordan, to take charge.[66] Jordan was a tall, dark-blond beanpole, a former SA man who swiftly fell into the spirit of Wewelsburg. He conducted a series of excavations in the countryside around the castle, digging ancient grave mounds and old village sites and collecting hundreds of artifacts, from Germanic urns and human skulls to iron knives, Roman coins, and Bronze Age needles. In his spare time, Jordan prospected local cliffs and construction sites for fossils of ancient sea life, designed dioramas, and pieced together ancient pots from shattered shards. To round out the Wewelsburg collection, he purchased some spectacular finds, including the complete fossil of an ancient marine reptile, a three-meter-long ichthyosaur.

The museum was located in the castle's western wing, down a few stone stairs from the courtyard. Upon entering, visitors were greeted by the sight of sleek steel-and-glass showcases and cabinets brimming with ancient Germanic artifacts. Several splendid dioramas decorated the walls. Perhaps the most important of these portrayed the simple bucolic life of a Germanic tribal family during the Roman era. For this, Jordan lovingly crafted a miniature German farmhouse complete with straw roof, similar in style to one excavated "in Oerlinghausen near Detmold."[67] For the farmyard, he created miniature models of men and women in rustic clothes and a small furnace for smelting iron ore. In the green fields beyond, tiny horses and sheep grazed contentedly on the grasses.

The diorama was clearly intended as a glimpse of paradise lost for those who visited the museum, an idyll of ancient Germanic simplicity. And in a "Nordic academy" devoted to contemplation and thought, it served as a template for an Aryan future that Himmler fondly envisioned for the Reich.

IF THE "EDUCATION OFFENSIVE" was truly to succeed, however, it needed a well-stocked arsenal of new ideological weapons. To re-create the lost world of the Nordic race, to capture primordial ways of thinking and beliefs, the SS had to push the boundaries of prehistoric research. Symbol experts had to find and decode the earliest written messages of the Nordic race. Experts in the ancient sagas and legends had to reconstruct the lost history and religion of the Aryans. Musicologists had to restore their music. Archaeologists had to dig their tombs and study their ancient treasures. Botanists had to rebreed their ancient seeds. In other words, the SS required an entire institute of elite researchers dedicated to reconstructing the lost golden age of the ancient Nordic past.

On July 1, 1935, in a spacious, sunlit office in the SS headquarters in Berlin, Himmler convened a meeting to discuss the new organization. Around the table sat five racial experts representing the interests of Darré, who had recently added the post of Reich minister for nutrition and agriculture to his duties as the head of RuSHA. Darré shared Himmler's enthusiasm for the new research institute. And the two men had agreed on inviting another party to the table, an elfin-looking scholar who spoke with a strong Dutch accent—Dr. Herman Wirth, one of the most famous prehistorians in all of Germany.

Those in attendance discussed at considerable length the structure of the organization, and at the end of their deliberations they agreed to found a brain trust that would, in effect, form a new department in RuSHA.[68] This organization was to be called "Deutsches Ahnenerbe" Studiengesellschaft für Geistesurgeschichte, or "German Ancestral Heritage," the Society for the Study of the History of Primeval Ideas—a hopeless mouthful that soon came to be shortened to the Ahnenerbe. Wirth assumed the position of president. Himmler, however, retained much power for himself, taking the title of superintendent and assuming control of the organization's board of trustees. The formally stated goal of the Ahnenerbe was "to promote the science of ancient intellectual history."[69]

5 · MAKING STONES SPEAK

A VISITOR ON OFFICIAL business in the late fall of 1935 would have found the new offices of the Ahnenerbe tucked away on one of the oldest streets in the most historic part of all Berlin. Brüderstrasse, as the street was known, dated back to the thirteenth century. It curved along a small island in the Spree, through a neighborhood that reminded some Berliners of old Amsterdam—what with its narrow four-story houses lining the waterfront and its working boats plying the river.[1] But Brüderstrasse was a particularly German street. At its northern end rose a stately old castle, the Schloss, where the electors of Brandenburg and the kings of Prussia had ruled in gilded majesty for nearly five hundred years.[2] And along Brüderstrasse's two-block length stood several famous homes, including Nicolai House, where publisher Friedrich Nicolai entertained Georg Hegel and other leading German intellectuals in the late eighteenth century.

The Ahnenerbe offices lay in the middle of Brüderstrasse, at number 29/30. They occupied two floors in a corner building owned by a Berlin department store magnate, Rudolf Herzog.[3] The floors were drafty and cold in the winter and the electrical wiring left a good deal to be desired. But the rent was modest and the space was large enough to accommodate the Ahnenerbe's staff of seven.[4] Inside, the offices were furnished with gleaming black Underwood typewriters and rows of filing cabinets, but there the similarity to most other offices in Berlin ended. Along the walls hung dozens of large framed photographs of strange symbols carved upon wooden staves. In the atelier, staff members consulted the plaster casts of ancient rune stones.

Nearly everyone who worked for the Ahnenerbe in late 1935 was involved, one way or another, in what the new Ahnenerbe president Herman Wirth fondly called script and symbol studies. At his behest, they took photographs of gravestones and church doors and the stenciled paintings that ornamented aging farmhouses. They collected baking molds, ceramic candleholders, and decorative wreaths, anything, in fact, that seemed to hold some ancient secret symbol or meaning. They wrote to museums scattered across Europe and the United States, offering to buy replicas of bone spoons, stone crosses, Mesopotamian cylinder seals, and carved boulders.[5] A young Finnish scholar, Yrjö von Grönhagen, who was studying ancient religious rituals for the Ahnenerbe, was pressed into gathering examples of traditional Finnish wooden calendars, carved with symbols.[6] All this immense effort was ostensibly devoted to preparing lectures and exhibitions for SS men and the staff of Darré's Ministry of Nutrition and Agriculture.[7] But in Wirth's mind, it served just one purpose: to further popularize his own peculiar ideas about the past.[8]

Wirth had an amazing effect on people. He was affable, energetic, enormously well read, and regarded by all as charming. He could quote long Sanskrit hymns from memory and discourse for hours about the meaning of spirit masks among the Yupik in Alaska or the inscriptions on remote dolmens in Ireland, and do so in such an engaging way that his listeners were convinced they had just glimpsed the heart of a mystery.[9] To his followers, Wirth seemed to combine enormous erudition with uncanny powers of perception. "Herman Wirth," noted one admirer, "is the artist who makes stones speak, he is the psychologist who feels hearts beating across the millenniums, he is the *Homo religiosus* who senses the deep eternal bonds of mind and spirit running through the whole of the human race."[10] Even his enemies could hardly find it in their hearts to dislike him.[11]

At the age of fifty in 1935, Wirth was at the peak of his fame. He was one of the most controversial prehistorians in all of Germany and he believed he was on the verge of a momentous discovery. He was convinced he had found an ancient holy script invented by a lost Nordic civilization in the North Atlantic many thousands of years ago. It was, he claimed, the world's earliest writing. He also believed that he could decipher this mysterious script, thereby unlocking the mysteries of ancient Aryan religion.

And he was convinced that if he could recover the meaning of these primeval sacred texts, he could reawaken Germany from its unhappy slumber and restore it to greatness again.

Entering the offices of Herman Wirth in the late fall of 1935 was to enter a world of chaos and confusion, a world where reason and rational argument had given way to wishful thinking, a world where scholarship had been abandoned for fantasy and dreams of glory. It was a realm that Wirth had inhabited for most of his scholarly career.

IN ALL LIKELIHOOD, Himmler first met Wirth sometime in the late spring or early summer of 1934, at the home of Johann and Gesine von Leers.[12] The von Leers were prominent in Nazi circles and took a keen interest in all things Aryan. Johann worked as a propagandist for the Nazi party, penning vicious anti-Semitic attacks in books with titles like *Jews Look at You,* as well as more theoretical writings on the Nordic race and German agriculture.[13] Gesine was inclined more toward the occult. She believed she was the reincarnation of an ancient German priestess, and frequently donned Bronze Age–style jewelry for evenings she hosted among similarly minded friends.[14]

Gesine had known and admired Wirth for years, and she had invited him to her home to discuss the research of Karl-Maria Wiligut, another of her personal mentors. She had also arranged for Himmler and Darré to attend. At first, Wirth stubbornly declined to go. He thought Wiligut an outright charlatan, and he seems to have feared him as a rival for the attention of Himmler and Darré.[15] But Darré insisted that Wirth come. He called the scholar up personally, offering to send around his private car, and so mollified was Wirth by this mark of favor that he agreed to put in an appearance.

Wirth was in his element on such occasions, and he made a fine impression on Himmler. He was a diminutive man, standing just five feet four inches tall, but he possessed thick locks of blond hair, sparkling blue eyes, and a finely chiseled handsomeness, all of which helped him considerably in passing the racial test that Himmler silently administered to everyone he met.[16] Wirth, moreover, was a polished showman who knew how to read and take command of an audience. It would not have taken him long to size up Himmler's interests.

The Reichsführer-SS, like many other prominent Nazis, had begun casting about privately for a system of spiritual beliefs that could eventually take the place of Catholicism and Protestantism in the Reich.[17] He found Christian doctrine immensely troublesome: It traced its origins to the deserts of the Middle East, rather than to forests of northern Europe. It presented the tribes of Israel, not the tribes of Germania, as the chosen people. It described Christ as a Jew. And it advocated charity, compassion for the weak, the brotherhood of men, and the equality of all in the eyes of God. All this, Himmler found abhorrent. He had attended Mass faithfully from childhood until his early twenties, but he had lost his faith at about the same time that he joined the Nazi party.[18] And since then, he had openly avowed Nazism as his new creed. "Our business," he explained to the readers of the party newspaper, *Völkischer Beobachter,* "is to spread the knowledge of race in the life of our *Völk* and to impress it upon the hearts and heads of all, down to the very youngest, as our German gospel."[19]

But Nazism needed a state religion. And it needed a god, or perhaps several gods, as well as suitable rituals to take the place of Mass and other Christian services. Himmler was strongly inclined to borrow these rituals, and he saw no better place to look for them than in the history of Germany's ancestors—the Germanic tribes and their Aryan forebears. He believed that the old pagan gods would be capable of weaning SS men from their Christian faiths. Indeed, as he later confided to his personal physician, he intended to make every *SS-Mann* drop his traditional church affiliation. "After the war," he explained, "the old Germanic gods will be restored."[20]

The trouble was that scholars knew very little about these old gods. Germany's tribes had left few written records of their sacred beliefs and practices before they converted to Christianity, and their Roman neighbors, who possessed a much earlier written tradition, took scant interest in the old Germanic religion. To recover its beliefs, Himmler realized that he needed to find new sources of information. Herman Wirth claimed to possess a rich trove of new details. The scholar believed he had discovered ancient sacred texts of the Nordic race, and through science, he intended to decipher them.

Wirth, moreover, was happy to yoke his research to the driving ambitions of the Nazis. Science, he believed, served a higher master than objectivity or truth. "The time is now past," he once explained to a German audience, "when science believed its task was to search for the truth, such as it is. Now the task of science is to proceed with its prophecy, to awaken. Like the morning dawn, it will light a new day."[21]

WIRTH HAD GRAVITATED naturally toward Nazi politics. Born in the Netherlands on May 6, 1885, he came from a background remarkably similar to that of Himmler. Wirth's German father was a schoolteacher and university lecturer in Utrecht, and he took great pride in his Teutonic ancestry.[22] At home, he drummed a love of German culture, history, and language into his children: Herman proved a particularly adept student. He pursued his father's interests, taking Germanic studies, philology, history, and music theory at the university in Utrecht, before moving to the University of Leipzig. There Wirth absorbed the extreme German nationalism that permeated many of the classrooms.

He came to loathe what he saw as the corrosive influence of cosmopolitan urbanism. He believed such thinking was responsible for the destruction of age-old German and Dutch folk culture—a priceless repository, in his opinion, of ancient Nordic thought.[23] In his Ph.D. dissertation, he lamented the disappearance of traditional folk songs in the northern Netherlands—portals, he believed, into the very soul of the ancient Nordic race.[24]

In 1909, Wirth landed a position as a lecturer in Dutch philology at the University of Berlin, but he craved a larger audience for his emerging ideas on folk music and the Nordic race. He turned performer, combining concerts of early Dutch music with lectures and magic-lantern shows. For these events, he dressed the part of a medieval musician, donning a flowing green velvet jacket. His fetching young lute player, Margarethe Schmitt, the daughter of a prominent German landscape painter, turned up at concerts in the robes of a fairy-tale princess, complete with a golden band in her hair.[25] She worshipped Wirth and complained loudly when anyone dared to criticize him, or refused to recognize him, as one friend observed, "as the coming messiah."[26] The two eventually married.

When the First World War began in August 1914, Wirth left his university job and enlisted in the army of his adopted homeland. Rather than sending the Dutch scholar to the front, however, German authorities dispatched him to a civilian post in occupied Belgium. While working in Ghent, Wirth became enamored of the Flemish independence movement, raising funds in Germany on its behalf. When the full extent of his political activism came to light, he was relieved of his post.[27]

Disillusioned by the war, he concluded in the early 1920s that all Western civilization was teetering on the verge of collapse.[28] The only hope for the future, as he saw it, lay in recovering the truths of the past. And it was at this time, while traveling through the northern Dutch province of Friesland, that he saw something that changed his life. On the gables of many old, neatly painted Frisian farmhouses, inhabitants displayed small wooden folk sculptures consisting of an assortment of carved shapes—curlicues, crescents, crosses, shamrocks, swan necks, stars, hearts, diamonds, and crowns.[29] To most passersby, they seemed little more than folk art, but Wirth was forever looking with reckless enthusiasm for hidden, secret meanings in the world around him. During his studies of Dutch folk music, he had glimpsed what he thought was the intellectual spirit of the ancient Nordic race. Now in the Frisian countryside, he thought he saw further remnants of a primeval Nordic civilization.

This sparked a great leap of logic. He concluded that the farmers' wooden carvings of hearts and diamonds and crowns were symbols from an ancient Aryan system of writing—a type of northern hieroglyphs.[30] Their significance, he concluded, had been all but lost to time. Not even the folk artists of the Frisian countryside realized anymore what their carvings truly meant. More bizarrely still, he inferred that the different shapes were remnants of the earliest writing system in the world and were therefore the mother of all scripts, from the Egyptian hieroglyphs to the Phoenician alphabet.[31]

These ideas flouted Western science and scholarship. For more than a century, linguists had examined ancient stone inscriptions from around the world, deciphering scripts and delving into the origins of writing. By the 1930s, they had made substantial progress, tracing the earliest appearance of the written word to one of two regions in the world: ancient Egypt or Mesopotamia. Inscriptions in both places, noted the editors of *Encyclopaedia Britannica* in 1938, went back to "an extremely early date; it is at present uncertain which is the earlier, but both show, before 3500 B.C. and possibly much earlier, a complete, organized system of writing which implies many centuries of development behind it."[32]

By comparison, the inhabitants of northern Europe were late bloomers when it came to writing. They had developed letters for a system of writing known today as the runes.[33] The runes were a type of script that blended letters of the Roman alphabet with several new inventions, and most experts in the 1930s agreed that the earliest runes had appeared around A.D. 250.[34] None of this, however, sat very well with Wirth and other nationalist scholars. If Nordic men and women were truly superior to all others, then it stood to reason that they had founded the world's first civilization in their ancient boreal homeland. It also stood to reason that they had invented the art of writing there. And since Nordic men and women were, in Wirth's opinion, the world's "intellectual sourdough starter," they had to have carried their writing system with them on their great migrations to Africa and Asia, thereby passing on the idea of writing to others, who devised cuneiform scripts, hieroglyphs, and the alphabet.[35]

Researchers had failed, however, to detect any proof of a primeval Nordic civilization in the north, one predating ancient Egypt and Mesopotamia. They had uncovered no ancient cities, no pyramids, no ziggurats of the north. "What are the facts?" asked prominent British scientist Julian Huxley in his famous book, *We Europeans*, in 1935. "The fundamental discoveries on which civilization is built are the art of writing, agriculture, the wheel and building in stone. All these appear to have originated in the near East, among peoples who by no stretch of imagination could be called Nordic."[36]

So, in the absence of evidence, Wirth let his vivid imagination run loose. Perhaps, he imagined, the Nordic race had evolved in an arctic homeland some two million years ago, then founded a civilization on a lost continent in the North Atlantic.[37] Perhaps strong Nordic women, who possessed "all-seeing capabilities," had ruled this primeval empire as a matriarchy for thousands of years, until a great cataclysm drowned their lands, sending survivors fleeing to northern Europe and North America.[38] And wasn't it just possible that this splendid Nordic homeland was the lost Atlantis of legend?[39]

WIRTH REALIZED THAT other scholars would be loath to toss away decades of scholarship in favor of ideas that rudely challenged the entire foundations of world history. He needed evidence to anchor his ideas to earth. To pursue the matter, he moved his young family—Margarethe and three children—to Marburg, which had a good library and a university willing to give him a teaching position. Three years later, in 1925, he took out a membership in the Nazi party, which he soon allowed to lapse, and began collecting a coterie of attentive young assistants—artistic, idealistic young men fascinated by Wirth's learned patter, his boundless enthusiasm, his knowledge of the ancient world, and his quest to revive ancient Nordic religion and culture.[40] They modeled themselves after him, dressing in short pants as he often did, and addressed him affectionately as dear father and Margarethe as dear mother.[41]

Wirth was in his element. He preached his own brand of clean living to his followers—he had given up meat and eschewed tobacco—and praised country living, far from the foul, polluting influences of the city.[42] A prophet of the past, he scorned universities and dreamed of founding a utopian community rather like the Artamanen, where young student scientists and farmers could live together and support themselves on their produce.[43] He frequently invited his young disciples to stay for dinner, and it was not unusual for fifteen people to crowd around the family table.[44] The meals could be strange and unpredictable. On one recorded occasion, Wirth's wife arrived at the table with a medieval-looking golden band in her hair and spent the entire meal in silence, while Wirth claimed to read her thoughts by means of telepathic communication. He also announced to his startled guests that she possessed clairvoyant abilities, seeing things that others could not.[45]

In private, Wirth rifled through the scientific literature, determined to prove that the Nordic race had evolved in the Arctic. He was delighted to learn, for example, that a Danish geological expedition had discovered fossil plants—most likely leaves of grapelike vines and oaklike trees—in Greenland's northern fjords. The finds were proof, he claimed, that the remotest parts of the Arctic had once been "a wonderful magical garden," eminently suitable for human evolution.[46] What he did not tell his readers, however, was that these plants had flourished in Greenland more than 50 million years before the first hominids walked the earth—a fact well known to contemporary geologists.[47]

He also pored over Plato's legend of Atlantis. He believed it to be an accurate description of the fate of his imagined Nordic empire in the North Atlantic, and he set about trying to prove this theory.[48] He combed through geological studies of the ocean floor in the North Atlantic, concluding confidently that the continent of Atlantis had occupied a region stretching all the way from Iceland to the Azores before it sank due to tectonic activity.[49] Only a few scraps of the ancient continent, he insisted, remained above water—namely the Canary Islands and Cape Verde.[50] But Wirth had little grasp of geological thinking. His ideas were based on antiquated, disproven theories scorned by contemporary geologists in the 1930s.[51]

All his talk of Atlantis and a lost Nordic civilization stirred a great deal of public interest, however. In the northern city of Bremen, Ludwig Roselius, the wealthy inventor of decaffeinated coffee, drew up plans for a stunning new building in honor of Wirth.[52] The edifice, dubbed Haus Atlantis, was to be part of a major urban development project known as the Böttcherstrasse. Blending modern expressionist architecture with themes from old Germanic mythology, Haus Atlantis would take visitors on a metaphorical journey of rebirth, from the dark blue depths of the ocean— where Atlantis languished—to the soft glow of the Sky Hall on the top floor. As part of the original design, Roselius created a lecture hall and exhibition space for Wirth.[53] He also displayed his own impressive collection of ancient artifacts—bronze swords, golden vessels, bronze musical instruments, fine jewelry, as well as casts of ancient rune stones.[54]

With the backing of such influential men, Wirth was propelled onto the national stage, much to the horror of many serious German scholars. His first major book in 1929—a huge rambling inchoate study of the origins of the Nordic race and their later migrations, all heavily larded with invented jargon—aroused intense criticism. "Only the feeling that the author has been taken by an almost holy insanity," observed Gero von Merhart, a prominent German prehistorian, "and the fact that he inspected a considerable mass of literature with unusual eagerness and diligence to support his delusions, which he considers science, restrain me from responding to this book . . . with rudeness."[55] Others were considerably less charitable. Taking firm aim at Wirth's risible powers of reasoning, another critic noted that the Dutch scholar "is unable to distinguish between probable, certain, possible and impossible."[56]

Outside Germany, discerning archaeologists were quick to pick up the extremist political tenor in Wirth's work. In Sweden, for example, archaeologist Nils Åberg published a warning about Wirth after attending one of his public performances in Germany. Wirth, he wrote, seemed harmless enough, but his visions of the murky past were intended to seduce unwit-

ting Germans into dangerous dreams of racial superiority and national ag-
grandizement, dreams that could fan the flames of war in Europe. As such,
his lectures were a thinly disguised call to action, in which Wirth "the magician, the Hitler of German scientists, captures his public."[57]

THIS WAS THE man that Himmler had placed in charge of the Ahnenerbe
in 1935 and made SS-Hauptsturmführer in RuSHA. And this was the man
who set about mounting, with the blessing and financing of the SS, the
brain trust's first major expedition abroad. Wirth's destination was the remote granite hills of Bohuslän in southwestern Sweden. Along the slopes
of these hills lay tens of thousands of ancient engravings. These, Wirth believed, were the sacred texts of the ancient Nordic race.

6. FINDING RELIGION

In 1936, BOHUSLÄN WAS a rugged land of small farms and fishing villages, isolated from the rest of Sweden by poor roads. Its residents were a stubborn, self-reliant people who possessed little that seemed to interest the outside world. The herring that had once flashed and roiled in the waters off Bohuslän each winter had departed some thirty years earlier, never to return.[1] And the quarries that once shipped Bohuslän stone halfway around the world—to Brazil and Argentina and South Africa—had fallen into idleness. Few builders, it seems, could afford fancy paving stones at the height of the Great Depression.

Even in its poverty, however, Bohuslän possessed one thing in abundance, one thing that still brought the occasional outsider to its doors. That was an astonishing wealth of primeval rock carvings. Northern Bohuslän counted more than five thousand rock-art panels teeming with images of the ancient world—ships, warriors, ploughmen, horsemen, acrobats, bulls, snakes, ducks, cranes, fish, trees, carts, axes, spears, battles, hunts, and rituals.[2] Their makers had engraved them in hard granite by pecking or grinding the rock surface to create furrowed lines. Some of their work was remarkably beautiful, expressive, and mysterious. Other figures were far rougher and cruder. Nearly all, however, dated to the Bronze Age, the era when bronze came into vogue as a material for weapons, tools, decorations, and jewelry. In southwestern Sweden, the Bronze Age began around 1800 B.C. and ended thirteen hundred years later.[3]

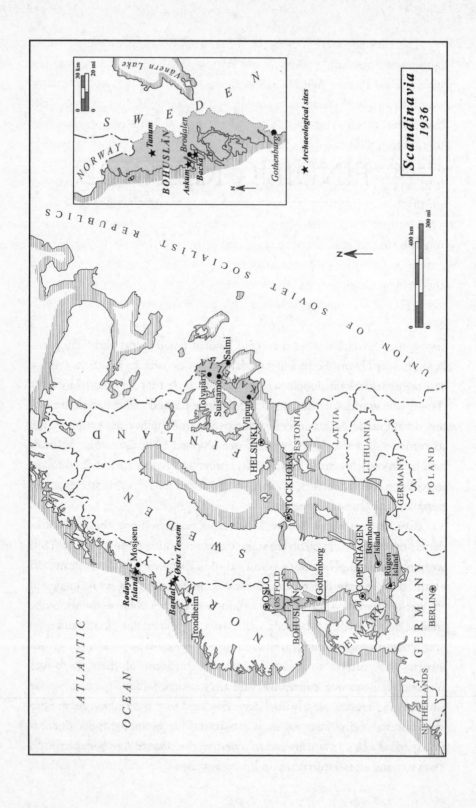

Scandinavia
1936

★ *Archaeological sites*

SWEDEN

NORWAY

BOHUSLÄN

Tanum ★

Brodalen

Askum ★ ★ Backa

Gothenburg

Vänern Lake

30 km
20 mi

N

UNION OF SOVIET SOCIALIST REPUBLICS

ATLANTIC

OCEAN

FINLAND

SWEDEN

NORWAY

Rødøya Island ★

Mosjøen

Østre Tessem ★

Bardal ★

Trondheim

OSLO

OSTFOLD

BOHUSLÄN

Gothenburg

DENMARK

Bornholm Island

COPENHAGEN

Rügen Island

GERMANY

NETHERLANDS

BERLIN

POLAND

GERMANY

LITHUANIA

LATVIA

ESTONIA

HELSINKI

STOCKHOLM

Tolvajärvi

Suistamo

Salmi

Vipuri

N

400 km
300 mi

The carvings were a point of pride among the farming and fishing families of Bohuslän.[4] For nearly two hundred years, Scandinavian scholars who studied Bronze Age art had beaten a path to the region, scrambling over hills and across fields to record the art and decipher its meaning. They had drawn careful sketches of the most important panels and published them in fine illustrated books, proving Bohuslän to be one of the world's great rock-art repositories.[5] No other part of northern Europe could touch it in terms of the sheer number of carvings or the immense range of motifs.[6]

Herman Wirth, who was captivated by ancient symbols wherever they might be found, read about the carvings of Bohuslän with enormous fascination. In his office on Brüderstrasse, he pored over the illustrations, with their enigmatic shapes and furrows. He was astonished and very, very excited. He felt certain he was looking at the script of an ancient Nordic holy text.

ON FEBRUARY 19, 1936, just a few weeks before Hitler violated the Treaty of Locarno and ordered German troops to cross into the demilitarized Rhineland, Wirth staged a private film screening at his home for Himmler and a small number of other guests.[7] As the projector clanked and sputtered, images of sunlit farm pastures and dark rock slabs whirred across a makeshift screen. Wirth, a newly minted SS-Untersturmführer, supplied a continuous running commentary. In the silvery light, he pointed out a succession of faint engravings outlined in chalk—circles intersected by a single vertical line, circles joined by a vertical line, and other equally cryptic markings. These, he announced proudly, were ideograms, symbols used to represent ideas and things, rather than sounds. They came, he claimed, from an old European script invented more than twelve thousand years ago by Nordic scribes inhabiting a lost North Atlantic homeland.

Wirth had filmed the carvings during a recent exploratory trip to Bohuslän. For nearly a month in the fall of 1935, he and a fellow SS officer, Wilhelm Kottenrodt, had bumped and scraped along the roads of Bohuslän, driving a shiny new Adlerwagen purchased by the Ahnenerbe.[8] They had roamed the countryside, sliding down ridges and trekking through shady forests of oak and elm, searching out the largest and most accessible rock-carving sites. Kottenrodt, who had trained as a sculptor, had assisted

Wirth in making plaster casts of selected engravings—the beginning of what Wirth grandly hoped would become the world's most important collection of primeval Nordic symbols and the nucleus of future museum exhibits on the superiority of the Nordic race.[9]

The autumn trip to Bohuslän had merely whetted Wirth's appetite, however. The scholar desperately wanted to return to Sweden, this time at the head of a major Ahnenerbe expedition. He was keen to broaden his studies, and put more men to work on the arduous task of collecting casts. For this, he needed the approval and financial assistance of the superintendent of the Ahnenerbe, Himmler.

As Himmler listened to the proposal, he considered the potential benefits to the SS and to his own personal ambitions. Much to his disappointment, Hitler had taken little interest in the Ahnenerbe.[10] In fact, the German Führer continued to complain about Himmler's passionate enthusiasm for northern European prehistory. "Why do we call the whole world's attention to the fact that we have no past?" he grumbled on one occasion to Albert Speer. "It's bad enough that the Romans were erecting great buildings when our forefathers were still living in mud huts; now Himmler is starting to dig up these villages of mud huts and enthusing over every potsherd and stone axe he finds."[11]

On this subject, however, Himmler was not to be denied.[12] Although he admired Hitler as a genius, he stubbornly believed that on the matter of the Germanic past, the Nazi leader had yet to see the light. Himmler sought avidly for new ways of illuminating the truth.[13] Almost certainly it occurred to him that a professional film production of Wirth announcing his discovery of the world's earliest writing system in Scandinavia might succeed where Himmler's other arguments had failed, sparking Hitler's imagination.

A film on Wirth's work, moreover, would be an invaluable weapon in the SS educational offensive, greatly bolstering Nazi claims of Aryan superiority. And the casts would become important assets in themselves. The Ahnenerbe could put them on public exhibit, and Himmler could use them himself as props in the speeches he gave to senior SS officers.[14] He could also present copies as gifts to Nazi party bosses. The previous fall, Wirth had made a cast of an ancient swastika—a symbol long associated with the Scandinavian god of thunder, Thor—in Bohuslän.[15] A replica of the ancient symbol would make a fine birthday gift for Hitler.[16]

Attractive as this sounded, however, Himmler had doubts about Wirth's ability to lead an expedition. He had discovered that the scholar was hopelessly profligate and wracked by massive debts. Wirth had run through a fortune over the years on his favorite expenses—secretaries, assistants, expensive books, private cars, palatial homes.[17] "In terms of money," observed one of the scholar's strongest supporters, "Wirth is as inexperienced as a seven-year-old: he can only spend it."[18] By the time the Ahnenerbe was founded in the summer of 1935, Wirth was tottering on the brink of bankruptcy or worse: a prison cell for debtors.[19]

When Himmler learned of this dismal state of affairs, he agreed to bail the scholar out. The SS chief wanted his officers to set an example in financial prudence, as in all other things: he did not want them to buy things for which they could not pay cash or to obtain goods by rent-to-own agreements. He most certainly did not want them beset by creditors.[20] So he arranged for two SS negotiators to settle Wirth's bills.[21] The two officers drew up a list of his principal creditors—princesses, merchants, industrialists, many of whom were prominent Nazi party supporters—and discovered debts totaling more than 84,000 reichsmarks, the equivalent of some $436,800 in today's dollars. The officers obtained the most favorable possible settlements, then arranged for the SS to pick up the tab.[22] But Himmler's largesse was not without strings. As a term of assistance, he seems to have demanded a pledge from Wirth guaranteeing that neither he nor his wife would ever again rack up debts.[23]

All this must have given Himmler, a man who liked making reichsmarks go a very long way, serious pause for thought. But a few months after the screening, he reached a decision. He ordered Wirth to organize and lead an official Ahnenerbe expedition to Scandinavia, departing in July 1936. The team would have a number of goals. It would study and make casts of ancient rock-art symbols in Sweden and along the western coast of Norway, from Stavanger to Trondheim.[24] It would also stop en route at local museums to photograph and make casts of old farm implements as well as household goods decorated with old patterns and designs.

IN THE AHNENERBE offices on Brüderstrasse, Wirth got down to work, hammering out plans for the expedition with the society's new managing director, Wolfram Sievers. At thirty-one, Sievers possessed an extraordinary talent for organizing complex projects and cutting through the tangles of red tape that ensnared so many ambitious schemes in the Nazi government. In the spring of 1936, Sievers, a tall, athletic man with a vulpine handsomeness that was attractive to women, was becoming indispensable to the operation of the Ahnenerbe.

The son of a church musician, Sievers had dropped out of high school as a teenager. He had resumed his studies on his own, concentrating on racial theory, German ethnology, German ethnic history, peasant customs, genetics, prehistory, and an examination of the stated enemies of the Nazi party, namely "Jews, Rome and Jesuits, Freemasons and Bolsheviks."[25] In 1929, Sievers joined the party, which welcomed him as a prime Nordic specimen. In 1932 he became Wirth's private secretary, and when Wirth moved to the Ahnenerbe, Sievers moved with him. He was accepted into the SS, and on his medical-examination form, a RuSHA official noted that "procreation is desirable."[26]

Working together, Wirth and Sievers estimated that the Scandinavian expedition would take two months and require a team of six. Wirth wanted to bring Kottenrodt, the SS sculptor who had accompanied him to Bohuslän the previous year. He also intended to recruit four other men for the expedition—two technical assistants to help make and transport the casts, a driver who could double as a photographer, and an experienced cameraman to shoot the SS documentary on the work, as well as two cultural films for the Ahnenerbe.[27]

Wirth drew up lists of necessary expedition gear, from workmen's clothing and rain boots to tents, steel-pipe beds, sleeping bags, folding chairs, and gas stoves. With the assistance of Bruno Galke, one of Himmler's most trusted aides, Sievers scrupulously budgeted the expedition costs down to the last pfennig, pegging the final tab at 12,590.76 reichsmarks, the equivalent of $65,000 today.[28] Galke then approached the Notgemeinschaft der Deutschen Wissenschaft, or the German Society for the Preservation and Promotion of Research, the most important German scientific funding agency of the day, for assistance.[29] The society had rejected Wirth's previous applications for funding. But in July 1934, Dr. Johannes Stark, a committed Nazi, had taken over the helm of the prestigious agency. Under his presidency, the foundation awarded Wirth 8,000 reichsmarks for the expedition.[30] Himmler made up the rest of the required funds from an SS account.[31]

By late July, Sievers had succeeded in solving nearly all the most pressing organizational problems. He had located a suitable cameraman for the expedition, thirty-three-year-old Helmut Bousset, an SS-Untersturmführer.[32] He had obtained an export visa to ship more than twenty tons of plaster of paris to Scandinavia and arranged for a small convoy of vehicles for the trip—most notably an Opel-Blitz truck from the SS.[33] And he had laid hands on the necessary foreign currency, no easy matter as Germany had quietly focused its economy on the re-armament of the Wehrmacht and dispensed with many of the exports that had once brought in foreign exchange.

But Wirth and his team still lacked something essential to the expedition: an official archaeological permit from Stockholm. Without this crucial piece of paper, they would be unable to conduct any research or make any casts of the Swedish rock art. The antiquities officials in Stockholm were in no hurry to issue the permit. They were furious at Wirth.[34] The previous summer, he had physically damaged some of the ancient engravings and neglected to clean others when he was done with them. At one site, an assistant he hired left several plaster casts still attached to the rock. This bizarre oversight had given grave offense.

Wirth swiftly penned a letter of apology, well aware that the entire expedition depended on making prompt amends.[35] But the Swedish officials dragged their heels on the question of the permit. Finally on July 29—halfway through the summer field season—they notified Wirth that they would issue a permit, but only for the work he wanted to do in northern Bohuslän. They did not want him ruining sites in lesser-known areas that had barely been recorded. Wirth had little choice but to accept their decision, but he stubbornly held out hope that he could still persuade the Swedish officials to relent.

Just before the team departed, Sievers received a list of detailed expedition rules drawn up by Himmler himself.[36] All participants, he instructed, were to conduct themselves in a comradely manner and leave behind all personal disagreements. Moreover, the expedition was to stay focused on the work at hand: its members were not to talk politics to outsiders or make any public display of the swastika. Nor were they allowed to drink or smoke during working hours. They were to spend expedition funds sparingly. So insistent was Himmler on this last point that he refused to give Wirth, the expedition leader, access to the team's funds. Instead he put another team member in charge of both the account books and cashbox.

The Ahnenerbe's first major expedition was to be all business. Germany's Reichsführer-SS wanted no embarrassing mistakes, no troublesome rows, no international incidents.

SOME THREE WEEKS later, on a soft green August morning, Wirth clambered up a gentle ridge of exposed granite known as Backa, not far from the Swedish village of Brastad in northern Bohuslän. On the slope below, his team started to unload the dusty trucks, carrying bags of plaster of paris, wooden planks, water buckets, and burlap to the base of the ridge. Setting down their loads, they stopped to watch as Wirth climbed up and down the rock, scouring its surface for ancient markings. In the early morning light, deep shadows accentuated the lines of the engravings: it was one of the best times of day to search for weathered furrows.

Wirth went about his work at a whirlwind pace, bursting with energy and exuberance. The previous summer at a training camp for Dutch youths, a newspaper reporter had observed him in action and marveled at his vitality. "No effort was too much for him. During those days, he was busy from nine in the morning until one at night. He didn't take any time for a meal. This man has the holy fire of a missionary in him."[37]

Already in the first two weeks of the expedition, Wirth and the team had covered an impressive amount of ground. Departing on August 4, just three days after the spectacular opening of Berlin's Olympic Games, the convoy had headed north to the Baltic coast and boarded a ferry for Rügen, a rugged island of white chalk cliffs and dark forests much loved by German holidayers. Over the next two days, they had cast an inscribed standing stone, then boarded a fishing surveillance cutter bound for the remote Danish island of Bornholm, some sixty miles to the west. Wirth believed that primordial Nordic migrants had chosen Bornholm as a stepping-stone on their journey south to Germany, and he was keen to find evidence of their passing. Under Wirth's direction, the team cast carvings at two sites on the island. Then they packed up and pressed on to Sweden. Lodged with friends in the village of Brodalen, Wirth was anxious to resume work at the famous site of Backa.

The scent of juniper rose in the morning air as Wirth bent down over the rock. Along one side of the outcrop, water trickled down in a thin dark ribbon, slicking the stone.[38] Between clumps of heather and patches of pale green lichen, dozens of engraved figures jostled across the rock. A five-foot-tall warrior, clearly a man of power, brandished a large erect penis and waved a giant battle-axe in his hand—in victory or menace, it was difficult to say. Off to one side, a tribe of lesser figures looked on. On the rock above, a fleet of sleek wooden ships rode invisible waves, while deer and cranes gamboled in thin air.

Backa was the first rock-art site in Sweden to be recorded by scholars. In 1627, a Norwegian university lecturer, Peder Alvsøn, passed through Bohuslän in a carriage and heard word of Backa's carvings. The ancient images so intrigued him that he made a freehand painting of some of the most obvious figures. It was a fine piece of early scientific illustration, but Alvsøn seems to have been a prudish man. He bowdlerized the largest and most obvious figure at the site. In place of the large man in proud tumescence, he modestly painted a large woman with small breasts.

Alvsøn's liberty was eventually corrected by more meticulous illustrators in the nineteenth century. And it was their work that stirred the first real scientific interest. What, wondered researchers in Stockholm, had the Bronze Age artists intended to portray? And why had they gone to so much trouble to engrave these figures in granite? In 1927, Oscar Almgren published a landmark book on the subject entitled *Rock Carvings and Cult Practices*. In it, the Swedish researcher proposed two main points. He suggested first of all that the strange jostle of figures on the Bohuslän rocks was the work of an ancient fertility cult that prayed for healthy children and abundant harvests, among other things.[39] Second, Almgren suggested that the rock-art panels often represented scenes of ancient sacrifices and fertility processions, in which people carried or pulled their great wooden ships.

Wirth had read and apparently admired Almgren's work, but he was temperamentally incapable of following in anyone else's footsteps.[40] What riveted his attention at Backa and at other sites in Sweden was not the scenes of human beings dragging or carrying boats, but the far more mysterious engravings of circles and lines—the ancient holy script of his imagination.[41] The key to reading this script, he firmly believed, lay in the far northern origins of the Nordic race. In Arctic lands shrouded in darkness for months each winter, Nordic men and women had felt a deep, abiding affinity for the sun, or so Wirth supposed. The movements of the sun across the horizon had eventually given rise to the awareness of a god, and the Nordic race had founded a great cult to worship this heavenly divinity. To mark the dates of their religious festivities and other events in the solar year, they had devised ideograms, Wirth believed, that evolved into a writing system.[42]

At Backa, Wirth scoured the rock for the symbols he believed he could read. He saw carved circles, disks, and wheels as ideograms for the sun and the annual cycle of life. He interpreted the figure of a man standing with raised arms as a symbol for a primeval divinity he called the Son of God, predating Christianity by thousands of years.[43] He believed the image of two circles linked by a vertical line represented the rebirth of life at the winter solstice, while a circle bisected by a vertical line indicated a year.[44] And so it went: Wirth's research had long ago abandoned reality and solid ground. It was soaring in a dense, impenetrable fog, no longer containing even the slightest fraction of science.

WIRTH INSISTED THAT his team cast the panels he found most important.[45] Whenever he found something suitable, he outlined its furrows in chalk and Bousset, the team cameraman, filmed the engraving. An assistant then hammered together a wooden frame to be fitted down upon the rock. With wads of damp clay, team members sealed gaps between the bottom of the frame and the uneven rock surface, and brushed a thin layer of soapy lubricant onto the engraving. They hauled heavy buckets of plaster of paris up the slopes and tipped the contents into the frames, then added a layer of burlap and iron bars for strength.[46] When the cast finally hardened, the team began the tricky work of prying it loose. They levered it upward with iron bars and mallets, praying that they would not break either the cast or the granite below.

That, however, was the easy part. A sweaty trek followed. Many of the casts were nearly ten feet long and weighed several hundred pounds. They were heavy and awkward to convey on trails over broken ground and through gnarly forests.[47] "The rock art does not just lie off the side of the road," explained Wirth in a later letter to Swedish officials, "but is often in hard to reach areas, where the work material (plaster of paris, wood, iron, water, etc.) had to be taken over quite a distance in a small cart or slid [on the ground] or carried on the back, and the plates had to be transported back with great difficulty."[48] Later at camp, team members packed the casts in sturdy handmade crates and hauled them by car to the nearest port, where they could be shipped to Gothenburg, and from there to Germany. It was not an expedition for shirkers.

Some local residents watched the goings-on with fascination, and two young men even went so far as to join Wirth's crew as assistants. But others adopted a more jaundiced view. The region was well known for its socialist sympathies, and most farmers and fishermen did not have much use for Nazis.[49] They were rather dismayed to see flocks of German tourists turning up in increasing numbers to visit the rock carvings in the area. "This is mainly due to a German professor declaring in Munkedal that he never felt more reverent than when standing in front of a rock carving," wrote one newspaper rather pointedly. "He must never have sat in front of the Führer then, for that is surely the greatest happiness for a Nazi."[50]

Wirth was well aware of the region's politics. He was pleasant and charming to the farmers and tried his best not to make enemies. From Brodalen, he and his team headed north up the Bohuslän coast. Pitching camp wherever they could, they toiled on sites from Askum to Tanum, collecting truckloads of casts. When they were done, they crossed the border into neighboring Østfold County in Norway, another region blessed with rock carvings. Wirth seemed tireless. "As soon as work got under way at a site, I had to immediately arrange things for the next. In between, material had to be bought and transported, sleeping quarters had to be found, local offices had to be dealt with and casts had to be done, which I did in work groups," he wrote in a final report.[51]

By the first week of September, the team had completed fifty-five massive casts: Wirth was in fine fettle. He believed that he and his team were doing something the Swedes could not be bothered with—making replicas of the engravings before quarrymen destroyed them. He couldn't resist crowing about this in a report from the field to Himmler. "This will be a collection that has no equal and never will," he boasted. "Whoever wants to study the Nordic race in relation to the rock art will have to come to Germany, to the 'Deutsches Ahnenerbe,' because it is only there that the main monuments will be collected together. No one in Sweden could see them for themselves because they are strewn far and wide. Soon the

Deutsches Ahnenerbe will have rock carvings that don't exist anymore in Scandinavia because they are being destroyed."[52]

Wirth was greatly exaggerating the situation. Swedish officials guarded the rock art vigilantly, well aware of their responsibilities to protect a national treasure. But the German scholar knew he had to put the best possible face on his research for Himmler.

ON SEPTEMBER 5, midway through the expedition, Wirth journeyed to Oslo to meet with Norwegian officials. He was eager to obtain advice and assistance for the next leg of the trip. He planned to lead the team northward along the western coast of Norway in order to cast several small scattered rock-art sites, one located just short of the Arctic Circle. He also planned to engage in some fund-raising in Oslo, although he did not mention this in his field report to Himmler. Specifically, he intended to sell replicas of the team's Norwegian casts to the antiquities museum at the University of Oslo—a transaction that would considerably undermine his claims about the exclusive nature of the collection he was compiling. In the museum offices, he met one of Europe's most respected archaeologists, Anton Brøgger.

At fifty-two, Brøgger was at the peak of his career. A generous, urbane man fond of the theater and music, Brøgger was a passionate advocate of Norwegian culture.[53] He had played a leading role in the excavations of the Oseberg grave mound, the richest Viking burial site ever found, and had written extensively about its many beautiful treasures. He had edited a popular journal, *Oldtilden,* or "Ancient Times," and inspired a generation of archaeology students in his lectures at the University of Oslo. He had also helped to host the meetings of the International Archaeological Congress, which had recently brought five hundred archaeologists from around the world to Oslo.[54]

Brøgger knew a great deal about German attitudes toward prehistory. And he had watched with alarm the way in which the Nazi party had twisted and distorted the past to advance its political agenda. "These days, prehistory and archaeology have become popular sciences across the entire world," he wrote a few months later. "Excavations, particularly those which take place on a grand scale, speak immediately and often dramatically to a nation's emotions and passions. Fantasy is set in motion. . . . Therefore, one must be cautious. These days, archaeological artifacts and results have an important meaning for nations. They are part of a people, a nation's spiritual capital. However, they should not be misused for nationalistic propaganda."[55]

Few were as guilty of misusing science and scholarship for political ends as Wirth. But Brøgger did not seem to know this, and Wirth, who was under strict orders from Himmler to avoid political discussions with outsiders, likely refrained from making his own position clear. By the end of the meeting he had struck a deal with Brøgger. In exchange for what he hoped would be 1,000 crowns, he agreed to provide the University of Oslo with a selection of casts of Norwegian rock art for its own research collections.[56]

Satisfied, Wirth went back to work, leading his team up the west coast of Norway. They journeyed along the winding highway north of Trondheim, past steep fjords and rugged cliffs, and stopped to make twelve casts of rock art at Bardal and another two at Østre Tessem.[57] It was spectacularly beautiful country, but the nights were growing increasingly cold and they could no longer camp out. So Wirth hunted each evening for whatever lodging he could find. In the darkness, after a long day of driving or casting, he found a farmer to give them shelter in a hayloft or looked for rooms in a local youth hostel.

At Mosjøen, a small fishing village just five hundred miles south of the Arctic Circle, Wirth and two assistants boarded a small boat bound for Rødøya.[58] The tiny island boasted a famous engraving of a very rare subject. It portrayed a human figure on skis and was thought to date back four thousand years.[59] Wirth was keen to cast it for the Ahnenerbe collection. So on a stormy day in mid-September, he and his team put out into the cold coastal waters. Rain and heavy wind lashed at the windows, and spray from the fierce gray waves drenched the deck as they crossed to Rødøya, but Wirth was not about to let the weather dampen his spirits.

He was closer than he had ever been before to the Arctic—the imaginary birthplace of the Nordic race he loved.

7. ENCHANTMENT

As THE DUTCH SYMBOLS expert and his team tossed in the stormy water off western Norway, a second Ahnenerbe researcher was putting the finishing touches on a very different research mission in the secluded forests of eastern Finland. Yrjö von Grönhagen had set out from Helsinki in June at Himmler's request to study the legendary sorcery of Karelia, a remote wilderness straddling the Russian-Finnish border. For weeks, Grönhagen and his two companions had journeyed to rustic log cabins, tape-recording elderly men and women as they chanted magical incantations, communicated with the dead, and sang songs of ancient wizards. Karelia, as one team member later observed admiringly, is a "land of witches and sorcerers."[1]

Certainly, Karelia clung stubbornly to the old ways. Blanketed by dense forest and threaded with dark bogs and cold lakes, the region was buried beneath thick drifts of snow for much of the year. And for many centuries, the extreme winter cold, the summer clouds of mosquitoes, and the rugged terrain had tended to discourage economic enterprise and other contact with the outside world. But in the late nineteenth century, forestry companies arrived to size up Karelia's timber. Loggers followed, thinning some of the dense stands of spruce and fir, while railway lines from the east and west began slicing through the taiga, transforming small towns into small cities. But modern life had yet to invade all corners of the remote fishing villages along the larger lakes. And in these isolated hamlets, some elders still passed the long winter nights singing songs of enchantment and casting ancient magic spells.

Grönhagen was a handsome young Finnish nobleman new to scientific research. He had become fascinated by Karelia after reading one of the most famous of all Finnish books, *The Kalevala*. *The Kalevala* was the work of a nineteenth-century country doctor, Elias Lönnrot, who believed that the old songs of Karelia were actually fragments of a lost northern epic that dated back thousands of years.[2] So Lönnrot set about collecting these songs, journeying across Karelia by boat and sled, writing down the words and knitting them together with snatches of his own writing, into a flowing narrative. This he first published in February 1835.

While a few Finnish intellectuals complained about the liberties that Lönnrot had taken, most Finns loved *The Kalevala*. And they particularly adored the hero of this boreal epic, a great sorcerer named Wäinämöinen.[3] In *The Kalevala*, Wäinämöinen and his assistant magically transform a treeless northern land into a vast verdant forest. And with his many enchantments, the sorcerer warms the sun and cleanses the land of pestilence by awakening the magical heat of a sauna. Indeed, at nearly every twist and turn of the plot, Wäinämöinen and his fellow characters rise above adversity with the help of their powerful spells. As one nineteenth-century English translator of *The Kalevala* observed, "Here, as in the legends of no other people, do the heroes and demigods accomplish nearly everything by magic."[4]

In 1935, Grönhagen published an article about *The Kalevala*—whose title literally means "The Land of Heroes"—in a Frankfurt newspaper. This act of patriotism had a very unforeseen consequence. In his office at the heart of the SS empire, Himmler picked up the article and read it with deep curiosity. He believed that the old legends and myths of Europe brimmed with clues to the primeval religion and technology of the Aryans. They needed merely to be studied and deciphered. Could the old songs of Karelia, he wondered, contain further valuable leads?

HIMMLER WAS NOT alone in this fascination with old legends. Many German ultranationalists were entranced by ancient epics such as *The Nibelungenlied*, with its tale of Prince Siegfried who slew a dragon and bathed in its blood to become invincible in battle. And they absolutely adored a famous collection of ancient Norse songs, the *Edda*.[5] First recorded on paper in Iceland during the thirteenth century, the *Edda* portrays an icy realm of gods and goddesses, kings and thanes, dwarves and wizards, sword-maidens and lordly heroes, magnificent beings who wield strange magical powers and fight to the death for their clan, their land, and their honor. Indeed, its tales had inspired J. R. R. Tolkien when he wrote *The Lord of the Rings*.[6]

But many German extremists considered these ancient myths to be much more than stories—an idea that a Munich newspaper editor named Rudolf John Gorsleben had done much to spread.[7] Gorsleben was closely connected to the Nazi party in the early 1920s and was fascinated with the history of the supposed Aryan race. The more he reflected on the *Edda*, the more he saw it, as one historian has noted, as "a distillation of Aryan religion."[8] According to Gorsleben, Aryan men and women had once possessed superhuman abilities and these were revealed in the *Edda*. He also believed that the Aryan race could regain its purported supremacy on the world stage if it could somehow activate these latent powers once again. Gorsleben was immensely concerned with the question of how to turn the switch back on. In 1925, he founded a study group, the Edda Society, to examine this matter. He described the *Edda* as "the richest source of Aryan intellectual history."[9]

Gorsleben published many of these ideas in a weekly right-wing newspaper he owned in Munich, *Deutsche Freiheit*, and they created something of a stir. Hitler himself seems to have incorporated some into his own thinking. At dinner one night in January 1942, for example, Hitler observed that "legend cannot be extracted from the void, it couldn't be a purely gratuitous figment. Nothing prevents us from supposing—and I believe, even that it would be to our interest to do so—that mythology is a reflection of things that have existed and of which humanity has retained a vague memory. In all the human traditions, whether oral or written, one finds mention of a huge cosmic disaster. . . . In the Nordic legend we read of a struggle between giants and gods. In my view, the thing is explicable only by the hypothesis of a disaster that completely destroyed a humanity that already possessed a high degree of civilization."[10]

Himmler seems to have taken Gorsleben's ideas more seriously still. As a child, Himmler had listened enthralled to his father and mother reading aloud from the German and Norse legends, and he had returned to these tales as an adult. In October 1923, just two months after he joined the Nazi party, he picked up a copy of *The Nibelungenlied* and read it from cover to cover, gushing later in his long-kept booklist about its "incomparable eternal beauty in language, depth and all things German."[11] Three years later, he gave a copy of the *Edda* as a present to a friend.[12]

Himmler cultivated the friendship of a man revered by the Edda Society, Karl-Maria Wiligut, the former psychiatric patient who helped locate Wewelsburg castle for the SS academy.[13] The two men frequently dined together and chatted for hours, and in time Himmler came to take a particularly literal view of the *Edda* and other ancient legends.[14] He believed they comprised a history, greatly fragmented but still intelligible, of northern Europe as it once was, before the arrival of Christianity. More important, he believed that the Aesir, the old gods of Norse legend—such as Odin, Thor, and Loki—were in fact beings of pure, undiluted Nordic essence, the earliest Aryans. As such, they were the possessors of superior knowledge, or so Himmler supposed.

With the help of Ahnenerbe scholars, Himmler hoped to recover the lost Aryan lore of the *Edda*, which seemed to span everything from superior weaponry to potent medicines.[15] He was particularly intrigued by stories of Thor and his lightninglike throwing hammer, Mjollnir, described as the strongest and most accurate weapon in the world. In a surviving letter to the Ahnenerbe, Himmler spelled out his particular interest in Thor and his hammer, without a touch of embarrassment or self-consciousness, as if it were a rational scholarly pursuit:

> Have the following researched: Find all places in the northern Germanic Aryan cultural world where an understanding of the lightning bolt, the thunderbolt, Thor's hammer, or the flying or thrown hammer exists, in addition to all the sculptures of the god depicted with a small hand axe emitting lightning. Please collect all of the pictorial, sculptural, written and mythological evidence of this. I am convinced that this is not based on natural thunder and lightning, but rather that it is an early, highly developed form of war weapon of our forefathers, which was only, of course, possessed by the Aesir, the gods, and that it implies an unheard of knowledge of electricity.[16]

If this advanced knowledge of electricity could be pieced together again from the old sagas, then German scientists could use it to develop a weapon capable of smashing the enemies of the Reich.

Moreover, there was no telling what other pearls of Aryan wisdom might lie buried in the *Edda, The Nibelungenlied,* or other ancient legends of northern Europe. Himmler was keen to find them and if possible put them to use in the Reich, and it was for this reason that he contacted Grönhagen, the author of the article on *The Kalevala,* in the autumn of 1935.

GRÖNHAGEN WAS A romantic idealist. He was born on October 3, 1911, in St. Petersburg, during a time when Finland was a grand duchy of the Russian Empire. His mother, Zina von Holtzmann, was a doctor of jurisprudence who traced her ancestry to the Finnish and Russian nobility.[17] His father, Karl von Grönhagen, was a Finnish officer and journalist whose family origins extended back to the Swedish, German, and Dutch aristocracy. Grönhagen grew up during a turbulent period in Finnish history. When the Russian Revolution began in 1917, Finnish patriots seized the moment and declared Finland's independence, unleashing bitter political rivalries. A bloody civil war followed, pitting Finnish communists and socialists against conservatives in the Finnish upper and middle classes. Grönhagen's family sided firmly with the conservatives. His older half brother fought as an officer in the White forces in Russia and was captured and executed in Moscow in 1920. All this made a lasting impression on Grönhagen. He grew up loathing communism and despising the Soviet Union.[18]

By his late teens, Grönhagen was a handsome lad of five feet seven inches, with delicate, even features, a pair of luminous eyes, a mop of thick dark hair, and a ready grin. He possessed fine manners and an aristocratic air. On leaving school, he landed a position as a trainee at the Finnish consulate in Paris, but he aspired to a career in the cinema.[19] According to a report submitted in 1937 to the Finnish Security Police, a French movie director offered him a role as a singer in a new film. Grönhagen accepted and signed the contract, but eventually turned the part down after the director made unwanted sexual advances.[20] Disillusioned, he enrolled in the Sorbonne in 1933. He dreamed of conducting anthropological fieldwork in India. But one of his professors suggested that Europe would be a better place for such research and Grönhagen heeded the advice.[21]

In the spring of 1935, he hatched a plan to travel by foot from Paris to Helsinki, studying "practical sociology" on the road.[22] He bought a journal so that people he encountered could write something they fancied in it—an old proverb perhaps, or a poem. He departed on June 27, crossing eastern France and trekking across the rolling hills and valleys of Belgium. He entered the Third Reich on August 1, and felt at home at once.[23] He had learned German as a child and was favorably disposed toward the Nazi government, which seemed to loathe communism and Bolsheviks almost as much as he did. The first person to sign his journal in Germany was a leader of a Hitler Youth group.

Soon after, Grönhagen met the editor of a Frankfurt newspaper, who agreed to publish his article on *The Kalevala*.[24] On October 1, not long after the Nazi government passed the Nuremberg racial laws prohibiting Jews from marrying German citizens or engaging in extramarital sexual relations with Germans, the young Finn sat down in a meeting with Himmler. The two men had an amicable discussion. Afterward, Himmler wrote an entry in Grönhagen's journal. It read: "Germans and Finns shall never forget that they once had the same fathers."[25] It was signed "H. Himmler."

From the beginning, Himmler took a personal interest in the affable young anthropologist, with his refined manners and his German pedigree. Moreover, Himmler obviously relished a conversation that he had had with Grönhagen on the ancestry of the Finns. For many years, European scholars had puzzled over the origins of the Finnish people. The Finnish tongue bore few similarities to German, English, French, or any other member of the Indo-European language group. Indeed, linguists had placed it, along with Hungarian, Estonian, Sami, and a group of lesser-known languages from Russia and Siberia, into a separate linguistic family—the Finno-Ugric languages.[26]

This linguistic classification raised a large red flag in the minds of many nationalist Germans. If the Finns didn't speak an Indo-European tongue, how could they possibly be Aryans? Some scholars suggested that they descended from the ancestors of the Hungarians; others speculated wildly that they sprang from the loins of Mongol invaders.[27] The short, round skulls of Finns, they observed, resembled those of Asians, and the reserved, reticent character of Finns seemed more Eastern than Western.[28] And they were fond of linking the Finns to an enigmatic tribe described by the Roman historian Tacitus—the Fenni, who had once lived somewhere in the northeast Baltic region.[29] The Fenni, noted Tacitus, lived in "unparalleled filth and poverty."[30]

But Grönhagen did not believe for a moment that he or his countrymen were of Mongol origins or that they were in any way members of an inferior race—quite the opposite. He was convinced, as he later wrote, that the primeval Finns arose from a vast wild land that stretched from Scandinavia in the north to the Black Sea in the south.[31] He had also persuaded himself that the Finns were blood brothers to the Aryans and therefore shared a common ancestry—a suggestion that would have shocked and dismayed most Finnish scholars of the day.[32] And Grönhagen wasn't above offering his own racial observations as evidence. "Even the outer appearance of the Finnish person is telling," he noted later. "Blond and tall, on average taller than the Germans, with mostly blue eyes or gray eyes, they belong racially to western or northern Europe."[33]

Himmler also relished what Grönhagen had to say on the subject of *The Kalevala*. As Grönhagen pointed out, the people of Karelia, whose songs were immortalized in *The Kalevala*, still preserved many of their ancient customs and religious beliefs.[34] Indeed, as Finnish folklorists had discovered, the remote territory was a refuge of paganism, a land where traditionalists still practiced a primeval form of shamanism. In Karelia, seers still recited magical spells to call upon spirits and gods, who in turn might help divine the future, cure chronic diseases, assist in childbirth, and protect families from harm. Many rural Karelians still revered and feared these magicians. "The entire life of the peasants is still filled with pagan customs, the use of magic and chants," observed Grönhagen later. "It is especially the power of the healing knowledge—which is full of secret chants—that is ingrained in the beliefs of the people."[35]

So impressed was Himmler by these ideas that he sent the young Finn to see Karl-Maria Wiligut, who was working in an office in RuSHA and whose relationship with Himmler had by then become the subject of much talk in the SS. One story going the rounds described an outing that Wiligut had taken with the Reichsführer-SS in his private car, a vehicle that bore, incidentally, the vanity plate of SS 1. As the two men were talking, Wiligut suddenly slumped in the leather seat, foaming at the mouth. Himmler immediately ordered his driver to stop. As the car screeched to a halt, Wiligut bolted from the seat and ran headlong into a field, his arms stretched out like wings. Himmler followed. Eventually, the old man dropped to his knees in the wet barley, instructing Himmler to dig at his feet. Himmler purportedly obeyed, dispatching a team of excavators to the site: they were said to have discovered an early historical settlement there.[36]

Himmler seems to have wanted Wiligut to assess the young Finn and his intriguing claims. So the two men—one paunchy, puffy-faced, and prone to psychosis, the other alert, curious, and eager to make a good impression—met several times.[37] Wiligut obligingly channeled memories and tales from his ancient ancestors and recited what he claimed were Gothic proverbs from the Black Sea region handed down in his family for centuries. Grönhagen listened attentively. After one such session, he reportedly told Wiligut, "Oh, I know this, too. I learned the same thing from my father."[38]

Grönhagen passed muster, and soon after Himmler offered him a job with the Ahnenerbe. The young anthropologist started work on November 1, 1935.[39] His first assignment was to return to Finland to conduct research in the folklore archives of the Finnish Literature Society in Helsinki, which held the original field notes of Elias Lönnrot as well as the manuscripts of many other prominent folklorists and experts on The Kalevala.[40] This research, Himmler and Wiligut seem to have hoped, would serve several purposes. It would prepare Grönhagen for future fieldwork in Karelia, collecting and recording ancient myths and magical chants. And it would also assist Wiligut in researching ancient Aryan religious ceremonies for the SS.

Himmler was immensely keen to replace Christian rites with rituals he deemed more Aryan. He believed—as Herman Wirth did—that the sun played a central part in the primordial religion of the Nordic race. In keeping with this idea, he wanted to create an SS summer solstice festival to celebrate life and a winter solstice festival to remember the dead and honor the ancestors.[41] He asked Wiligut and Professor Karl Diebitsch, a cultural advisor in the SS, to research and draw up suitable rituals.[42] Already, the two men had developed an SS naming ceremony to replace the traditional christening ceremony for a baby.

From Finland, Grönhagen corresponded with both Himmler and Wiligut, keeping them posted on his work.[43] The two men sometimes passed his letters back and forth and discussed his progress. On April 19, 1936, Himmler wrote a chatty letter to the young anthropologist, inquiring solicitously after his health—Grönhagen had recently suffered a bout of jaundice—and instructed him to extend his stay in Helsinki until the summer.[44] As soon as weather permitted, he wanted Grönhagen to take a journey through Karelia to photograph the sorcerers and witches and record their songs and incantations for study.

Grönhagen must have been delighted, but he could see potential problems. He was uncertain whether the elderly Karelians would agree to pose for their pictures, so he proposed to recruit a Finnish illustrator, Ola Forsell, for the journey.[45] A more serious problem lay in Grönhagen's lack of musical training. The fieldwork really required an ethnomusicologist familiar with diverse folk-music traditions in Europe and abroad. For this, Wiligut recommended Dr. Fritz Bose, a musicologist from the University of Berlin.[46] Bose was a member of the Nazi party and a self-described expert on music and race.[47] He firmly believed that the world's diverse musical styles reflected racial traits, rather than cultural influences.[48] It was a theory that appealed greatly to the new Nazi regime, and Bose capitalized on it whenever he could. In 1934, at the age of twenty-eight, he had become head of the Berlin Acoustics Institute, after his mentor was sacked for being Jewish.[49]

Himmler must have hoped that Bose, with all his theories on race and music, could determine whether the ancient chants and incantations of Karelia were Aryan or something very similar. And Bose, for his part, seems to have been delighted with the assignment. He joined the SS, where he was assigned to RuSHA and to Himmler's personal staff.[50] He then set about preparing for the trip, searching for suitable sound recording equipment. While other scholars of the day generally employed disc cutters, a device that captured sound on acetate-coated metal discs, Bose hunted for more sophisticated equipment. He chose a piece of audio gear that had debuted only a few months earlier at the Berlin Radio Fair.[51] It was the magnetophone, a prototype of the modern tape recorder.[52]

IN JUNE 1936, Grönhagen, Bose, and Forsell set out from Helsinki by rented car to Viipuri, the principal city of Karelia.[53] Viipuri was a busy cosmopolitan port on the Gulf of Finland, a place that comfortably mixed old medieval stone churches and a thirteenth-century castle with modern theaters and cafés. It was a far cry from the northern wilderness that Grönhagen yearned to explore, so he and his companions soon set off to the east. Along the north shore of Lake Ladoga, they found a land of dark fens and gentle ridges topped by pine forests, and there, among scatterings of small log cottages, they began scouting out sorcerers and witches to record and photograph.

It was so close to midsummer that the sun hardly set, and they often worked late into the evening. One day, in the woods that stretched beyond the small village of Suistamo, they visited Timo Lipitsä, a traditional singer well known among Finnish folklorists.[54] On their arrival, they saw an elderly man dressed in a snowy white tunic and dark work pants. Lipitsä's hair was white and his eyes, as Grönhagen later wrote, gleamed with "a far away look."[55] He had never attended school. He could neither read nor write, and he had never so much as laid eyes on a train. Grönhagen was delighted. Lipitsä, he later recalled, "knows nothing of civilization or technology."[56]

Grönhagen introduced himself and his teammates, explaining that he was collecting material for a department of Finnish folklore he planned to establish in Germany.[57] Then, after gaining the old man's confidence, he asked if he would sing some verses from *The Kalevala*. This rather confused Lipitsä, for he did not know what *The Kalevala* was, but he offered to sing for his visitors just the same. He called over his son, and the two men sat across from each other by the hearth. Holding each other's hands so that the "power becomes stronger," they began to rock back and forth.[58] As they moved, they sang an old story about the creation of the world, a story that closely resembled one published in *The Kalevala*. This amazed Grönhagen. "They sing the language of *The Kalevala* without even suspecting the existence of a scientific collection of song."[59]

Bose diligently recorded the songs, fussing over the dials to get the best sound.[60] He was fascinated by the performance. The old man, he later noted, seemed to enter a trance state. "His focus is fixed out into the beyond and he no longer senses his environs. It is almost impossible to interrupt him. He does not take kindly to any attempt to halt his sacred singing which is also an enormous and difficult artistic effort."[61] Finally, at the end of the recording session, Grönhagen asked the elderly singer to pose for a photograph in the bright sunlight outside his cabin. Lipitsä graciously obliged, squinting into the light.

The trio resumed their travels. In the small village of Tolvajärvi, they photographed and recorded a prominent Karelian musician, Hannes Vornanen.[62] Vornanen was a master of the kantele, a traditional Finnish string instrument made from birch wood. The kantele and its murky origins interested Bose greatly. Some scholars of the day speculated that the kantele traced its roots to Asia. They observed that it had originally possessed just five strings and bore a resemblance to an old Chinese zither known as the *guqin*. Intriguingly, the *guqin* was also reputed to possess magical powers—making snow fall in summer and flowers blossom in the cold and darkness of winter.[63]

Bose, however, discounted the Asian theory, pointing to a myth in *The Kalevela*. According to one ancient song, the wizard Wäinämöinen had created the first kantele from the jawbone of a common species of fish in Finland—the pike. Other clues, he noted, pointed to the Finnish origin of the instrument. The kantele shared a tonal system with the songs of *The Kalevala*, which suggested that their histories were intertwined. Morever, Bose believed the kantele was far older than its Asian counterpart. Its tonal system was more primitive and its overall construction was much simpler, lacking as it did the arched cover of the *guqin*.[64] Based on this, Bose concluded that the kantele was a thoroughly European instrument—a scholarly opinion that must have pleased Grönhagen, who was anxious to refute any suggestion that the Finns were of Asian origins.[65]

As they continued their travels, the team learned of an elderly soothsayer living in the woods near Salmi. The local people called her a witch and advised Grönhagen against visiting her, but the warnings merely whetted his curiosity. He went off to find her. When he arrived at her hut, it was empty. He waited for her to return. In the distance, he heard someone talking. It was ninety-two-year-old Miron-Aku. She was picking mushrooms. She looked up at him and stared into his eyes. Grönhagen was riveted by her first words. She told him that she had dreamed about him three days earlier. "You came to me in my sleep and wanted to take away my secrets. Since then I have been sick and will die soon. What do you want of me?"[66]

Grönhagen tried to befriend her: he wanted to learn more about her reputed ability to see into the future. He paid her several visits. She brewed him a bitter drink from local plant roots and talked of an old god that people in the region had once worshipped before they embraced Christianity. She also described how the spirits of her ancestors resided with her in her hut.[67] By summoning them ritually, she could divine future events. This information fascinated Grönhagen and Bose. After a long negotiation, they persuaded her to perform the ritual for the camera and tape recorder.[68] She reluctantly gave them the demonstration, but she was terribly dismayed when Bose later played back the tape. She told him that "she would never be able to practice magic again," because she had committed a sacrilege when she performed a spell simply to satisfy others' curiosity.[69] Worse still, this error in judgment was now preserved on the tape.

The team moved on to the next village, and as the summer stretched on, they broadened their research. Grönhagen diligently recorded all the folklore he could concerning an ancient Finnish tradition—the sauna. Karelian healers believed that the sauna's combination of water and fire-warmed rock generated a powerful force capable of warding off many forms of sickness.[70] To better avail themselves of the mystical powers of steam, healers often took the sick into their own saunas, where they performed sacred ceremonies.

Bose, for his part, kept adding to the collection of tapes. He recorded old Karelian dance songs with titles like "Boys Travelling" and "Children on a Cliff."[71] He recorded ancient lullabies, work songs, and patriotic tunes. He also recorded women's songs of lamentation, which impressed him greatly. "As a kind of magical priestess," he observed later, "the mourning singer helps the souls of the dying and the dead to cross from this side to another."[72]

In all, the team taped more than one hundred songs, chants, and instrumentals, and by the end of the summer, Bose was convinced that he had successfully adduced the racial origins of the otherworldly songs that made up The Kalevala. The improvised nature of the lyrics suggested to Bose that these songs were very old, much older in fact than the earliest relics of Nordic poetry. The Kalevala, he suggested, "must be a very early Nordic cultural level," possibly even dating back to pre-Aryan times.[73]

BACK IN BERLIN, Grönhagen sent Himmler a note about the trip and a copy of the photograph he took of Timo Lipitsä, the old man who had sung about the creation of the world. He then set to work writing a short article for Germanien, a monthly archaeological and anthropological magazine partially owned by the Ahnenerbe.[74] As his subject, Grönhagen chose Karelian magic chants. He described with journalistic flair the age-old rituals that Karelian women performed to fend off evil spirits in cemeteries and to heal the sick in the cleansing heat of the saunas. The editors of *Germanien* quickly published the piece, listing the author as Georg von Grönhagen.[75] The change of name made the young Finn sound more German and hence a good deal more acceptable to the magazine's nationalistic audience.

In Helsinki, Finnish researchers were growing suspicious of the Ahnenerbe team. Grönhagen had met several important folklorists the previous winter in the Finnish capital, and some, such as Martti Haavio, had received reports about his field activities from Karelian friends.[76] Haavio, the director of the folklore archives at the Finnish Literature Society, was greatly concerned about the use that Grönhagen's work might be put to.[77] He was loath to see Finnish folklore falsified or manipulated in some fashion in order to advance the Nazi cause. So in the fall of 1936, he advised at least one of his trusted Karelian sources against cooperating any further with Grönhagen.[78]

Grönhagen seems to have been oblivious, however, to the controversy he was stirring in his homeland. On January 27, 1937, he and Bose attended a private meeting at Himmler's home in Berlin.[79] The two researchers were ushered into a study, and as they glanced around, they were startled to see their photo of Timo Lipitsä hanging like a treasured icon above Himmler's desk.[80] They had arrived bearing other mementos of their travels—an album of Karelian photographs and a kantele they had acquired—which made quite an impression. Himmler and his wife Marga flipped through the photographs together, then Bose produced the kantele and gave Himmler a short impromptu lesson on it. Himmler was so delighted with the instrument that the two researchers ended up giving it to him. He immediately placed an order for ten more for the SS.

The SS leader was keen to hear Bose's tape recordings and he wanted to know how old the Karelian songs were and whether they were related to music of the ancient Germanic tribes. After listening intently to Bose's replies, he mentioned the possibility of a grant for further research on these subjects. He and Wiligut, it transpired, were planning to use ancient Nordic musical instruments in SS solstice ceremonies, and they hoped to find examples of authentic ancient Nordic music.[81] Bose's ideas on the origins of European music so impressed Himmler that he later asked the scholar to help cast replicas of another ancient Scandinavian instrument: the *lur*.[82] Bronze Age artists had depicted this wind instrument in their rock carvings in Sweden.

Grönhagen also shone that evening with his colorful stories of Timo Lipitsä and the old seer, Miron-Aku.[83] Less than a month later, Himmler named the young anthropologist head of the Ahnenerbe's brand-new department of Indo-Germanic-Finnish studies, a prestigious appointment for a young Finn of twenty-six who did not possess a graduate degree.[84] The mission of the new department was to search for parallels between the Aryans and the Finns in order "to establish their shared origins."[85] The top two priorities were studies on the metaphysical meaning of *The Kalevala* and research on magic, witches, and rune singers.

Himmler intended to make use of this Finnish lore, too. Grönhagen's description of the mystical healing powers of the sauna, with its combination of fire and water, had particularly aroused his interest. He suggested to Dr. Ernst-Robert von Grawitz, the chief physician of the SS, that SS men would benefit from using traditional Aryan methods of body cleansing. So Grawitz's staff dutifully launched a joint study with the Ahnenerbe on "Germanic bathing" in 1937.[86] As part of this project, Sievers instructed Grönhagen to share his research on Finnish saunas with the SS medical staff.

8. THE ORIENTALIST

THROUGHOUT THE EARLY SPRING of 1937, Himmler mused with pleasure upon the new research from Finland. It conjured up for him visions of a simpler, more radiant Europe, a land of powerful gods and magic, a countryside undimmed by the modern evils of racial mixing and industrialization. Moreover, it seemed to open up new possibilities for strengthening the SS and preparing its men for life in the agricultural settlements of the future. But as delighted as Himmler was with Grönhagen and Bose, as content as he was to listen to the sound of the kantele and gaze at the portrait of the old Finnish singer that hung over his desk at home, he fretted over the future of the Ahnenerbe. He had yet to convince Hitler of the importance of the research institute or the value of its studies—quite the contrary. When Hitler finally deigned to take public notice of the Ahnenerbe, it was only to tear a strip off the institute and its wayward president, Herman Wirth.

Hitler had delivered this rebuke during one of the most important events in the annual Nazi calendar—the great party rally in Nuremberg. Amid all the goose-stepping and saluting, all the spectacle and speeches in September 1936, Hitler had given a stirring address on the state of German culture. As thousands of adoring faces gazed up at him, Hitler had expressed his views with customary forcefulness, then issued a kind of fatwa against two prominent heretics. The first was Ludwig Roselius, the coffee merchant who had built a spectacular lecture hall for Wirth along Böttcherstrasse, a major urban development project in Bremen. The second was

Wirth himself. "We have nothing to do," Hitler thundered, "with those elements who only understand National Socialism in terms of hearsay and sagas, and who therefore confuse it too easily with vague Nordic phrases and who are now beginning their research based on motifs from some mythical Atlantean culture. National Socialism sharply dismisses this sort of Böttcherstrasse culture."[1]

Hitler had clearly taken great exception to the modernist architecture of the Böttcherstrasse and had finally lost all patience with Wirth. The two had first met at Nuremberg in 1928, and over the years Wirth's grandiose talk of reawakening German society had come to irritate Hitler.[2] More recently, the scholar's rabid attacks on Christianity had added to Hitler's exasperation. While Hitler dreamed of exterminating Christianity from the Reich, he was not yet ready to take on the Catholic and Protestant churches, and he did not want a loose cannon like Wirth complicating matters.[3] Furthermore, Hitler despised Wirth's vision of ancient Nordic civilization as a matriarchy ruled by priestesses and female seers. It was a far cry from the Nazi notion of German women as compliant baby factories. Hitler had no objection to twisting and distorting the past—indeed he encouraged it, provided that the results fit with his own political agenda. The last thing Hitler wanted was a new feminist world order.

His denunciation of Wirth at the party rally presented a serious problem for Himmler. The SS leader was inordinately fond of Wirth and his eloquently expressed theories of the past.[4] But he could not ignore Hitler's scathing criticism, particularly if he wanted to salvage the Ahnenerbe. Moreover, he had come to realize that Wirth was entirely the wrong man to direct the affairs of the institute and build it into a strong arm of the SS. Wirth was incapable of managing his own life, much less taking charge of others. So Himmler decided to remove him from his prominent post as the Ahnenerbe's president. He demoted him to the position of a department head, and then proceeded to silence him publicly by prohibiting any further books or lectures.[5]

WIRTH, WHO WAS in Norway at the time of the Nuremberg rally, had little idea of the storm brewing back in Berlin. On arriving home, he was stunned to discover doors closing to him everywhere. He could not quite grasp that he was finished as a public figure in the Reich. His mind still percolated with expensive plans for a new open-air exhibition of his casts and a series of public lectures. He even wrote to Himmler asking for yet more money to help him through a difficult patch.[6] Himmler exploded. As Reichsführer-SS, he ordered his trusted aide Galke to impress upon the prodigal scholar his precarious position. "I will not participate in politically foolish things which will also bring us to financial ruin," he wrote to Galke. "It should be made very clear to Prof. Wirth that without the financial settlement of his affairs by the SS, which today is unwished for but a fact, he would be sitting in jail, if not in the penitentiary, because of his serious disregard for the law in several matters."[7]

Himmler had no intention, however, of putting a halt to the studies of sagas and myths that had so exasperated Hitler. As deeply as Himmler admired and respected the Nazi leader, he regarded the Ahnenerbe and its studies as far too essential to his plans to be jettisoned: the future of the SS in his view depended on a thorough understanding of the past. But the incident at Nuremberg convinced Himmler of the wisdom of putting the Ahnenerbe on a firm academic footing, where it could deflect all criticism. This meant finding the right person to take charge of the institute, someone with impeccable academic credentials and a comprehensive understanding of Nazi doctrine.

Already, Himmler had met someone ideally qualified for the job. Dr. Walther Wüst was an authority on the ancient literature and religions of India, with a particular research interest in the migrations of the Nordic race. Small and solidly built, with hooded eyes, dark hair, and a round face adept at concealing his innermost thoughts, Wüst was a dean at one of Germany's most prominent academic institutions, Ludwig-Maximilian University in Munich. Just thirty-six years old, the scholar was an admirer of Adolf Hitler and a member of the Nazi party.

In nearly every imaginable respect, Wüst was the polar opposite of Wirth. He was cautious and calculating, calm and orderly, discreet and circumspect. He was polite but not pleasant, reserved but not shy, admired but not particularly liked. He was financially solvent and meticulously organized, and accustomed to keeping his own counsel. Those who thought they knew him, did not really know him at all. He was and had been for some time a *Vertrauensmann*—informer and spy—for the SS Security Service.[8]

On February 1, 1937, Himmler appointed Wüst the new president of the Ahnenerbe.

WÜST WAS A very unconventional kind of Nazi. He was born on May 7, 1901, in the Palatinate city of Kaiserslautern, the son of a city official who was, as Wüst later observed, "of pure Aryan background."[9] At the age of nineteen, the young Wüst enrolled at Ludwig-Maximilian University, and over the next five years, he studied broadly, covering subjects as diverse as comparative religion, anthropological geography, and the origins and distribution of races in Asia. Under the tutelage of a prominent Indologist, Wilhelm Geiger, he learned to read Sanskrit and puzzle over the meanings of the Vedas, the ancient scriptures of Brahman priests. He was fascinated by the connections he saw between the old songs and legends of Europe and the ancient literature of the East.

After completing his studies, he found a teaching position at Ludwig-Maximilian University, first as a private lecturer in Sanskrit in 1926 and later as a professor. He had a dull, pedantic manner in the classroom, as one student later recalled, and his seminars tended to attract few students.[10] But his research on Sanskrit and Old Persian philology was well received, and he began to make a mark at the university. In 1927, he married a young Munich woman, Bertha Schmid. She had apparently given birth two months earlier to a daughter, and in conservative Catholic Bavaria, this likely gave rise to some tongue-wagging and gossip.[11] Wüst seems to have ignored this, however: he did not have much use for Christianity.[12]

Like many of his Nazi colleagues at the university in the early 1930s, Wüst had developed a strong research interest in the Nordic race, but he was in no rush to join the hoi polloi by taking out a membership in the party.[13] In 1933, however, after the Nazi seizure of power, he changed his mind. It had become plain to many German professionals that party membership was a necessary stepping-stone for future advancement.[14] Hundreds of thousands applied for memberships, swamping the party offices. Wüst was one of them. He received membership number 3,208,696 on May 1, just squeaking in before the party closed its membership rolls.[15] To some old party comrades, this made Wüst suspect as an opportunist.[16] As a means of erasing their doubts, he agreed to spy on his university colleagues and students for the SS Security Service.[17]

It was Wolfram Sievers, the Ahnenerbe's astute managing director, who first brought Wüst to Himmler's attention. Sievers had met the scholar in the offices of Bruckmann Verlag, a prominent Munich publishing house with important connections to the Nazi party.[18] He was deeply impressed by the breadth of Wüst's research interests, his soothing manner, and his apparent grasp of ancient history. A visit to Wüst's office, he declared to a colleague, was like spending time in "a German cathedral where one gains insight and reflection."[19] So in May 1936, when SS officials were scouting for new staff for the Ahnenerbe, Sievers approached Wüst and persuaded him to join the organization as a corresponding member.

A few weeks later, Sievers wrote to Himmler requesting that he take the time to meet Wüst in person. The Orientalist, he declared, "understands how to make the most incomprehensible scientific research findings understandable to the simplest man, as his lectures before the district [Nazi] party group and the National Socialist Teachers' Association have demonstrated. A short speech by Professor Wüst with the theme, 'The Führer's *Mein Kampf* as a Mirror of Aryan Worldview' in the large auditorium at the University of Munich resulted in fifteen minutes of applause."[20] All this intrigued Himmler greatly. Although he was deeply mired in work— restructuring most of the police forces in Germany, stamping out political opposition, and developing the network of concentration camps—he took time out from his pressing duties to meet Wüst. Indeed, he invited the scholar down to his alpine chalet in Gmund, a small town on Tegernsee, one of the most beautiful lakes in Bavaria. Wüst arrived for the meeting carrying a copy of one of his favorite books, the *Rig Veda*.[21]

The *Rig Veda* is the oldest of the Sanskrit scriptures. Composed around thirty-five hundred years ago, it consists of hymns addressed to the gods, and as such it marks the starting point of recorded Hinduism. Wüst, however, saw the *Rig Veda* as considerably more than a Hindu holy book. He believed it was also an important document of the Nordic race. Indeed, he had persuaded himself that members of the Nordic race had written it as they swept eastward out of Europe and colonized the deserts of Iran, the high mountain valleys of Afghanistan, and the rich river plains of India.[22] He also believed the *Rig Veda* contained clear traces of an ancient sun religion from Europe—the same religion that Wirth talked about when he described the rock art from Bohuslän.[23] Wüst brought the *Rig Veda* with him to Gmund in order to share these thoughts with Himmler. In the quiet of his host's study, he read aloud from its hymns, translating the passages into German.[24]

Himmler listened attentively. He was impressed by the Orientalist's erudition and eloquence. But he was riveted by Wüst's theories on the racial significance of the *Rig Veda*. Other extremist German scholars had speculated at length about primeval Nordic migrations to the east, but Himmler had never paid much attention to them. Indeed he had shown little real interest in Asia or Asian religion.[25] But Wüst's recitation from the *Rig Veda* and his earnest talk of an ancient blond-haired, blue-eyed ruling class in the Far East seems to have electrified Himmler, giving birth to a strange new passion for Asian religion.[26]

As the two men talked after the reading, Himmler considered other possible advantages to hiring Wüst. The SS was beginning to extend its tentacles into many diverse sectors of German life, creating what many historians and writers would eventually call "a state within the State."[27] Himmler intended to add Germany's universities to his empire. By placing SS officers in key academic positions, he planned to obtain control one day over everything taught in university classrooms. In this way, the Nazi version of history, prehistory, literature, genetics, and biology would replace authentic scholarship.[28] Sitting in his study at Gmund, it must have occurred to him that Wüst, a dean at one of Germany's most important universities, would make a fine new SS bridgehead.

Wüst, for his part, soaked up Himmler's admiration. For a pedantic Sanskrit professor accustomed to dispensing wisdom to a few bored students each year, the opportunity to mould the minds of many thousands of SS men must have seemed irresistible. And he felt genuine excitement about the Ahnenerbe's commitment to fieldwork. He had never traveled himself to India or Iran—indeed he had never set foot out of central Europe.[29] But he had avidly followed the results of German expeditions to Central Asia since 1929 and he felt certain that fieldwork in the region would yield new and important insights into the Nordic race.[30] A German researcher working in the Hindu Kush region of Afghanistan, for example, claimed to have discovered a high percentage of people with blond hair and blue eyes: Wüst considered them to be "people of our own blood."[31]

So Wüst did not hesitate to accept the new post at the Ahnenerbe. He explained to Himmler, however, that he wanted to remain in Munich so that he could continue his work at Ludwig-Maximilian University. Far from being offended, Himmler was delighted by this prospect. He agreed to furnish Wüst with an office and two assistants.

WITH HIMMLER'S STRONG backing, Wüst began taking charge of the Ahnenerbe, reading through the thick stacks of correspondence and documents that Sievers regularly dispatched from Berlin and familiarizing himself with its operations. By the fall of 1937, the Ahnenerbe boasted thirty-eight workers, eleven of whom were senior researchers, mostly folklorists and symbols scholars hired to assist Wirth with various aspects of his studies.[32] Wüst intended to expand the scope of the Ahnenerbe's research and greatly polish its scholarly and scientific reputation. The first step was to distance the organization from any out-and-out crackpots. One obvious target was Wiligut. Wüst thought him a crank, and saw to it that there was little further direct cooperation between the Ahnenerbe and Wiligut's office in RuSHA.[33]

Wüst also intended to weed out those he deemed scholarly upstarts. To do this, however, he needed to take the measure of his new staff, so on October 25, 1937, he convened a meeting of the senior researchers at the new Ahnenerbe offices in the exclusive neighborhood of Dahlem in Berlin. The brain trust—along with Wirth's mammoth collection of rock-art casts from Scandinavia—had outgrown its quarters on historic Brüderstrasse. So members of Himmler's personal staff had located temporary digs just a few blocks away from where Himmler himself lived.

Sievers called the meeting to order, while a secretary began recording the minutes. The staff had already heard a great deal about their important new head, and Wüst did his best to reassure them that he had merely called them together in order to become better acquainted with their research. The purpose of the meeting, he explained, was "to discuss concerns, experiences and successes with each other. Each co-worker in the Ahnenerbe was to consider himself to be in a close working relationship with his colleagues, and their guiding principle had to be achievement."[34] He then asked each researcher in turn to give a short report on his work.

Yrjö von Grönhagen took the floor first. The handsome young Finn had just returned from a second summer of fieldwork in Karelia. He had spent nearly five months shooting film footage of things that fascinated Himmler—Karelians taking part in traditional midsummer celebrations, sorcerers performing magical spells, seers conjuring up spirits, mourners conducting old heathen rituals for the dead in graveyards.[35] In addition,

Grönhagen had purchased nearly two hundred books on Finnish culture for the Ahnenerbe library. But his sense of satisfaction with the summer's work was rather short-lived.

In the space of just ten minutes, from 9:50 to 10:00 in the morning, the young Finn traveled the humiliating public arc from golden boy to has-been. In the process, Wüst let everyone know who was now in charge. He listened to Grönhagen's short report about his research, then proceeded to grill him mercilessly in front of the others on arcane linguistic connections between the Finns and the Aryans. Grönhagen could not answer a single one of his questions.

When the meeting was over, Wüst sat down and wrote a letter to Himmler, informing him of what had taken place. Grönhagen, he observed, "has shown that he does not unfortunately possess the necessary knowledge required of a department head. He does not recognize important parts of his study area, and he was unable to answer questions I had on fundamental publications that have recently appeared on Finnish-Indogermanic cultural connections."[36] The only solution for the problem, in Wüst's opinion, was to send him back to the classroom.

Himmler's new craving for academic respectability outweighed his fondness for the young Finn. Just as he had dumped Wirth as the Ahnenerbe president when it had been expedient to do so, he now agreed to Wüst's plan. A few days later, the Orientalist delivered the bad news to Grönhagen: he would be stripped of his new position as department head, reduced to the position of researcher, and sent back to the classroom. The young Finn was furious, throwing the entire Ahnenerbe offices into an uproar.[37] But he gradually accepted Wüst's verdict, particularly after the Ahnenerbe agreed to subsidize his university tuition.[38] In the end, Grönhagen continued to perform Finnish research for the Ahnenerbe, but his days of intimate meetings with Himmler had ended.

It was the beginning of a new era in the Ahnenerbe. Henceforth, Wüst intended to recruit scholars who would be much like him—bright, ambitious, and highly regarded in their respective fields. They would be scientists and scholars who would not mind bending the truth to fit the new political realities of the Reich. And they would be individuals who could be called upon reliably to perform a variety of important services for the SS, from indoctrinating recruits to conducting human experiments and secretly gathering intelligence for the Reich.

9. INTELLIGENCE OPERATIONS

ON MARCH 15, 1938, Adolf Hitler strode out upon the balcony of the old imperial palace in Vienna, surveying with satisfaction the restless, swarming host at his feet. In the late-winter sunlight, some two hundred and fifty thousand bystanders stood in Heldenplatz, chatting comfortably with their companions and craning their necks toward the old palace, the seat of the Habsburg dynasty for nearly six centuries. Glimpsing Hitler's familiar figure at the railing, they let out a loud, resounding cheer. Just three days earlier, the Gestapo had seized control of the Vienna airport and German Panzer divisions had pushed across the Austrian border, followed by convoys of security police. Overnight, Austria as a nation had disappeared from the map, absorbed into the Third Reich. Vienna, the gilded capital of Franz Schubert and Gustav Klimt, Johann Strauss and Sigmund Freud, had become little more than a provincial German city.

Far from resenting this news, however, many Viennese rejoiced. Church bells pealed joyously, while bands of brown-uniformed storm troopers, barely old enough to shave, roamed through once peaceful city quarters, belting out *"Sieg Heil"* and *"Heil Hitler"* at the top of their lungs.[1] Jeering mobs attacked Jewish men and women on the street, beating them viciously or forcing them to their knees to scrub sidewalks with buckets of soap and cold water. In the Heldenplatz, the crowd strained for a glimpse of the man many had worshipped from afar. Dressed in military uniform, Hitler announced publicly what most in the square below longed to hear.

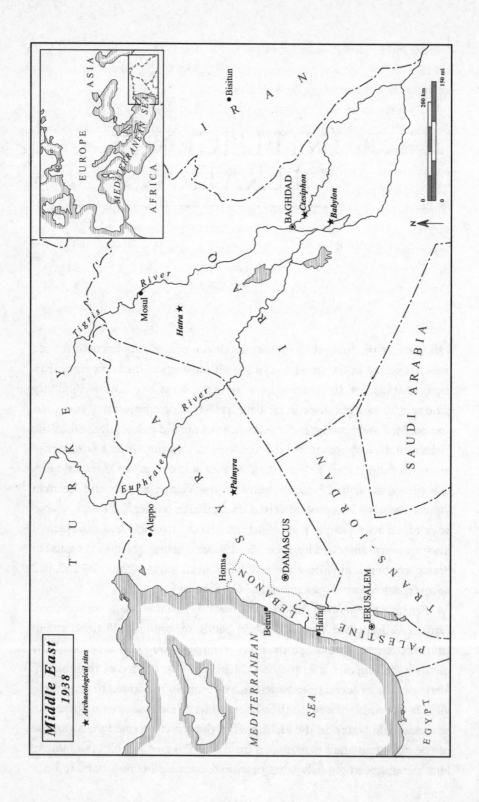

"As the Führer and chancellor of the German nation and the Reich, I now declare before history the incorporation of my native land into the German Reich."[2] A deafening cheer followed.

To prepare for this triumph, Himmler had departed Berlin two days earlier with a large entourage and an entire SS company in two Junkers Ju-52 transports. As the leader of the SS and the Gestapo, Himmler was responsible for eliminating all opposition to the new Nazi regime in Austria. To assist in this grim work, he called upon one of his most trusted officers, thirty-three-year-old Reinhard Heydrich, the head of the SS Security Service or SD. Heydrich, a tall, raptorial-looking man with pale blue eyes preternaturally alert to the world, but dead to most emotion, was a great admirer of the British secret intelligence service, MI6. He believed, without any foundation, that the entire English aristocracy served pro bono as secret agents for it. So he had set out to emulate MI6 in the early 1930s, personally recruiting university professors, business leaders, and government officials to work as unpaid SS agents.[3] From each he had demanded exact and, in some cases, exhaustive information on opponents to the Nazi party—details of their business affairs, finances, lovers, family secrets, character flaws, and daily habits, as well as up-to-date photographs, addresses, and telephone numbers, and the names, addresses, and telephone numbers of their friends and relatives.[4] His staff stored all this information on cards, then filed them away in an index in one of several categories, from Jews and Freemasons to "Political Catholics," "Bourgeois Conservatives," and "Nobility hostile to National Socialism."[5] Often they classified individuals according to their future fates in a Nazi regime. Some opposition members were destined clearly for arrest and execution; others were slated for intimidation in order to terrify them into submission.[6]

Heydrich and his officials had been slaving away feverishly for weeks on their Austrian card index in anticipation of the *Anschluss*, or union with Austria.[7] Soon after German tanks crossed the frontier, they drew up lists and dispatched special police task forces to make the desired arrests. Armed with the SD information, the squads had little difficulty locating their targets. They arrested and frequently murdered leading opposition figures before they had time to find hiding places.[8] The newly deposed chancellor of Austria, Kurt von Schuschnigg, for example, received a knock at the door just a few hours after his forced resignation. He was bundled into a car and taken off to a prison cell on the fifth floor of the Hotel

Metropol, the new Gestapo headquarters in Vienna.[9] There, for nearly seventeen months, Gestapo officers starved Schuschnigg and forced him to clean their toilets with his own wash towel, before finally dispatching the emaciated former statesman to a concentration camp. He did not see freedom again until the end of the war.

Satisfied that they had crushed the most serious Austrian opposition, Himmler and Heydrich returned to Berlin. It must have been obvious to Himmler, however, that the homeland of the Habsburgs was only a trial run. Hitler intended to unite all ethnic Germans in Europe under the flag of the Reich. Czechoslovakia, with its large German population in Sudetenland, was next on his lengthy list of desirable real estate, and it was conceivable that such aggression could push the Western democracies, wavering as they were, to war. In order for Germany to triumph, a wide range of intelligence was needed. It was time to recruit yet more agents and cast a broader net of espionage across Europe and as far away as the Middle East.

The Ahnenerbe, with its bright young staff and its expeditions and research trips to foreign lands, was an obvious place for Himmler to look for potential spies. A few of its researchers undoubtedly possessed the necessary resourcefulness, intelligence, and discretion for a good agent.

FOUR MONTHS AFTER the *Anschluss*, Himmler read a promising letter from Wolfram Sievers. The Ahnenerbe had received a proposal from two German researchers for a major new trip to the far borders of the old Roman Empire in the Middle East. Wüst had given the proposal his scholarly approval and passed it on to Sievers, who requested financing for it.[10] The researchers in question were a prominent classical scholar, Dr. Franz Altheim, and his collaborator Mrs. Erika Trautmann. Together they proposed a two- to three-month-long journey by train and car, from Eastern Europe all the way to the Middle East—Beirut, Baghdad, and Damascus.

Altheim was an expert on the origins of Roman religion and the history of the Latin language. At the age of thirty-nine, he had already built a major international career as a classical scholar, counting the German historian Oswald Spengler and other prominent intellectuals among his friends. Clever and irreverent, Altheim chafed at conventional armchair research. For years he had spent his summers wandering Italy's classical ruins in search of weathered inscriptions and worn statuary that others had over-looked. He now wanted to extend his studies to the far frontiers of the Roman Empire.

Trautmann, a rock-art researcher, was Altheim's lover. The daughter of a once wealthy estate owner in eastern Prussia, she had impeccable Nazi connections. Hermann Göring had attended cadet school with her older brothers, and according to one story, he had fallen in love with her as a young woman and asked to marry her, a proposal she declined.[11] Be that as it may, Göring never forgot his friends' little sister. Two decades later, as the portly Reich minister for aviation, he took her under his wing, inviting her to sit in his box at the theater and introducing her to his prominent Nazi colleagues.[12] His second wife, Emmy, looked a great deal like Trautmann, with her blond hair, oval face, high cheekbones, and thin lips.

The two researchers' proposed itinerary to the Middle East took them through several strategically important areas. The Third Reich was keenly interested in Eastern Europe and the gulf states, particularly in their oil fields. Germany possessed very little crude of its own, relying for much of its supply on the Ploesti fields in Romania.[13] "Without Romania," Göring is reported to have declared in 1937, "we cannot start a campaign."[14] Romanian oil producers made a tidy fortune from their exports to Germany, but the political situation in Bucharest had become unstable with the emergence of a new ultranationalist movement. Turmoil in Romania needed careful watching. Nothing could be allowed to disrupt German oil supplies.

Other major European powers were even more dependent on foreign reserves. Great Britain, for example, had turned Iraq into what amounted to a British colony after the First World War and helped develop its vast reserves. In 1934, Iraq joined the ranks of the world's major oil producers, shipping its black crude by a new pipeline to the Syrian port of Tripoli on the Mediterranean. From there, tankers ferried much of the production to Britain. It was a long, tenuous supply line, particularly in the event of a war in Europe. If German forces could seize control of the oil fields near Kirkuk or sabotage the pipeline, then they could sever one of Britain's vital arteries.

A couple as clever and charming as Altheim and Trautmann, with their easy entrée into academic and government circles, could be extremely valuable intelligence assets. The two could readily locate and befriend Nazi sympathizers, individuals who could later be recruited as local SD agents. In trains and restaurants, hotels and museums, the witty scholar and the statuesque blonde could easily glean useful bits of intelligence from casual conversations with foreign nationals and local inhabitants. Moreover, they might even succeed in contacting Bedouin leaders on the excursions they planned into the deserts of Iraq and Syria, gathering valuable information on their predisposition toward Germany.

All in all, Himmler and the members of his staff must have taken a keen interest in recruiting the pair. But they must also have devoted some thought to Altheim's reaction to such a proposal. The scholar was not a member of the SS. Nor had he ever taken out a membership in the Nazi party or in any other of a multitude of Nazi organizations. He had never expressed the slightest desire to awaken Germany to some glorious future or resuscitate some mythical Aryan past. He was a leading European intellectual, a member of the café society that thrived on ideas and took matters such as science and scholarship seriously. He was perfectly capable of seeing through the web of Nazi deceit. Worse still, Altheim was one of the least conventional of men.

The scholar had inherited his bohemian ways from his father, Wilhelm Altheim, a talented artist whose sketches and paintings of country life earned much critical acclaim. The elder Altheim liked nothing better than to shock the local bourgeoisie. He dressed, as the mood took him, in cowboy gear or a hussar's uniform and kept a menagerie of animals about the house—dogs, donkeys, monkeys, and even a bear. Inordinately fond of weapons, he shot holes through the curtains in the house and fired slingshots at his neighbors as they gardened. He turned up for galas in Frankfurt dressed in an expensive suit and riding a donkey. As one biographer later noted, he was "an eloquent, imaginative jokester."[15] When he sold his paintings, he often disappeared on drinking binges, sometimes for weeks at a time. Despairing of all this, Altheim's mother eventually left her husband and the artist sunk into depression, finally shooting himself on Christmas Day 1914.[16] His son could never again bear to celebrate the holiday.[17]

The young Altheim joined the German army in 1917 and was dispatched to a school for translators, then posted to Turkey.[18] He seems to have been fascinated by his first experience of Asia. After Germany's defeat in the war, he began casting about for a career. He had hoped to follow in his father's footsteps, becoming a sculptor, but he lacked artistic talent. So he went to university, majoring in classical philology, archaeology, and linguistics, and worked in a bank to support his studies. Each year, he journeyed to Italy with grants from the German government, soaking up all he could of the classical world. "His intellect was restless and his thirst for knowledge was . . . immeasurable," recalled one writer.[19]

Never handsome, he was by then a short, slim man with a round, doughy face, a shiny shaven pate and a wry sense of irony and humor. Despite his physical shortcomings, he had a definite way with women. He had married a fellow classicist, a teacher of Latin and Greek, but the marriage quickly soured and he soon moved on to other conquests.[20] "All women loved him," observed one of his closest companions and lovers half-jokingly later. "It was terrible."[21]

His professor, historian Walter Otto, recommended him for a post as a private lecturer at the University of Frankfurt in 1928. To supplement his income, Altheim began dealing in art on the side. He had an eye for beautiful things and a talent for buying low and selling high.[22] As Altheim settled into teaching, Otto introduced him to Leo Frobenius, one of the most celebrated European anthropologists of the day. Frobenius, a tall, bluff man who peered out at the world through a fussy monocle, was an Africanist who specialized in the study of ancient art and who had traveled the world studying it. The two men became fast friends, and it was likely Frobenius who brought Altheim to the attention of the deposed German kaiser, Wilhelm II. Once a year, Wilhelm invited a select group of archaeologists, historians, and art collectors to his home in the Netherlands to dine royally, drink prodigiously, and present papers on a variety of scholarly subjects. Wilhelm asked Altheim to join them. The young classicist loved these occasions and apparently bragged about them to friends, showing off an engraved silver trophy Wilhelm had given him.[23]

Socializing with royalty, spending summers in Italy, and cultivating a circle of influential international scholars, Altheim seems to have paid little attention to the rapid spread of Nazi ideas among the students and faculty of German universities. He shrugged off suggestions that he join the party and avoided serious political discussions. Many of his colleagues had no idea where he stood politically. They struggled to define his views in official reports, using phrases such as "no *Homo politicus*," "focused on scientific and aesthetic fields," or more simply "impenetrable."[24] When pressed for an expert opinion on Altheim's reliability, a certain Dr. Cordes, head of the lecturers' organization, was forced to take a stab in the dark. "In the political respect," he reported, "[Altheim] has not been very active before or after 1933. His whole political attitude is probably impeccable, not very active and certainly harmless."[25]

Increasingly, however, such indifference to politics carried a price in academic circles. In 1935, an influential colleague wrote a damning letter about Altheim to the Ministry of Education, which was considering promoting him to a much-coveted post as professor in Frankfurt. The poisonous letter took firm aim at Altheim's reluctance to incorporate Nazi doctrine into his work. "He belongs," noted the writer disdainfully, "to those intellectuals who still try to write ancient history as if the problems of 'race' did not exist. This pale aestheticism will not do in a field where there are especially great tasks for research in a National Socialist spirit."[26] After a year of cautious deliberation, the ministry decided to overlook Altheim's political shortcomings and appointed him professor.

It must have dawned on Altheim, however, that his academic future hung by a thread: little pressure would be needed to snap it. A growing number of Nazi students took their seats in the classrooms, recording their professors' comments and judging lectures on the basis of their political content. In addition, secret SD informants attended faculty meetings, taking note of any untoward remarks of their colleagues. When Altheim delivered a lecture in London in 1936, for example, he chose to speak in English, rather than German. This was observed disapprovingly by an informant and duly recorded as a black mark in his file at the Ministry of Education.[27]

By 1936, Altheim could no longer shun politics if he wanted to travel abroad and conduct his research unhindered. He needed to join the Nazi fray or find a solid connection to influential Nazis: quite likely he had to do both. And it was right about this time that he began his love affair with Erika Trautmann, a very unconventional woman with superb Nazi contacts.

TO A GOOD NAZI, the most important part of a woman's anatomy was not her brain, but her womb. "The mission of woman is to be beautiful and to bring children into the world," explained future propaganda minister Joseph Goebbels in 1929. "This is not at all as rude and antiquated as it sounds. The female bird pretties herself for her mate and hatches the eggs for him. In exchange, the mate takes care of gathering the food, and stands guard and wards off the enemy."[28] By these Nazi standards, Erika Trautmann was a failure. In 1936, she was thirty-nine years old and childless. She had little apparent interest in prettying herself for her mate and even less in staying home, accompanying as she did scientific expeditions to France and Spain each summer.

She had been born into a wealthy family, the Nehrings, just outside Konitz, in what was then a province of Germany known as Posen. Her father was a descendant of Johann von Nehring, a celebrated seventeenth-century Prussian architect who designed part of Charlottenburg Castle in Berlin, as well as nearly three hundred other buildings.[29] Trautmann grew up on the family's 538-acre country estate with five brothers and one sister.[30] Her older brothers were away a good deal at cadet school, and Trautmann and her younger sister received their lessons from a governess and a local minister. It was "a dreamy, lonely but quite intense life and experience," recalled her sister later.[31] Even so, the Nehrings were a large and sociable family: the two girls spent their summers visiting back and forth with relatives.

The Great War shattered this rather idyllic childhood. Two of Trautmann's older brothers died at the front; a third was taken prisoner and sent to Siberia.[32] In December 1918, her mother died, possibly a victim of the influenza epidemic, and when the Weimar government reluctantly handed Posen back to Poland, the Polish government confiscated the family estate. Trautmann and her father and sister fled to Germany. She moved in for a while with an older brother, a physician in Magdeburg, but eventually she escaped to Berlin to attend a women's school, the Lettehaus.[33] In her mid-twenties, she was "a very striking blond woman with whom you could have an intelligent conversation," recalled her cousin Walther Nehring, who became a general in the Wehrmacht during the Second World War.[34]

Trautmann possessed a talent for sketching and enrolled in the state arts and crafts school in Berlin.[35] It was the height of the Jazz Age. The school, located in a bright, airy building at 8 Prinz Albrechtstrasse—the very place that was to become Himmler's headquarters nearly a decade later—bubbled with laughter, gaiety, and high-spirited escapades. The students delighted in throwing costume balls and masquerades, transforming the school's corridors into Arab harems and Turkish minarets, bars and buffets, all pulsing as a band pumped out the latest music.[36] Trautmann spent a year and a half at the school. She took classes in life drawing and fashion design, studying under one of Berlin's best-known couturiers, Otto Ludwig Haas-Heye.[37]

She stayed until the spring of 1924, struggling no doubt to pay her tuition in a time of rampant inflation. A year later, she married a young German civil engineer, Bernhard Trautmann.[38] The couple moved to Frankfurt. After the Pernod-laced gaiety of art school, however, the subdued life of a bourgeois matron seems to have held little appeal for Trautmann. "Her deep interest in the intellect," noted her sister in later years, "as well as an amazingly strong artistic sense spanning all eras, defined Erika's life."[39] In 1933, she landed a job as a scientific illustrator for the Research Institute for Cultural Morphology, founded and directed by Leo Frobenius. Trautmann was one of half a dozen or so women employed in this work.[40]

Housed in one of the loveliest old buildings in Frankfurt, near the elegant Thurn und Taxis Palace, the institute specialized in the study of ancient art. It brimmed with colorful, eccentric people who were quite willing to pack their bags one day for Timbuktu, the next for Khartoum. As part of the institute's extensive research program, it regularly dispatched small teams of brilliant young artists and scholars to far-flung corners of Africa and Europe. Often living out of tents and traveling rough, the teams recorded on film and in watercolor the mysterious cave paintings and rock carvings of ancient human cultures. One prominent expedition member, Count László Almásy, guided team members through the deserts of Libya in 1933 and later entered modern popular culture as the tragic hero of Michael Ondaatje's novel *The English Patient*.[41]

Frobenius chose Trautmann for the institute's 1934 expedition to the famous Paleolithic caves of France and Spain.[42] As a country girl raised amid the freedom of a great estate, Trautmann seems to have reveled in the experience. She went about the work with an intensity that one of her companions still recalled nearly seventy years later.[43] And she blended seamlessly into expedition life. The team spent the summer crawling on their bellies one by one, down narrow subterranean passageways, and painting by the flickering light of kerosene lanterns. They bathed in goat stables and lounged on the stone seats of the Roman amphitheater in Nîmes. They attended a bullfight in Barcelona and chatted late into the night over wine and dinner.[44]

Trautmann had a fine, observant eye for detail and a knack for discovery. Frobenius made her a regular on the summer expeditions. In August 1936, he dispatched her and two other artists to Val Camonica in northern Italy under the direction of a classical scholar whom he had befriended and recruited for the institute—Altheim.[45] Italian researchers had made a major discovery at Val Camonica. Beneath thick layers of moss and sod lay large sprawling panels of antique rock carvings dating back to the Bronze Age, if not earlier. Frobenius hoped to display detailed paintings of the new finds in a forthcoming exhibition. In all likelihood, Altheim hoped for a new glimpse into the ancient cultic practices of Rome's ancestors.

Sprawled side by side together on the sun-warmed rock, Trautmann and Altheim spent weeks tracing the outlines of the weathered carvings. Far from her engineer husband and her conventional life in Frankfurt, the artist fell under the spell of the restless scholar, with his bohemian ways and seductive charm. Late at night, as they dined out in the little country restaurants, she listened in delight as Altheim talked of the ancient world with an eloquence and wry humor that she had never encountered before.

A few months after her return, Trautmann obtained an official separation from her husband. Many of her Nehring relatives were scandalized by her behavior. They disowned her, refusing to speak to her or of her for decades.[46]

RELUCTANT TO RETURN to her old life, Trautmann began skillfully campaigning to advance her lover's career. There was one clear avenue of opportunity. Prominent Nazis, who fanatically touted all things German, had long struggled mightily to explain why it was that Romans, not Germans, had colonized and conquered much of the known world, forging a magnificent empire that commanded awe even two millennia later. Despite all the reverential talk in Germany of ancient Teutonic valor and manliness and bloodlust, many Nazis envied the Italians the glory of their ancestors.

The only acceptable way of explaining Roman civilization for German extremists was to furnish it with a Nordic pedigree.[47] The grandeur of Rome, they claimed, could be traced back through distant time to blond-haired migrants from the north.[48] Italian scholars loathed such talk, bristling, quite naturally, at any suggestion that Rome was at heart a Teutonic civilization.[49] But German researchers, particularly those eager for advancement in the Third Reich, didn't care whose feathers they ruffled. They kept an alert eye open for any scientific evidence that might support such contentions. At Val Camonica, Altheim and Trautmann thought they had discovered something that would help clinch this theory—carvings of deer, chariots, carts, warriors in battle, and sacred rituals that looked similar to the Bronze Age engravings Wirth had studied in Sweden.[50]

Together, the pair published a paper on the Val Camonica art in a German historical journal.[51] Trautmann, well aware of the importance of such an article, passed it on to Göring, who proceeded to introduce the couple to Himmler.[52] The SS chief was delighted with the pair's research, particularly with its connection to the Swedish rock art that Wirth had studied. He agreed to finance another season of fieldwork, and in the summer of 1937, Altheim and Trautmann departed for Italy and the Adriatic coast of Dalmatia in what is now Croatia to find more evidence of Nordic migrants.[53] They dispatched regular reports from the field, outlining their new finds. On their return, Himmler instructed Sievers to contact the couple and recruit them for the Ahnenerbe.[54] Any concerns Sievers may have had about Altheim's commitment to the Nazi cause were soon set aside. The couple readily agreed to join the SS organization.[55]

Increasingly, Altheim began to take on the coloration of a German ultranationalist. In March 1938, for example, he wrote to a German colleague who had been highly critical of his work, and insisted that he had mended his ways politically. "The completed work [at Val Camonica] shows that the topic, 'The Nordic race and its migration to Italy and South-East Europe,' is at the heart of my work."[56] He also found a way of injecting the subject of race into a major new research project he was planning along the frontiers of the Roman Empire.

For some time, Altheim had felt restless in his work. He worried that he was merely following the same deeply worn ruts as his colleagues, who concentrated exclusively on the political, social, and economic factors at work within the Roman Empire, rarely looking at the influences which came from beyond. Altheim dreamed of taking a radically different approach.[57] At Rome's height, the law of the Caesars stretched all the way from the rocky shores of the Atlantic in Spain to the deserts of the Persian Gulf. Altheim desperately wanted to investigate the complex relations between Rome and the various civilizations and tribes of Asia and Africa. To examine physical evidence of their interplay, he yearned to travel to the empire's most remote frontiers, examining the distant caravansaries, entre-pôts, outposts, and battlefields where Rome had greedily sought to expand its might.

He was particularly keen to journey to the old imperial frontiers in Eastern Europe and the Middle East, taking Trautmann along as a collabo-rator and photographer. But such a research trip would be costly and re-quire significant sums of foreign currency—something in very short supply in the Reich.[58] After the seizure of power, the Nazi government had forced many factories and mills to turn their production toward manufac-turing armaments. As a result, German exports had shrunk dramatically, bringing in only tiny dribbles of foreign exchange—too little to pay for the vital raw materials the Reich imported.[59]

In search of foreign currency, Trautmann chose to first approach her old friend Göring, rather than her new associates at the Ahnenerbe. Hitler had placed Göring in charge of the Reich's Four-Year Plan in 1936, a posi-tion in which he wielded enormous economic clout in Germany. But the former aviator—an astonishingly rapacious man who was in the process of renovating his country home to include an old-style beer hall, a bowling al-ley, a lion kennel, a fifty-seat cinema, a personal museum, and a large gallery hung with Old Masters and fine Gobelin tapestries—was prepared to part with only 4,000 reichsmarks in foreign exchange, the equivalent of some $20,900 today.[60] So in late June 1938, Altheim and Trautmann wrote to the Ahnenerbe, requesting another 4,000 reichsmarks for the trip.[61]

Altheim carefully cast the project in racial terms. The purpose of the trip, he explained, was to examine evidence for a great power struggle that wracked the racially diverse Roman Empire during the third century A.D. This struggle pitted "Indogermanic peoples of the North" against "the Semites of the Orient" for control of the empire.[62] Such an examination, he added, would break important new scholarly ground. American and British scholars had viewed this tumultuous period largely from a Marxist point of view, focusing on class warfare within the empire.[63] Altheim, however, fa-vored a very different approach. He and Trautmann proposed to study the conflict "between peoples and races" and determine how it was that the Nordic race triumphed in the struggle for power over the empire.[64] This, they promised, would be a significant contribution. "Here we have a great opportunity to present the meaning of race in the writing of history."[65]

Wüst read the proposal with great interest, noting at the top of his copy, "Agree very much."[66] It was precisely the kind of research that he wanted to encourage in the Ahnenerbe. And quite possibly it was Wüst—an SD informer himself—who first hit upon the idea of putting Altheim and Trautmann to work gathering intelligence on their travels. European scholars, after all, had proven remarkably adept agents in the Middle East. Nearly a quarter of a century earlier, a young Oxford graduate, T. E. Lawrence, joined a crew of British archaeologists at the famous site of Carchemish in Syria.[67] The lanky young scholar appeared to be little more than a keen student of the past, joining the excavators with enthusiasm. But in reality the future Lawrence of Arabia was working for a British intelligence agency, maintaining surveillance on a German railroad line under construction from Berlin to Baghdad.

As attracted as Altheim may have been to intelligence work in the Middle East—and indeed he later spoke with great admiration of Lawrence of Arabia[68]—he could not have been blind to the moral corruptness of the Nazi regime. Altheim, for all his scholarly preoccupation, must have noticed the repressiveness of the Nazi state and the increasingly brutal treatment accorded to Jews. Two and a half years earlier, the Nazi government had summarily stripped all Jews of their citizenship and prohibited them from working in any public office. Since then, Nazi officials had greatly stepped up the pace of persecutions, barring Jews from tutoring Aryan children, studying medicine, practicing medicine (except upon Jewish patients), receiving university degrees, playing music by Beethoven or Mozart at Jewish cultural events, or even bathing with Aryans in public baths and spas.

Altheim and Trautmann, however, overcame any qualms they may have had about collaborating with Himmler and his scholars. Outfitted with the right official documents, signed and stamped by the proper authorities, and equipped with the necessary sum of foreign currency, neatly arranged by the Ahnenerbe, they headed east.

A WAVE OF INTENSE dry heat greeted the couple as they stepped out of the central train station in Bucharest in August 1938. Along the city's spacious tree-lined streets, young women in elegantly tailored summer dresses strolled arm in arm, eyeing the colorful bouquets of flower sellers. Young officers in gold-braided uniforms ambled past old men in suits shiny with wear. Along rows of haberdasheries and bookshops and furniture makers, peasant men in high straw hats and colorfully embroidered chemises laid out plump melons in baskets on the sidewalks, urging passersby to stop and buy.[69] The two travelers, dusty and weary from their train trip, searched for their hotel.

For the last few days, Altheim and Trautmann had tramped the remote forested hills of northern and southwestern Transylvania with Romanian scholars, searching out the moss-covered ruins of Dacian kings.[70] The ancient Dacians, whom Altheim regarded as a warlike Nordic people, had founded a magnificent kingdom with a capital near the modern village of Hunedoara. Finally defeated by the Romans in the early second century, the Dacians had been absorbed into the Roman Empire. Altheim had been eager to see their fallen fortresses. But as he and Trautmann journeyed by car along the rutted country roads of Transylvania, some little more than cart tracks, they could not help but notice how preoccupied their Romanian hosts were with the political events of the day. Indeed, turmoil seethed in nearly every village and town, stirred by the feverish supporters of a Romanian paramilitary organization known as the Iron Guard. "Everything," noted Trautmann and Altheim in a later report, stamped SECRET, "was overshadowed by the events regarding the Iron Guard and the government."[71]

The paramilitary group was the creation of a certain Corneliu Codreanu, a handsome, charismatic Romanian lawyer whose hatred of Jews—a major segment of the country's population—rivaled that of Hitler.[72] Codreanu had first encountered the Nazi party in 1922 while he was pursuing his studies in Berlin. He admired Hitler's anti-Semitic platform. On his return to Romania, he set out to build a strong nationalist movement. He dressed in a white peasant costume and rode a snowy white horse through the countryside, preaching a bizarre blend of Christian mysticism and rabid nationalism, all wrapped in a vicious form of Jew-hating.[73] He urged his followers to dig earth from the former battlefields of Romania—where the country's heroes had shed their blood—and tuck it inside little cloth bags around their necks. His speeches stirred the hearts of Romanian peasants, some of the poorest in all Europe, who came to regard Codreanu as an emissary of the Archangel Michael.[74]

Indeed, so successful was the former lawyer in rallying the countryside that he and his followers began to alarm Romania's king. Carol II was a grandson of Queen Victoria, an imperious playboy far more interested in seducing beautiful women and perfecting his bridge game than he was in the tedious business of governing the impoverished country he had inherited.[75] He was venal and greedy and he broke his promises with gay abandon. All this, however, Romanians may have been willing to overlook. But they could not forgive his choice in mistresses. They hated his favorite— Helen Wolff, popularly known as Madame Lupescu, *lupescu* being the Romanian word for wolf. She interfered terribly in political affairs, but her chief crime was something beyond her control. She was Jewish.[76]

In the spring of 1938, Carol had seized dictatorial powers for himself, and soon after he ordered the arrest of his charismatic rival Codreanu and other key members of the Iron Guard. The countryside blazed with unrest. In the old historic capital of Transylvania, Altheim and Trautmann learned that two prominent scholars, Professor Dumitrescu and Professor Christescu, had mysteriously disappeared. "Their 'whereabouts were unknown,' i.e. they had been arrested," noted Altheim and Trautmann in their report.[77] But the couple had met someone whom they considered potentially useful to the Reich—Constantin Daicoviciu. Daicoviciu was the director of the Institute of Classical Studies in Cluj-Napoca. As one of Romania's leading classical scholars, he had kindly taken the Germans on a personal tour of one of his Dacian excavations. They had found him "an impressive personality."[78] He was not a fascist. Indeed, as they noted in their report, "his political position is neutral. However, he is a former reserve officer in the Austrian-Hungarian empire, has an encompassing grasp of German scholarship and possesses one of the most capable minds."[79]

After thanking Daicoviciu for his help, the pair journeyed on to the Romanian capital. Rising up from the dusty Wallachian plains, Bucharest was a surprisingly worldly city that looked more to the West than to the East. Wealthy city fathers had spent a fortune hiring French architects to bring a note of Parisian elegance to the city. They had laid out leafy parks and broad boulevards and streets of opulently decorated buildings that would not have looked at all out of place in the 16th arondissement. They had opened chic cafés and decadent nightclubs with names like the Alhambra. Indeed, so deep was the love of all things French in Bucharest, that architects had even built a plaster replica of the Arc de Triomphe, complete with a Romanian version of the Champs-Elysées. By the late 1930s, travelers had begun to refer to Bucharest as the "Little Paris of the Balkans."[80]

Rested and refreshed, Altheim and Trautmann began their official work in the city, poring over museum collections of old Roman inscriptions and texts, and chatting with high-ranking museum officials. They took a keen interest in the aristocratic director of the Municipal Museum. Grigore Florescu was a member of Romania's privileged class, the *boyar*; his wife's kin included two prominent leaders of the Iron Guard, General Cantacuzene and Prince Cantacuzene. Altheim and Trautmann found him fascinating. "We learned an excellent series of facts from Florescu," they noted later in their report. "He described the adventurous flight of the two Cantacuzene in disguise. He told us of the battle between the Iron Guard and the role of the Jewish Madame Lupescu. Not only does she possess the 'heart' of the king (many have followed after her), but she still has an inexplicable hold on him and determines his political actions. The position of the large Jewish banks in Bucharest is more powerful than ever."[81]

The two scholars then went on to evaluate Florescu's future usefulness to the Reich. "Mr. Florescu speaks fluent German, is related to almost all the big Romanian families (Bratianu, Bibescu, Ghika, Cantacuzene, etc.), and because of this and his access to the court and other high positions, he is full of information. Should his position become endangered, then he would like to flee to Germany, where he owns land because of his German relatives. We would like to point out that through him, we could win 1.) a supporter of the Iron Guard, 2.) a friend of Germany, 3.) a sharp and well-informed mind."[82]

Already Altheim and Trautmann had slipped smoothly into their assignment as spies.

1938

DURING THE SWELTER of late summer, Altheim and Trautmann journeyed by train from Bucharest to Istanbul and Athens, then boarded a steamer bound for the Middle East. In Beirut, where men passed their evenings languidly smoking nargilehs and drinking cups of bitter Turkish coffee, they eased themselves into the Middle East. Alert to their political assignment, they quickly cultivated a helpful friend, a businessman from Northern Ireland who was clearly no stranger to German politics. "A constant companion and loyal helper," they reported, "especially in the first days, was the Irishman JAMES H. JACQUES, (Belfast, Ireland, 35 HADDINGTONS GARDENS). He is an enthusiastic supporter of the NSDAP, knows *Mein Kampf* and the most important publications and has them constantly with him. He arranges for these works to be distributed in his home, Ireland, and on his worldwide trips."[83]

Perhaps it was Jacques who helped them make arrangements for the next stage of their trip to Damascus. In the Syrian capital, where Roman administrators had once plucked fruit from local fig and pistachio trees, muezzins chanted and sang, calling the faithful to prayer. Citified Arabs, tribal Bedouins, Kurds, Turks, Circassians, and Armenians crowded the narrow passageways of the souks. Beneath the noisy bustle, however, the atmosphere was tense. Syria's largely Muslim population chafed under the colonial-style rule of the French. Moreover, recent events in neighboring Palestine had added to their discontent. With the rise of the Nazi party in Germany, European Jews had begun flooding into Jerusalem, eager to escape further persecution. This had incensed Palestine's Muslims. To keep the peace, the British proposed dividing Palestine into two states—one Jewish, one Arab—a prospect which enraged much of the Middle East.

All the Arab unrest had alarmed French officials in Syria. They suspected German designs on the region, and according to Altheim and Trautmann's secret report, the local police kept a watchful eye on the pair. "The constant surveillance of the 'Sûreté Général' limited our movement," reported the couple later. "We did not make any connections with French scholars."[84] Ordinary Syrians, however, offered the two Germans an effusive welcome: many seemed to think of Hitler as a natural ally against the Jews, something that the two Germans found intriguing. "The conflict in Palestine resulted in a deeply felt reaction, and each word of the Führer on this subject made a great impact, reaching from Syria into the deepest corners of Iraq. A flyer was given to us by a Mr. GEORGE I. SAAD (Firma IBRAHIM I. SAAD & Fils, Beirut, POSTAL BOX 66) with ten commandments against the Jews written on it. He brought it back from Iran, where thousands of examples are distributed. (We will be sending an original one in the next few days.) Herr Saad and his wife (a Belarussian) are, like all Syrians, strongly anti-French and anti-Jewish."[85]

From the intrigues of Damascus, the pair headed south and east. As their train car clattered and clanked over the rails, they listened to the unfamiliar cadence of Arabic among their fellow passengers, and gazed out the window at the tiny mud-brick villages and the herds of woolly sheep grazing on little more than sand and the swaying figures of camel riders in the distance. After crossing hundreds of miles of sand and dune, sparse oases and plantations of date palms, they arrived at last in Baghdad, a city of some four hundred thousand. In the scorching heat of midday, men in fezzes and long white cotton *thoubs* ambled down the main thoroughfare, past an odd mélange of small flat-roofed shops and British colonial buildings. No one seemed in much of a hurry.

After centuries under the thumb of the Ottoman Turks and more than a decade of British rule, Iraq had become fully sovereign in 1932. Germany was keen to cultivate it as an ally. The German envoy to Baghdad, Dr. Fritz Grobba, worked diligently to spread Nazism to the Middle East. His office in Baghdad produced a daily newspaper, the *Al-Alim al-Arabi*, which had, among other things, published an Arabic translation of *Mein Kampf*. His staff handed out Nazi pamphlets and organized showings of anti-Semitic films.[86] Grobba himself did his best to capitalize on a growing interest in fascism in Iraq. He arranged for the head of Hitler Youth, Baldur von Schirach, and fifteen assistants to visit Baghdad in 1937 to lend their assistance to a new fascist-style youth organization, Al-Futuwwa.[87] The following year, Grobba arranged for the leaders of Al-Futuwwa to travel to Nuremberg to attend the annual Nazi party rally.[88]

Grobba was delighted to see the two German researchers. He introduced them to his wife and his circle of friends in Baghdad, which included Dr. Julius Jordan, a prominent German archaeologist who had founded a local branch of the Nazi party.[89] And with Grobba's assistance, Altheim and Trautmann hired a driver and car and set off to photograph and study southern Iraq's historic ruins. Motoring along the eastern bank of the Tigris, they stopped at Ctesiphon, the ancient winter capital first of Parthian kings and later of Persian monarchs, whom many German scholars of the day considered to be Nordic.[90] Roman armies had attacked the city five times in the second and third century A.D., racking up four victories. Only in the final battle did the Persian forces defeat them. While Altheim mused over the reasons for this victory, Trautmann photographed Ctesiphon's famous towering arch. Then the two pushed on to Babylon.

After considerable wandering, they returned briefly to Baghdad, then headed north with a car and private driver. At Assur, they met someone very important in the Arab world. Sheikh Adjil el Yawar was a leader of the Shammar, the most prominent Bedouin tribe in northern Iraq and Syria. He was an astute statesman and a fine soldier who commanded the northern Camel Corps. After the fall of the Ottoman Empire at the end of the

First World War, he had skillfully represented the interests of the Sham-mar during international deliberations over the formation of the new Arab states, greatly impressing one knowledgeable European. "From the first moment I saw him," noted British advisor and intelligence agent Gertrude Bell in a private letter in 1921, "I reckoned him foremost of the Shammar sheikhs in character and influence. He is 6 ft. 4 in. odd, a powerful magnif-icent creature; not an ounce of spare flesh on him, hands you would like to model, not too small but exquisitely shaped. Under his red kerchief, four thick plaits of black hair fall to his breast."[91]

El Yawar had helped shape the boundaries of Iraq, but like many other Arabs in the Middle East, he had grown restive under British and French rule. He was searching for new allies. In 1937, noted Altheim and Traut-mann in their report, the Bedouin leader had traveled to Berlin at the invi-tation of the Ministry of Foreign Affairs, which had given him a warm welcome.[92] In Assur, he insisted that the two researchers stay as his guests. The German scholars and the elegant sheikh clearly hit it off. They dined leisurely and talked through an interpreter for two days about politics. "We had to tell him continuously about Germany and the Führer," observed Alt-heim in his report. "For him, it was enough to hear from us that the Führer would never allow for a Jewish state in Palestine."[93]

Almost certainly, Altheim made no mention of certain Nazi racial the-ories that classified the Arabs as a Semite people, greatly inferior to the Aryan race. Nor did he likely make any mention of the stated goals of his own research—namely to study the great battles of the northern Indo-Germanic people and the inferior peoples of the Orient. Instead he paid close attention to his host's political views and aspirations. The Shammar sheikh, he noted, seemed very keen on following the example of Saudi Arabia's new Bedouin king, Abd al-Aziz ibn Saud. With only a small force of men, Ibn Saud had captured Riyadh, uniting the entire country under his kingship. The Shammar sheikh hinted at doing the same in Iraq—but as the couple pointed out, he needed foreign support. "The Sheikh is partially independent and receives money from the Iraq government. We estimate the Sheikh's army to be around 4,000–6,000 warriors: they suffer from older weapons (German guns from the World War) and insufficient am-munition. The example of Ibn Saud is in their minds; from our point of view, we think that he hopes for a European war, which will allow him to declare independence. He identifies with Germany and sets his hopes on it."[94] Indeed, the Bedouin leader asked the two researchers to send him literature on the Nazi government, preferably in picture form.

To ensure the safety of his honored guests, the Iraqi sheikh arranged for an armed guard to accompany them on the next leg of their trip—a desert trek to the ancient ruins of Hatra. His brother Sheikh Mesch'an agreed to serve as their guide. Before the pair departed, however, the Bedouin leader conferred on Altheim the title of honorary sheikh of the Shammar—a mark of distinction that the scholar later included proudly on his business cards.[95] Thus suitably feted, Altheim and Trautmann bid farewell to their host and set off by car across the rolling desert for Hatra.

With their Bedouin escort, the two German scholars explored the sprawling ruins of Hatra. Lying along an ancient boundary between the Roman and Persian Empires, Hatra had blended both the East and the West into its distinctive religion and culture. Along its honey-colored stone walls, Altheim admired what remained of the ancient statuary, spying with delight a carved mask of the Greek goddess Artemis. To commemorate their visit, the two researchers posed for a series of photographs in front of the toppled walls, wearing borrowed Arab headdresses and robes. Trautmann gamely attempted to clamber atop one of the gentler Bedouin camels, and Altheim later joked that the Sheikh Mesch'an was so enamored with his companion that he offered to pay fifty camels for her.[96]

When they had finished their explorations, the two German researchers returned to Baghdad and from there headed west. They stopped at the ancient city of Palmyra, a spectacular place, and chatted to the local people, gleaning what snippets of political information they could. "The Bedouins," they concluded in a later report, "are always ready to attack at any opportunity. They told us that in the event of a European conflict, things would immediately erupt. In addition to IBN SAUD, they speak the names 'Hitler' and 'Mussolini' as if they are holy. . . . The routes of the oil pipelines are hidden but every Arab and Bedouin knows exactly where they are. Time and time again we were told that the PIPELINE ran here or over there."[97]

BACK AT HOME in Germany, Altheim and Trautmann diligently wrote a long secret report by hand, appending several personal recommendations to the end. Among these was a proposal to strengthen their blossoming friendship with el Yawar by a follow-up trip to Iraq. "The connection to Sheikh Adjil will thereby be maintained," they observed. "He himself has urgently invited us. We recommend sending an SS officer along also. There are many possibilities for scholarly studies, as our recent results have shown. A present of a hunting gun and sufficient ammunition would serve to deepen the friendship."[98] In addition to this, they advised sending the sheikh the requested propaganda material.

Sievers sent the report to be typed and promptly passed it on to Himmler, who in turn forwarded it to the staff of the SS Security Service.[99] On May 12, 1939, two intelligence officers from the service met with Sievers to examine the report's recommendations.[100] The two officers were clearly impressed by the information that Altheim and Trautmann had gleaned, particularly in Iraq; contacts with the restive Bedouins would prove very helpful down the line if Germany needed to cut off British oil supplies in the Middle East. To further this strategy, the two intelligence officers agreed to the couple's plan to cultivate el Yawar, and they approved Altheim's request to return to Iraq. They also arranged to dispatch the requested Nazi propaganda to Baghdad by diplomatic pouch, so that it would not be intercepted.

Sievers relayed the substance of this meeting to Altheim, who was at work on a new book, *The Soldier Emperors,* which drew on the new Middle Eastern research and described the clashing empires of East and West that had made the third century such an important turning point in history. The scholar was delighted by the prospect of a return trip to Iraq. Again, he and Trautmann did not hesitate to volunteer their services as agents of the Reich. "We plan on keeping a diary for the trip to Iraq and central Arabia that will include not only scientific results, but also everything that is important in the ethnological, economic and political respect," they noted in a later letter to the Ahnenerbe. "The talks with Sheikh Adjil el Yawar shall be recorded in his characteristic wording as much as possible."[101]

10. CRO-MAGNON

At his desk in Munich, Wüst waded through thick stacks of Ahnenerbe letters and reports, doing all he could to put a new, more scholarly face on the organization. In between recruiting new staff and reining in the old, he fielded phone calls and letters from Himmler, who had a seemingly endless supply of new research projects for the Ahnenerbe. The SS leader wanted folklorists to search for references in ancient myths to the healing springs of Helgoland, a tiny German island in the North Sea.[1] He desired saga experts to study the role of the human heel in early European tales and runologists to decode certain inscriptions he had come across on a sculpture in Florence's archaeological museum.[2] And he was eager to put Germanists to work studying the sexual practices of ancient Germanic tribes. He had read that in some tribes, men and women had engaged in lovemaking only at midsummer so that their children would be born in the early spring. Himmler wanted to verify this idea by research—presumably so he could develop guidelines for SS men on the most propitious times for sexual relations.[3] But he offered no suggestion to Wüst as to how this study might be carried out.

As irrational and foolish as Himmler's requests often were, Wüst diligently relayed them to his staff.[4] He understood that a certain amount of compromising, of abandoning academic standards in favor of pursuing political ends, would be required of him and his staff. And he was prepared to make that sacrifice: he believed that he was serving a greater good, laying the building stones of a strong new future for Germany. And he was fascinated by his new proximity to the corridors of power. Since taking over the scientific leadership of the organization, he had managed to cultivate Himmler's friendship.[5] The two men shared a deep, unshakable belief in the majesty of the mythical Aryan race, and Himmler greatly relished conferring with Wüst on arcane points of research, talking as one scholar might to another. Some Ahnenerbe staff observed in private that Wüst was becoming Himmler's "Father Confessor."[6]

Indeed, so strong was Himmler's confidence in Wüst that the Ahnenerbe had begun to mushroom in size, consuming its potential rivals. It made short work, for example, of something called the Excavations Department in the SS. Himmler had created the department in 1935 to sponsor or direct archaeological digs at major sites in Germany. He intended these excavations to be exemplars of German research, places where SS men could be trained in the science of recovering the ancient Germanic past from the ground.[7] In three years, the department had mounted eighteen SS excavations, from an ancient hill fortress at Alt-Christburg to a major Viking trading post at Haithabu in northern Germany, not far from the Danish border.[8] But in February 1938, Himmler transferred the department lock, stock, and barrel into the Ahnenerbe.

Before this transfer, the Ahnenerbe had largely confined its studies to poring over ancient written texts, rock engravings, and folklore. But the Excavations Department brought a new scientific acumen to the brain trust. Its staff consisted of dirt archaeologists trained in analyzing bits of ancient stone, bone, and ceramics and interpreting things such as pollen samples and geological layers. This new expertise in the natural and biological sciences, Himmler and Wüst hoped, would assist in reconstructing the lives of Germany's ancestors before the first written histories and greatly extend knowledge of the mythical Nordic race.[9]

Few German ultranationalists had paid much attention to the era that preceded 5000 B.C. Indeed, some pointedly referred to this period as the Pre-Nordic Age.[10] But the Ahnenerbe's new scientific staff was by no means daunted by the task of proving the primacy of the imaginary Nordic race even at such a distant time. Indeed, one of its most ambitious young researchers claimed he could trace its origins all the way back to the Paleolithic era in Germany, when woolly mammoths and cave bears wandered the chill tundra.

DR. ASSIEN BOHMERS was a young Dutch national with a remarkable talent for dissembling, scheming, and self-advancement. According to documents he later supplied to the Ahnenerbe, he was born on January 16, 1912, into a Mennonite family in Zutphen, an old merchant town in the eastern Netherlands. He boasted that his father was the personnel director of the largest psychiatric hospital in the country and came from a proud line of sea captains based in the Frisian town of Harlingen.[11] The truth was rather more modest, however. Bohmers's father worked on the hospital wards as a psychiatric nurse and his Frisian lineage seems to have been limited to a Frisian grandmother.[12]

Nevertheless, Bohmers reveled in all things Frisian.[13] As a boy he sopped up stories of the Frisians—farmers, fishermen, and seafarers who inhabited the coastal regions of the Netherlands, Germany, and Denmark, as well as the Frisian Islands curling along those shores. In his youth, Bohmers learned to speak and write the Frisian language and he later made a personal study of "the symbols and customs in the area."[14] He took a keen interest in Frisian politics. The Frisians were a people without a nation, rather like the Basques. Over the centuries, some had nourished a bitter grievance against Holland's wealthy nobility, coming to see themselves as an oppressed minority. By the 1930s, a few Frisians looked longingly toward the German Nazi party, with all its flattering talk of the racial superiority of peasant farmers. Bohmers sympathized with these Nazi converts. Indeed, he later claimed to be a leader of a small Frisian Nazi party, although it is doubtful that such a party ever existed.[15]

At the age of seventeen, Bohmers enrolled at the University of Amsterdam, studying geology, petrology, and paleontology. He became an expert on soil and sediment analysis and on interpreting the complex stratigraphy of caves. He also took a strong interest in racial research, conducting what he called "some pretty detailed studies" on the Nordic race.[16] He traveled widely, doing geological and archaeological fieldwork in southern France and northern Spain, Sweden and Norway, and in 1936, at the age of twenty-four, he graduated with a doctorate degree. By then, he was an immensely confident man with a driving sense of ambition.

Superbly trained as a geologist, he had his pick of jobs in the European oil companies.[17] But in April 1937, he decided to apply for a post as a Paleolithic archaeologist in the SS Excavations Department.[18] Almost certainly his political views and his sense of opportunism influenced the decision. He dearly hoped that Hitler would unite all the former Frisian lands into one territory in the Greater Reich. If that day were to come, he wanted to be properly positioned to become the Gauleiter, or head of the new Nazi regional district of Friesland.[19]

Bohmers sent off his curriculum vitae, as well as a two-page description of his preferred excavation techniques and a small head-and-shoulders photograph.[20] The latter, clearly intended to demonstrate his racial suitability, showed a young man with a thatch of blond hair, deep-set eyes, large ears that angled out conspicuously from his head, and the clear, ruddy good looks of someone who grew up with plenty of fresh air and wholesome food. Even into his mid-twenties, Bohmers could have passed as a farm boy fresh from the country.[21]

Bohmers's application arrived at a rather opportune time in the SS headquarters in Berlin. A German scholar, R. R. Schmidt, had just begun excavations at Mauern caves in Bavaria.[22] The site consisted of a network of five caverns, and near the eastern entrance of the middle cave, Schmidt and his diggers had found something very important—a bed of soil stained red with a mineral pigment known as ochre. Spread out along this red layer was the skeleton of a woolly mammoth blanketed with handmade ivory beads and beautifully fashioned stone weapons and tools.[23]

The ivory beads and the finely worked stone tools were the product of the Cro-Magnon, a Paleolithic people who hunted reindeer and other big game. The Cro-Magnon had often chosen red ochre as a pigment for their cave paintings and for the ceremonies they performed for the dead. In all likelihood, the mammoth skeleton at Mauern had been part of some ancient Cro-Magnon religious rite. This news seems to have excited local Nazis greatly, for one of most prominent racial researchers in Germany, Hans F. K. Günther, had declared that the Nordic race arose from the earlier "Cro-Magnon race," based on physical similarities that the racial scholar somehow perceived in the two groups.[24]

In August 1937, the SS Excavations Department took control of the dig at Mauern. The head of the department, Dr. Rolf Höhne, became the site inspector and a unit of the Austrian SS—an underground organization at the time—began preparing the area for further excavations.[25] By September, however, Höhne realized that he needed an experienced Paleolithic excavator to take charge of the ancient cave site, someone capable of retrieving every possible scrap of evidence concerning the "Cro-Magnon race."

He offered the position to Bohmers.

SCIENCE FIRST LEARNED of the Cro-Magnon in 1868, when workmen laboring on a local railway line in the small French town of Les Eyzies-de-Tayac stumbled upon a grave in a rock-shelter, a huge horizontal groove in a cliff face that provided protection from the elements.[26] The grave contained the skeletons of five humans laid to rest with various body parts of extinct Ice Age creatures such as the woolly mammoth, as well as an assortment of stone blades and knives and pierced seashells. These discoveries immediately stirred the attention of two scholars, Louis Lartet and Henry Christy. Together, the two men excavated what remained of the grave and reassembled the bones that the railway workers had badly disturbed.

The five skeletons looked very different from an ancient human unearthed in the Neander Valley in Germany twelve years earlier. The Neandertal man possessed a low, receding forehead, with a thick bony ridge looming above the eyes and a distinctive bony bulge jutting from the nape of the neck. The humans of Les Eyzies, by contrast, were clearly modern *Homo sapiens*, with rounded crania, steep foreheads, short faces, and jutting chins. But they had obviously died a very long time ago, buried as they were among the bones of long-extinct animals. Indeed, later studies showed that they had roamed the region in the midst of the Ice Age thirty thousand years ago, when dirty gray ice sheets sprawled over much of northern Europe. Scientists dubbed these modern humans the Cro-Magnon, after the rock-shelter where they were found, Abri Cro-Magnon.

Later excavations at a scattering of sites revealed much about the Cro-Magnon. They had descended into subterranean caves to paint achingly beautiful murals of horses and bulls, ibexes and mammoths. They had sat about hearths carving tiny bosomy Venus figurines from the ivory tusks of mammoths. They had bedecked themselves in clinking shell necklaces and soft fur robes and invented new and ever more lethal stone and bone weapons to bring down the game they hunted. They had also lived side by side with the Neandertal in Europe for several thousand years, until finally the Neandertal—who neither painted great murals nor carved beautiful figurines—disappeared under mysterious circumstances. The Cro-Magnon clans had then helped themselves to the sun-warmed rock-shelters and caves of their former Neandertal neighbors.

European researchers interested in questions of race soon turned their attention to these Paleolithic finds. In 1878, a pair of prominent French researchers, Jean Louis Armand de Quatrefages de Breau and his assistant Ernest Jules Hamy, interpreted these finds according to then-popular ideas on the Aryans. The Neandertal, they claimed, were members of an ancient race vanquished in most of Europe by a superior immigrant race—the primordial Aryans.[27] Their notion that fossil humans represented diverse ancient races, as opposed to diverse species, made a deep impression on some researchers, particularly in Germany. By the 1930s, scientists in England and America had abandoned this idea, but racial theorists in Germany stubbornly clung to it.[28] Hans F. K. Günther, for example, insisted upon describing the Cro-Magnon as "the ruling race in central Europe," a cold-loving people who handily conquered all of its contemporaries.[29] In an effort to prove that they had spawned the Nordic race, he pointed to similarities in facial shape between the Cro-Magnon and some modern Scandinavians.[30] Other German racial theorists attempted to prove that Cro-Magnon men and women were blond-haired.[31]

But while Nazi researchers happily claimed the artistic, inventive Cro-Magnon as their own, they were somewhat uneasy about the origins of this primeval "race." Contemporary research suggested that the Cro-Magnon had arisen somewhere in Asia or even Africa, but these places were not terribly acceptable to German ultranationalists who posited that all important events in human history took place in northern Europe. In 1938, however, Dr. Assien Bohmers advanced a far more politically palatable theory for the origins of the Cro-Magnon, based on his finds at Mauern. He contended that these ancient clans had arisen directly from the barren tundra of Ice Age Germany.[32]

THE SITE OF Mauern lies in the southern part of the Jura Mountains, between the two old Bavarian cities of Ingolstadt and Donauwörth.[33] Its caves have been hollowed out by nature on a high sloping valley wall and look down upon a pretty stretch of small farmhouses and fields. In the fall of 1937, Bohmers arrived at Mauern to take control of the SS excavations.[34] He studied the lay of the land, examined the cave floors, and scrutinized the exposed stratigraphy. Then he ordered his team to expand upon Schmidt's excavation. As the days grew shorter and cooler, the team dug deeper into the sediments near the entrance to the middle cave. Not far from where Schmidt had worked, they struck more fossilized bone and ochre. It was the skull and several vertebrae and ribs of a woolly mammoth.[35]

Bohmers pored over the osseous debris. Someone had broken off the mammoth's long ivory tusks and scattered precious carved-ivory pendants and a collection of drilled animal teeth—fox, wolf, bear, and caribou—all ready to be hung on a necklace, near the mammoth bones.[36] Then he or she had scattered red ochre nearby, and lit a fire, for the ground was littered with bits of charcoal. Close by, Bohmers and his team collected seventy burins—small stone chisels for shaping and decorating bone and ivory—as well as a rich assortment of other stone and bone tools. Bohmers was fascinated. He concluded that the area was a cultic site.[37]

To explore the site further, he divided his team, dispatching some men to a lower cave entrance, others to an upper. As the excavators dug downward, they encountered ancient hearths and the refuse of a different group of visitors: the Neandertal. In these layers, Bohmers's team found hand axes, scrapers for cleaning animal skins, and spearheads suitable for hand-held spears wielded like bayonets. There was nothing terribly unusual or striking in any of this—except for one curious find. In the upper portion of these layers, Bohmers and his team unearthed thirty-three leaf-shaped points, tapered at both ends.[38] These points resembled the streamlined spearhead of a technologically advanced weapon: the throwing spear or javelin favored by much later Cro-Magnon clans.[39] Armed with this weapon, a human hunter could slaughter dangerous prey—animal or human—from afar rather than closing in and risking serious injury. Bohmers considered the discovery immensely significant.

He diligently collected soil samples from the dig and sent them out for pollen and mineral analyses. The results allowed him to visualize local vegetation cover and climate at the time the Paleolithic clans had visited the cave. Bohmers then compared these conditions to sweeping, well-documented environmental changes that had taken place in Europe over the last three glacial ages, as vast ice sheets advanced and retreated in a global minuet. On the strength of this comparison, he assigned rough dates to the cave's occupation, a standard technique in an age before more accurate methods, such as radiocarbon dating, were invented. Bohmers was terribly excited by the results. According to his calculations, which later proved to be quite wrong, the curious tapered leaf points dated more than seventy thousand years ago, to a period between the last two glacial ages.[40]

The young researcher chose to interpret this finding in a very dramatic way that would appeal to the SS. Since he believed inherently in the superiority of the "Cro-Magnon race," he assumed that only Cro-Magnon hunters were intelligent enough to invent leaf points and throwing spears. Therefore the presence of these projectile points in the stratified layers at Mauern proved that Cro-Magnon clans stopped in there more than seventy thousand years ago.[41] By this reasoning, Mauern was the earliest Cro-Magnon site in the world—by tens of thousands of years.[42] It was a very bold interpretation for a young prehistorian, for the business of dating cave floors was a particularly difficult one. Moreover, in hammering these ideas together, Bohmers had ignored another major possibility—namely that the "subhuman" Neandertal had created a superb new weapon.

Bohmers proudly announced his results in a handwritten letter to the Ahnenerbe, which had just absorbed the SS Excavations Department. "Your assignment," he declared in July 1938, "has brought me face to face with a conclusion that is extremely important for all future Stone-Age research and especially for all areas of study which are concerned with the predecessors of the Germanic tribes—the so-called Cro-Magnon race. Until now almost every German, and without exception every foreign investigator, has assumed that the race migrated to Europe from somewhere in the East. The excavations at Mauern and Ranis [a second German site where excavators had noticed leaf points in early layers] have revealed for the first time the key that proves that the Cro-Magnon race must have developed in greater Germany."[43] This news, he knew, would please Himmler mightily, for it neatly eliminated another potential source of foreign blood in the pedigree of Germans' ancestors. Moreover, it seemed to confirm another favorite boast of SS school leaders: that all important advances took place first in northern Europe.[44]

Bohmers did not describe in this letter how he envisioned this evolution to have taken place. But he had become interested in Aimé Rutot, a maverick Belgian geologist and archaeologist who contended that a very early genetic mutation gave rise to a single person with an immensely large brain.[45] Such an individual, Rutot surmised, was capable of founding "a race of the strong."[46] Rutot also suggested that primeval climate change—such as the advent of an Ice Age—had triggered human evolution, sparking intense clashes among existing humans for food and shelter. From these pitched battles, a new, more advanced human race emerged.

Such ideas would surely have appealed to Bohmers. He believed that the Cro-Magnon evolved in Germany from the Neandertal populations just as the earth cooled, the forests died, and immense ice sheets in the north and in the Alps began advancing and gnawing away at the land.[47] And he must have thought it natural that in such a hostile environment, the very first invention of the ur-Nordic race would be a new weapon, a javelin that could be heaved through the air to wipe out more slow-witted and primitive rivals for scarce supplies of food. Violence and progress marched apace in Nazi political doctrine, and now Bohmers thought he had discovered an important example of this in the Paleolithic world, too.

When word of the young prehistorian's work at Mauern reached Himmler in late July 1938, he was fascinated. Although he was mired in work, engineering a major new anti-Jewish campaign in Germany and drawing up plans for covertly undermining the government of Czechoslovakia, he asked to read Bohmers's papers even before they were published.[48] Bohmers was well aware, however, that the European scientific community would be considerably less enthralled by these ideas. Indeed, his foreign colleagues would need serious convincing, for Bohmers knew that many European scholars scoffed at the science of the Third Reich. "Everywhere outside of Germany," he noted in a letter to Sievers, "German science is seen as chauvinistic, without foundation."[49] To deflect such criticism, Bohmers wanted to talk with his foreign colleagues and examine their collections of early Cro-Magnon tools. He also wanted to tour the painted caves in France—one of the best surviving records of the Cro-Magnon and their religious rituals, a subject that clearly intrigued him.

New as he was to the staff of the Ahnenerbe, he did not hesitate to request funding for a research trip to the Netherlands, Belgium, France, and England. And the powers at the Ahnenerbe did not dawdle or drag their heels—as they often did with others—in approving his plans. But there seems to have been little interest in recruiting Bohmers as an intelligence agent. A few months earlier, in the spring, Sievers had warned the young archaeologist by letter to be more discreet while conducting research among the Frisians.[50] Senior staff at the Ahnenerbe had discovered that Bohmers lacked judgment and circumspection when it came to political matters.

None of this, however, interfered with the plans for his trip. In mid-October 1938, as the field season at Mauern closed down, Bohmers packed his bags and papers. He caught a train for the old Frisian capital of Leeuwarden and then on to Groningen and its important archaeological department, before pushing on into Belgium.

AS ONE OF the great crossroads of Europe, Brussels was in an anxious mood. For weeks, the city's residents had followed the intricate dance of promises and betrayals that had led to the Czechoslovakia crisis. Hitler had sown violent unrest among the Germans living in Sudetenland, then demanded that Czechoslovakia hand over the important territory to the Reich. Stunned by his rapaciousness, the Western democracies had searched for some face-saving way of abandoning Czechoslovakia in its hour of need. By the end of September, both France and Britain had capitulated to Hitler's demands. The Sudetenland passed effortlessly into German hands. The immediate crisis was over, but those reading the newspapers in Brussels cafés suspected that Hitler was not done, and as Bohmers wandered the city streets, a certain sense of foreboding, of living on borrowed time, hung in the air. If the young Dutch researcher noticed this, however, he made no mention of it in his surviving reports to the Ahnenerbe. He was preoccupied by the work at hand.

He presented himself to the Royal Museum of Natural History, part of an ambitious complex of educational and research facilities grouped around Parc Leopold. The museum had won international renown for its collection of fossil iguanodons, large plant-eating dinosaurs that once waddled contentedly across the Belgian countryside. In 1878, Belgian coal miners struck a fossil layer of the creatures near the village of Bernissart, and researchers from Brussels had salvaged twenty-nine of their giant skeletons, reassembling them for public display in an immensely popular gallery.[51]

Bohmers almost certainly stopped to admire the giant skeletons—he had been trained, after all, as a geologist and paleontologist—but he was keen to get on with the study at hand. He had traveled to Brussels to examine the large collection of flint tools and pebbles gathered by Aimé Rutot.[52] Among other things, Rutot had championed before his death a bizarre theory of human evolution. He believed that ancient apelike humans had fashioned simple stone tools in Europe during the Tertiary Age, which spanned a vast period of time, from the extinction of the dinosaurs to the beginning of the great Ice Age some 1.8 million years ago. This radical theory had initially attracted droves of converts, but Rutot had failed to win lasting scientific acceptance. Many of his prize flint tools turned out to be little more than the products of natural forces. Bohmers was eager, however, to check through Rutot's collection, searching, it seems, for something that resembled the leaf points from Mauern.

When he was done, he moved on to Paris and the Institute of Human Paleontology on rue René Panhard, a short stroll from the Museum of Natural History and the laboratory where French researchers first discovered radioactivity in 1896.[53] The Institute of Human Paleontology was housed in a formidable-looking building, a modern fortress in yellow brick and dull gray stone. Founded by the Prince of Monaco in 1920, the institute had become a mecca of Paleolithic studies, attracting researchers from around the world. Bohmers spent two weeks there, sifting through the institute's collection of flint tools.[54]

In the hallways and offices, he introduced himself to the institute's renowned staff. He chatted amicably with physical anthropologist Raymond Vaufrey and presented his findings from Mauern to seventy-seven-year-old Marcellin Boule, an authority on the Neandertal. He also paid his respects to a small, round-faced man in a worn black cassock, who always made time for young researchers. Abbé Henri Breuil was the world's foremost expert on Paleolithic cave art.[55]

In 1901, Breuil and two companions had discovered a wealth of ancient cave paintings and engravings in France's Dordogne region, and it was Breuil who helped convince European prehistorians that the cantering horses, charging bison, leaping stags, and lumbering cave bears were indeed the work of Paleolithic artists. Since then, the adventurous cleric had spent nearly forty years wriggling through narrow subterranean passageways and clambering up steep cavern walls. He showed no signs of slowing down or losing his enthusiasm. He had just returned from a month in the

Pyrenees, copying "the frightful tangle of engravings" at Tuc d'Audoubert and an assortment of painted figures at Les Trois-Frères, work he dearly hoped to finish before war brought cave-art research grinding to a halt.[56]

Despite these worries, however, Breuil took time to see Bohmers. Almost certainly the genial priest rolled and lit a crumbling cigarette—he was seldom to be seen without one—as Bohmers described the leaf points, red ochre, ivory beads, and mammoth bones of Mauern. Then Breuil, a generous-spirited man who could not possibly have guessed the extent of Bohmers's Nazi ambitions, helped to arrange a rare honor for the young Dutch researcher—permission to enter one of the most important French caves, Les Trois-Frères. The ancient site lay on a private estate, closed to the public. The owners permitted just sixteen or so people a year—archaeologists for the most part—inside.[57]

Before availing himself of this honor, however, Bohmers made a quick trip across the English Channel. He toured museum collections in London and Cambridge, examining their fossil human remains and stone tools. "I am very happy with the success," he reported to the Ahnenerbe. "I have been able to study all of the worthwhile collections for our study, and in England and France, I was able to talk to all the important researchers. They were all very enthusiastic about the finds and geological results from Mauern."[58] Bohmers, however, had clearly mistaken politeness for acceptance of his interpretations.

He returned to France at the end of the first week of November and traveled by train south to Les Eyzies, the little market town in the Dordogne where railway workers had discovered the first Cro-Magnon skeletons seventy years earlier. Since then, curious residents and researchers had combed the surrounding countryside, turning up dozens of painted caves and prehistoric rock-shelters in the neighboring limestone cliffs, and for decades journalists had publicized their finds. Les Eyzies had become synonymous with archaeology, and throngs of tourists in knickerbockers and boots and heavy woollen sweaters turned up at the railway station each year, ready to tour the caves. Bohmers checked into the Hotel des Glycines and almost certainly paid a visit to the local museum of archaeology, founded in 1910 to preserve the region's Paleolithic treasures after a collector was found to be buying and shipping off artifacts to German museums.[59]

Bohmers had a clear idea what he wanted to see in the surrounding countryside.[60] He squeezed down the long, narrow meandering gallery at Les Combarelles, past dank walls encrusted with stalactites and thick dollops of calcium-rich moon milk, and gazed in wonder at the vast Paleolithic bestiary on the walls—stags, mammoths, cave lions, bears, fallow deer, reindeer, ibex, wolves, and a profusion of horses, one often superimposed upon two or three others. All had been engraved by firelight by ancient artists equipped with flint tools. Breuil had detected what he believed to be four ancient breeds of horses at Les Combarelles—Lybian, Celtic, Tarpan, and what he called the Nordic horse, "with an arched profile from ears to nostrils with a long saddle back."[61] Bohmers struggled to find them: such depictions, he later noted in a memorandum for the Ahnenerbe, could prove useful to science, giving "important information on the development of modern horses."[62]

When he was done there, he toured another of the region's great treasures, La Font de Gaume, less than a mile away across a patchwork of small tobacco fields and groves of walnut. Cro-Magnon artists had adorned the cave with spectacular paintings of Ice Age bison, oxen, and horses, many delicately colored in shades of red and brown and all observed with the acuity of the hunter's eye. Their sheer beauty astonished Bohmers: it overshadowed, in his view, the artistic achievements of many of the Cro-Magnon's successors. The engravings and paintings of the Dordogne caves, he later conceded, "are at an artistic level only seldom reached in the history of the Germanic tribes."[63] He then set off to lesser-known caves of the region, Teyat and La Mouthe.

From the caves of the Dordogne, he pressed on south to the rugged Pyrenees and the small French village of Montesquieu-Avantès, where one of the great mysteries of the ancient world, Les Trois-Frères, awaited him. The cave had haunted Breuil for years. The abbé believed that it had long served as a subterranean shrine, a place where generations of sorcerers and enchanters came to pray to their gods and perform their ancient magic. As a spiritual man, Breuil felt certain that only initiates were permitted to venture into its dangerous subterranean passages and enter the chamber he called the Sanctuary.

Even after a decade of meticulous study, Breuil still marveled at this innermost chamber. Along its walls, human hands had incised and painted a strange phantasm of creatures. In the darkness, a swirling maelstrom of superimposed bison, horses, ibex, rhinoceros, and a host of mythical or masked creatures, half-human, half-animal, capered and pranced. Some danced, others played what appeared to be an early musical instrument.[64] A few gazed out mischievously toward the viewer. The Sanctuary teemed with life and vital energy, and above all this tumult, beyond all the charging, rearing, swaying, someone had engraved and painted the lord of the chamber: a tall antlered man with a stag's ears and a horse's or wolf's tail. Breuil dubbed this surreal figure "The Sorcerer." He theorized it was a primeval divinity of the Paleolithic world—the spirit that controlled the fertility of game and the success of the hunt.

As the light of his lanterns flickered across the walls of Les Trois-Frères, with their strange unearthly visions, Bohmers gazed at the tumult of engravings. He had arrived at the center of a mystery, at a place where Himmler and where so many other Nazis had long dreamed of standing—in the shrine of the ancient dead, in the dark embrace of the ancestors.

THAT WINTER, BOHMERS basked in the warmth of Himmler's favor. The SS leader lapped up his theories, so much so in fact that he asked him to report on his findings at a conference of the SS-Gruppenführer, the major-generals who would faithfully carry out orders for the Holocaust.[65] Himmler also found time to take Bohmers aside at a gathering to convey his personal views on the subject of human evolution. It must have been an instructive conversation. As Bohmers later reported, Himmler dismissed outright, for example, the current notion that the human race was closely related to primates. He was also outraged by an idea proposed by another German researcher that the Cro-Magnon arose from the Neandertal.[66] To Himmler, both these hypotheses were "scientifically totally false."[67] They were also "quite insulting to humans."[68]

Bohmers must have felt stunned. He, too, believed that the Cro-Magnon had evolved from the burly Neandertal, based on the finds from Mauern, and he was unaccustomed to taking scientific direction from amateurs. He had clearly reached a major turning point in his career. He could continue gathering up evidence for his ideas on evolution, pursuing his own lights and paying the inevitable price for his stubbornness: scholarly obscurity. Or he could quietly drop all talk of evolution and build an important career in Germany with the help of the Ahnenerbe. He chose the latter. He wrote to Sievers in March 1939 stating that he "always agreed with the Reichsführer's and the Ahnenerbe's position" on matters of human evolution.[69]

Bohmers had already discussed plans with Sievers to create a satellite Ahnenerbe research center for Paleolithic studies in Munich. He believed himself to be the natural choice for its head, and he confidently predicted to Sievers that the new center would soon outshine the Institute of Human Paleontology in Paris and "become the leading research site in the world."[70] For this, Bohmers needed a professional staff of four young German prehistorians, whose specialties would nicely complement his own. He also wanted a team of ten experienced excavators and his own anthropological journal, "where all our investigators are to be published."[71] This would help build an international reputation for the center. In addition, he requested editorial control over all articles on prehistory and anthropology that appeared in the Ahnenerbe journal, *Germanien,* and asked for funds for a new research trip to Hungary, Czechoslovakia, Poland, and Yugoslavia.[72]

Indeed, his plans for the future seemed to be growing by the day, and they even encompassed something that would have astonished Abbé Breuil and his colleagues in Paris. Bohmers, it seems, dreamed of taking a leading role in French cave-art research. The ancient murals and reliefs, he noted in an official memorandum, could shed much light on subjects dear to the hearts of the Ahnenerbe staff, such as "symbol studies," the "history of hunting methods, i.e., wild horse, bear and mammoth hunting," and "knowledge about the magical customs of the Cro-Magnon race, i.e., hunting magic, initiation rites, etc."[73] The French, in his view, had shamefully neglected this priceless legacy. He thought their scientific publications on the caves scanty and inadequate, and their preservation methods a disgrace. He noted that both he and Erika Trautmann had observed during their visits "that the named caves are owned by farmers, or are managed by farmers, who don't have the faintest idea about the cultural value of the rock art."[74] As a result, American and French tourists had swarmed in droves through the caverns and run amuck, scratching the walls with penknives.

To Bohmers, the solution was perfectly obvious. If the legacy of the ur-Nordic race was to be protected and preserved for future generations in Europe, then his new institute would have to take the lead in cave-art research. "The Ahnenerbe," wrote Bohmers, "would serve the whole culture well if it were to succeed in preserving these valuable documents from their eventual destruction by photographing, drawing or creating casts for the future."[75]

Bohmers, it seems, already dreamed of a day when the swastika would fly over France.

11. THE BLOSSOMING

In the early spring of 1939, the Ahnenerbe was blossoming like some strange hothouse flower, sending out twisting new shoots and twining tendrils into German academic life. It was becoming a place that courted bright young scholars and catered to their research, a place that advanced academic careers and arranged coveted university appointments, a place that financed dreams of foreign research, then sliced effortlessly through the hopeless tangle of red tape that restricted most travel outside the Reich. It was a place where the spirit of National Socialism, such as it was, ruled and where the *Führerprinzip,* or "leadership principle," meaning absolute allegiance to those higher up in the pecking order, was put into practice daily. It was a place where SS commissions were doled out liberally. As such, the Ahnenerbe had little difficulty in hiring its pick of young Nazi scholars.

Indeed, since the beginning of 1937—when Walther Wüst assumed the presidency—the institute had more than doubled in size, from thirty-eight workers to more than one hundred.[1] For its new headquarters in Berlin, the SS had acquired a sprawling mansion in Dahlem, a district of quiet leafy streets and tall hedges, wrought-iron gates and servants' quarters. Dahlem was a neighborhood of choice for those who had profited from the corruption of the Nazi regime. Those who hadn't sometimes called it by another name—Bonzopolis—after a popular English cartoon character famous for spineless greed.[2] The new Ahnenerbe headquarters was set in a lovely green estate at 19 Pücklerstrasse. It stood just five minutes away from Grunewald, a former royal hunting preserve turned parkland with horseback trails and lakes and swimming beaches.[3]

Like many other properties in Dahlem, the mansion and its attendant buildings had recently been "Aryanized," meaning that the Nazi government had succeeded in forcing its previous Jewish owners to flee the country and give up their spacious home at a bargain price.[4] Certainly, the Ahnenerbe had profited handsomely from the situation. A real-estate appraiser valued the Pücklerstrasse estate in October 1938 at 675,000 reichsmarks, or some $3.5 million today, after inflation is taken into account.[5] But the Ahnenerbe seems to have snapped up the property for less than half this price—300,000 reichsmarks.[6] And almost certainly, the unfortunate Jewish owners received a good deal less than this, for the government customarily forced Jewish emigrants to pay 25 percent of their assets for the *Reichsfluchtsteuer*, or Reich flight tax, as well as additional amounts for other forms of official extortion.[7]

The newly acquired estate consisted of three buildings. The largest was built in 1910 of ivory-colored sandstone and an abundance of expensive wood paneling and stained glass. It boasted a conservatory, dining room, library, laboratory, microscopy room, photographic studio, darkroom, and some two dozen offices furnished with comfortable leather chairs, Persian rugs, antique-looking German paintings, and reproductions of Wirth's rock-art casts.[8] The second building housed a large workshop for the Ahnenerbe's sculptors and other technical staff, while the third and smallest structure served as a private residence for Wolfram Sievers and his family, a convenient arrangement that allowed the young administrator to go horseback riding in Grunewald before starting work on summer mornings.[9]

Spacious as the complex was at 19 Pücklerstrasse, however, it merely served as the Ahnenerbe headquarters. Many of the staff, including Wüst himself, toiled in what were called teaching and research sites—smaller satellite institutes scattered across the Reich, from Munich to Vienna. These centers varied in size from one or two scholars to nearly a dozen researchers.[10] Ahnenerbe officials planned to house the most important of these in select "Aryanized" properties, too.[11] For Wüst and his department of Indo-Germanic-Aryan linguistics and culture, for example, they were planning to create a major bureau at an elegant address in Munich, somewhere close to the university. In January 1940, they scooped up a prize piece of "Aryanized" real estate at 35 Widenmayerstrasse, a lovely terrace house situated across the street from the tree-lined Isar River.[12]

All in all, the Ahnenerbe was rapidly becoming a rather large and lavish operation. Rather than order cost-cutting measures, however, Himmler encouraged further expansion. He regarded the Ahnenerbe's research as essential to the future of the Reich—even if Hitler had yet to take much notice of it. To keep this large and expensive research organization afloat, the SS leadership was increasingly forced to resort to some rather unconventional sources of financing.

IN ITS EARLY days, the Ahnenerbe had received nearly all of its money from just two sources—the Deutsche Forschungsgemeinschaft, or German Research Foundation, the country's preeminent scientific funding agency, and the Reich Agricultural Organization. The latter was a Nazi cartel responsible for planning and organizing Germany's agricultural affairs in order to render the country self-sufficient. Germany's heavy dependence on foreign food before the First World War had contributed to its defeat in 1918, and Hitler was determined to avoid that mistake.[13] As a first step, he had appointed Richard Walther Darré, the Reich farmers' leader, to take command of the Reich Agricultural Organization. Darré in turn had combined these appointments with his continuing work as head of RuSHA. So he had taken it upon himself to quietly funnel money from the agricultural cartel into the Ahnenerbe.[14]

Such finangling certainly helped pay the bills. But Himmler was eager to find new sources of tax-free money that were independent of any political rival. So he set up something called the Ahnenerbe Foundation, and under his direction, the SS leadership began searching for donors in private industry. They landed one big fish, Emil Georg von Strauss. Strauss sat on the supervisory board of the Deutsche Bank, as well as on the boards of several major German manufacturers. The prominent financier agreed to milk his business contacts for donations for the Ahnenerbe Foundation. In 1937, Strauss succeeded in rounding up 50,000 reichsmarks from a short list of corporate donors that included Daimler-Benz and Bayerische Motorwerke, better known in the English-language world today as BMW.[14]

Such piecemeal donations, however, would cover little of the Ahnenerbe's operating expenses. So the SS leadership drew up a more ambitious scheme. This involved something that bore little relation to pre-history: a reflector pedal for bicycles. The pedal was the brainchild of one of Hitler's former chauffeurs, Anton Loibl, an old party comrade. Loibl was a machinist and a driving instructor by profession.[15] By avocation, however, he was an inventor. In his spare time, Loibl tinkered away in his shop in Berlin, designing new and improved mechanical devices, such as carburetors. While mulling over a way of making bicycles more visible at night on the road, he stumbled upon the idea of fastening small pieces of glass to the pedals in order to reflect headlights from approaching cars.

Eventually, word of these inventions reached the SS leadership, and in 1936, the SS formed a joint company with the machinist to market his inventions. Himmler intended to use part of the revenues to fund the Ahnenerbe. As it turned out, however, Loibl's design was not nearly as original as it had first seemed. Indeed, another German inventor had devised a similar safety device and applied for a patent. But this competitor lacked something very important: the SS as a business partner. His patent application was buried. Loibl's sailed through, and in 1938 Himmler used his supreme authority as the head of the German police to pass a new traffic law. This required all new German bicycles to be equipped with Loibl's reflective pedal.[16]

German manufacturers suddenly had no choice but to use Loibl's design and pay licensing fees to the company owned by the inventor and the SS. Those who initially refused to pay the fee soon saw the wisdom of doing so.[17] Often it took no more than a single letter, signed by Himmler personally, to convince a reluctant manufacturer to obey the new law.[18] And before long, the revenues were flowing into the Ahnenerbe Foundation. In 1938 alone, the foundation received a tidy 77,740 reichsmarks from the bicycle-pedal proceeds.[19]

The Loibl scheme was insufficient, however, to meet the Ahnenerbe's growing stack of bills. So in the absence of other legal channels, SS aides began to indulge in some very shady accounting—a common practice in the offices of many senior Nazis.[20] Although Himmler demanded the appearance of financial probity from his SS men, he encouraged fraud when it came to financing the SS itself. Senior officers on his staff juggled large sums of money from one secret account to another, deftly evading taxes and leaving behind a paper trail so tangled and snarled that no auditor could hope to unravel it.[21] They also signed for massive bank loans that the

SS had no intention of repaying. Himmler and his staff simply assumed, with good reason, that in due time the SS would become so powerful and fearsome that no bank director in his right mind would dare to call in one of its loans.[22]

SUCH FRAUDULENT FINANCIAL maneuvers permitted the Ahnenerbe to take on a much broader range of scientific and scholarly research, a plan that Himmler clearly endorsed. From the very beginning, he had regarded the Ahnenerbe as the source of a glorious new history of the Aryan race, a history that could be used to teach SS men and their progeny to act and truly think like Aryans. Moreover, he had long hoped that such knowledge of the ancestors could be used to convince SS men to turn their backs on the moral decay of Germany's cities and return to the countryside to take up the simple life with their families in special SS farm colonies. Such colonies, he hoped, would give birth one day to a new golden age, replacing industrial Germany with an agricultural paradise and creating a fertile breeding ground for the Aryan race.

The SS command, however, had made little progress in achieving this rural utopia. Land, after all, was at a premium in Germany. SS officials had to first locate, then purchase large tracts that could be readily divided into settlement farms—an expensive proposition. In 1937, for example, SS officials decided to found a model farm colony at Mehrow, east of Berlin. For its land base, they bought part of a thirty-two-hundred-acre estate from the financially struggling daughter of a Berlin industrialist.[23] The official sale price was reportedly 1,000,000 reichsmarks.[24] SS authorities then proceeded to slice the property up among just twelve SS families. The largest block of land—some one hundred acres—was reserved for an SS doctor. Smaller parcels were then distributed to men of lower ranks.[25]

Drained by the cost of these purchases, officials founded only a handful of the colonies in Germany between 1937 and 1938. Himmler, however, was not discouraged. He held high hopes for the future, especially after Hitler engineered the *Anschluss* with Austria and the annexation of the Sudetenland. As German armies marched farther and farther east, he reasoned, the SS could easily scoop up confiscated country estates for its men. To prepare for this day, RuSHA officials drew up detailed plans for new SS settlements in the east.[26]

At the heart of each of these colonies would be a distinctive outdoor amphitheater known in Nazi parlance as a *Thingplatz*.[27] The idea was borrowed from the old Scandinavian *Thing*, an assembly of free men who met in a field or village common to elect chieftains and resolve disputes. The Nazi *Thingplatz*, however, was considerably more contrived and far less democratic. Architecturally, it was intended to blend into a natural setting that possessed some prominent feature—a hill, perhaps, or a lake or an archaeological site—of legendary or historic importance.[28] There SS families would be encouraged to hold torchlight rallies, stage SS solstice celebrations, or present their own propaganda plays.

In addition to the *Thingplatz*, each colony would possess other key features. It would have a shooting range where the men could sharpen their marksmanship and a distinctive graveyard where the living could suitably honor and remember the dead. It would have buildings to house local branches of the Nazi party, the SS, and Hitler Youth, as well as a variety of Nazi women's organizations.[29] And last but not least, it would have a *Sportplatz*, where young men and women in the community could receive physical training in a wide range of sports and gymnastics. Hitler himself had stressed the importance of such training. Sport, he had noted in *Mein Kampf*, would "make the individual strong, agile and bold" and "toughen him and teach him to bear hardships."[30] Such training, he further opined, would produce both defiant men and "women who are able to bring men into the world."[31]

The colony's farmhouses would be spacious and solidly built, as befitting homes of a master race. SS planners favored a primeval housing style known as *Wohnstallhaus*, which dated back to at least the Roman era in Germany—and possibly earlier.[32] One basic design called for a long, narrow wooden building of nearly 9,500 square feet that combined both the family home and barn under one roof.[33] The front half of the spacious building featured a downstairs parlor and a roomy kitchen, where several small children could run about freely, as well as a number of bedrooms upstairs. The back half housed the family's stable, as well as a barn for chickens, pigs, and cattle. But the design was very flexible. SS men could add on more space as new babies arrived.

Each family in the settlement was expected to observe SS doctrine. Simply stated, this meant maintaining the purity of their Nordic bloodlines at all costs and producing as many children as possible. To prove the purity of its lineage, each family would keep a detailed genealogical chart of its ancestors, as well as a copy of its *Sippenbuch,* or clan history.[34] Moreover, settlers would be encouraged to research and display their clan symbols and family coat of arms.[35]

Such was the cookie-cutter pattern that Himmler intended to stamp out across Eastern Europe, and he spread word of these plans among his men, knowing that the promise of free land and a spacious house in the country would motivate many to obey even the most odious order.[36] The proposed SS settlements were a far cry from the cluttered urban apartments that many of Himmler's subordinates called home.[37] Moreover, these country estates were a world apart from the cold, disease-infected barracks that Himmler was creating in the concentration camps. These crowded miniature cities were the places that German and Austrian Jews were now increasingly being funneled into.

TO ASSIST FUTURE SS colonists, the Ahnenerbe greatly expanded its field of research. It assigned heraldry experts to pore over the history of clan symbols and scholars of comparative religion to retrieve details of ancient Aryan rites. It hired philologists to collect German place-names and folklorists to record German legends and fairy tales. It employed archaeologists to excavate primeval settlements, engineers to study ancient German building styles, and landscape specialists to uncover possible astronomical alignments of old villages. It also appointed botanists to search out potential new food plants and zoologists to chart the evolutionary history of the horse and other farm livestock, in order to help restore old "Germanic" breeds.[38]

In addition, Ahnenerbe researchers worked very hard to recover the lost history of the fictional Aryan race. Linguists studiously reconstructed words in the ancestral Aryan tongue, and classicists hunted for evidence of the Nordic founders of classical Greece and Rome, while Orientalists searched for proof of the Nordic roots of Near Eastern civilizations. Geologists mapped traces of prehistoric life in German caves, while biologists conducted racial studies on SS men and on Jews, and astronomers tested a popular Nazi theory on the origins of the solar system, a theory that neatly explained the supposed demise of the first great Aryan civilization.[39]

Research was only half the battle, however. The Ahnenerbe strongly encouraged staff to popularize their work. Some gave lectures to SS officers and officer trainees and wrote articles for the SS training magazine, *SS-Leitheft*, and the Nazi party newspaper, *Völkischer Beobachter*.[40] Others organized scientific conferences that attracted reporters from the popular press.[41] Most wrote for the Ahnenerbe's own publications, thereby avoiding impartial peer review of their work. The institute possessed its own press, the Ahnenerbe-Stiftung Verlag, which produced both popular and scholarly books, as well as three journals on diverse subjects.

The best known of the journals was *Germanien,* a monthly magazine whose covers generally ran to German folk art and colorful historical paintings of Nordic farmers. The issues blended articles of interest to SS planners—"Around the *Thingplatz,*" "On Nordic Storage Buildings," and "Wooden Construction and Farmhouses in Norway"—with more arcane pieces intended for historians of the Aryan race, such as "The Atlantis Problem" and "Nordic Runes and House Marks in Chinese Script." In addition, the Ahnenerbe staff produced two other magazines of interest to SS officials—one devoted to research on German family names, the other to studies of clan symbols.[42]

Senior Ahnenerbe scientists also assisted in the making of SS training films on subjects ranging from the ancient symbols in Swedish rock art to traditional Finnish ceremonies for burying the dead. Other Ahnenerbe staff labored in the sculpture studio to create plaster reproductions of the cryptic rock art from Scandinavia.[43] The Ahnenerbe intended to display these enigmatic inscriptions in new SS "celebratory sites and cemeteries," as well as on "monuments and items of daily use."[44]

And of course, the SS staff in Dahlem continued to labor over their towering stacks of paperwork, planning major expeditions abroad. Himmler had begun to fix his gaze to the east, far to the east, and at the beginning of 1939, he waited impatiently as one of his favorite researchers led a scientific team across the remote mountain kingdom of Tibet in search of lost Aryan tribes.

12. TO THE HIMALAYAS

SINCE HIS FIRST MEETING with Walther Wüst, Himmler had grown increasingly curious about Asia. It was as if the Orientalist had thrown open a door to a secret chamber that promised rare burnished treasure: Himmler could not resist exploring it. Privately, Himmler began to carry a copy of the Hindu religious poem, *Bhagavadgita,* wherever he traveled.[1] He browsed the speeches of Buddha and pored over the biography of Genghis Khan, the ruthless Mongol leader whose empire stretched from southern Manchuria to the Caspian Sea in the thirteenth century. Himmler then arranged for copies of the biography to be distributed to all SS leaders.[2] He also cultivated the friendship of Oshima Hiroshi, Japan's military attaché to Germany and later its ambassador.[3] The two men talked of many things, but Himmler was particularly keen to learn more of the samurai, Japan's warrior nobility.

This was not an idle curiosity. Himmler had come to believe, as had Wüst, that the elites of Asia—the Brahman priests, the Mongolian chiefs, the Japanese samurai—all descended from ancient European conquerors. It was a bizarre idea. But many German ultranationalists willingly swallowed it because of Hans F. K. Günther, one of the Reich's most famous racial scholars. In his immensely popular books on race, Günther claimed that the primordial master race had first launched its assault on Asia some four thousand years ago.[4] A few of its tribes, he suggested, took a northern route. They swept across Inner Asia and successfully fought their way to

China and even Japan, where they became large landowners and nobles.[5] Even in modern times, he noted, the Chinese and Japanese aristocracy often displayed discernible Nordic traits—"a decidedly long skull and an almost white skin, sometimes combined with handsome European features."[6]

In Günther's view, however, most of the Nordic invaders had taken a more southerly route. After sweeping through the Caucasus, they fought and slaughtered and pushed their way east to the fertile plains of India.[7] So awed were the native inhabitants of the subcontinent, alleged Günther, that they described the invaders as gods in their legends. Along India's fertile river plains, the flaxen-haired foreigners made themselves at home, becoming the lordly Brahmans. They instituted a strict caste system, prohibiting their children from marrying anyone other than a fellow Brahman. In this way, they zealously protected the purity of their bloodlines, and over time they extended their power to northeastern India and Nepal. There, explained Günther, a wealthy young Nordic couple gave birth to a prince—Buddha.[8]

As Günther saw it, these powerful lords were ultimately unseated by something beyond their control. Pale Nordic men and women, he suggested, were simply unable to bear the tropical sun, originating as they did in the dreary, drizzly, snowy lands of northern Europe. Their infants sickened and died at the height of summer, and the few that survived, he contended, had little choice but to marry into native families, giving birth to generations of babies with dark hair and brown skin. Even so, Günther was convinced he could discern certain Nordic traces in the aristocratic Brahmans. "Their height," he noted, "is six to nine centimeters taller than the rest of the population. In addition to their generally lighter skin, they have a narrower face and nose, and brown hair that they share with other high castes."[9]

All these far-fetched claims for Nordic overlords in Asia made a deep impression on Himmler.[10] He was keen to unearth hard archaeological proof of these golden-haired conquerors—the farther east the better. Such finds would provide key new evidence of the primacy of the Aryan race through time. Moreover, they would almost certainly rivet Hitler. The German Führer had taken little, if any, interest in the Ahnenerbe's expeditions and research in northern Europe, much to Himmler's despair.[11] But Hitler greatly admired Günther's work, even taking the time to attend the

As a young man, Heinrich Himmler took a keen interest in three things: extremist politics, ancient history, and archaeology. *USHMM, courtesy James Blevins*

A nineteenth-century illustration of ancient Germanic warriors. Himmler saw them as "high, pure and capable."

Himmler (third from the left) poses with his extended family, including his father, Gebhard (far right), and his wife, Marga (center). As a young Nazi official, Himmler hoped to build a new Germany that mirrored a mythical German past. *USHMM, courtesy James Blevins*

Das deutsche Gesicht

1. Nordische Kernrasse

2. Fälisch-nordische Kernrasse

3. Westischer Einschlag

Nazi theorists classified Europeans into five disparate races, publishing posters to educate Germans on these fictional groups. The Aryan or Nordic race was said to be tall, blond-haired, bold, creative, and valiant in war—a natural master race. *USHMM, courtesy Hans Pauli*

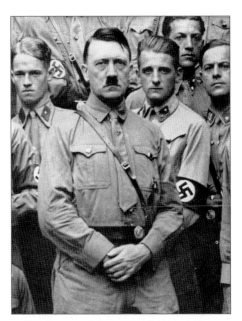

1. Schutzstaffel-
Appell
der Gruppe Ost in Berlin
11., 12., 13. August

Hitler poses with a group of SS men. In 1929, he placed Himmler in charge of the SS, founded as an elite personal bodyguard. *USHMM*

To attract the right recruits, Himmler honed the sinister image of the SS, redesigning its uniforms and launching recruitment campaigns. *Mary Evans Picture Library*

SS men march at a public rally. Himmler wanted to transform the SS into a racial elite of blond Nordic men. He instructed his officials to grade each SS applicant on a five-point scale, from "pure Nordic" to "suspected non-European blood components." *ullstein-SV-Bilderdienst*

To teach SS officers to think like ancient Aryans, Himmler recruited a former psychiatric patient, Karl-Maria Wiligut. *Bundesarchiv, SSO Weisthor, Karl-Maria*

On Wiligut's advice, Himmler leased Wewelsburg Castle for a senior SS academy. Its furniture and china soon sported bizarre runic designs. *Heather Pringle*

Himmler, Wiligut, and a group of SS officers examine purported runic symbols in a quarry. In Nazi circles, Wiligut took the name Weisthor, meaning "Wise Thor" and claimed to channel the wisdom of his forebears in trances. *Bildarchiv Preußischer Kulturbesitz*

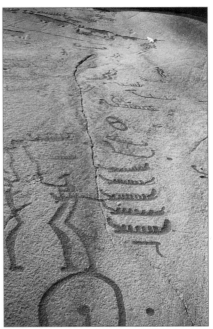

In 1935, Himmler founded an SS research institute, the Ahnenerbe, to re-create the lost world of Germany's ancestors. The first president was Herman Wirth. *ullstein–ullstein bild*

Wirth claimed in one wild theory that Bronze Age rock art in Sweden (above) represented the world's earliest writing system. *Heather Pringle*

In 1936, Himmler sent Wirth (far right) and an SS expedition to Sweden to make massive casts of the rock art. As one prominent German critic noted with derision, Wirth was "unable to distinguish between probable, certain, possible and impossible." *Landsarkivet i Lund, Schlyterska samlingen*

Yrjö von Grönhagen was a young Finnish nobleman traveling on foot through Europe when he first met Himmler. *Juhani Grönhagen*

Himmler believed that remote Karelia in Finland remained a stronghold of Aryan lore. He hung this photo of a Karelian musician in his private study. *Juhani Grönhagen*

Grönhagen chats with an elderly Karelian couple. A year after the founding of the Ahnenerbe, Himmler dispatched Grönhagen and a German musicologist to Karelia to record with a magnetophone the magical spells of sorcerers and witches. *Juhani Grönhagen*

As the Ahnenerbe's managing director, Wolfram Sievers was on top of every file, meticulously carrying out Himmler's orders. *Eike Schmitz*

Walther Wüst lectures on Aryan philosophy to SS officers in March 1937. A former SS informer, Wüst taught Sanskrit and Aryan studies at the University of Munich. In 1937, Himmler handpicked him to replace Wirth as the new president of the Ahnenerbe. *Bundesarchiv Koblenz*

Historian Franz Altheim and photographer Erika Trautmann clown for the camera in August 1936. Clever and immensely charming, the couple wrote a secret intelligence report for the SS while conducting research for the Ahnenerbe in Eastern Europe and the Middle East. *Frobenius-Institut*

Erika Trautmann casts a backward glance (far left) while recording rock art in Spain. An old family friend, Hermann Göring, introduced her to Himmler. In 1936, she used her important Nazi connections to advance the career of her lover, Franz Altheim. *Frobenius-Institut*

scholar's inaugural lecture at the University of Jena in 1930.[12] An expedition to Asia in search of the golden-haired master race would raise the Ahnenerbe's profile dramatically in Hitler's eyes. All that Himmler needed was to find the right scientist.

ON NOVEMBER 2, 1935, a muscular young German SS-Mann turned up at the German consulate in Chung-King with an impressive tale of survival and resourcefulness.[13] Ernst Schäfer was a twenty-five-year-old zoologist, the son of a prominent German industrialist. Volatile and headstrong, he had set off from Shanghai in early July 1934 with a close friend and another companion on an American-financed expedition to the wilderness that straddled the border between Szechuan and Tibet. Schäfer and his colleagues had planned on collecting bird and mammal skins and other specimens for the Academy of Natural Sciences in Philadelphia. But the remote region, rife with robbers and warring tribes, tested their endurance to the limit. The trio battled off murderous brigands, faced down a mutiny among their porters, and ferreted out treacherous spies in their midst. But during a perilous siege mounted by the machine gun–toting troops of a Chinese warlord, Schäfer's friend, Brooke Dolan, decided to flee for help disguised as a Mongolian merchant. He never returned. Schäfer finally managed to talk his way out of the siege, but his second American companion departed the expedition soon after.

Left in charge of the native porters and what remained of the caravan, Schäfer stewed over his fate. "I was totally alone," he later wrote, "just as the hardest part of the expedition was beginning and the most important research lay ahead of us."[14] The route ahead led deeper into a war zone, but Schäfer resolved to continue collecting zoological specimens. He set off with the porters, and when he arrived back in civilization seven months later, his caravan was laden with the carcasses of some three thousand exotic animals, including at least two species unknown to science. He had taken nearly seven thousand photographs of the countryside and its people and shot rare film footage of the remote Inner Asian wilderness—all of which would be useful for magazine articles, books, and public lectures. He had also charted the remote region in maps and scribbled his impressions on nearly three thousand double-sided pages.[15]

In Shanghai, Schäfer coolly plotted his future. He planned to spend a few months in Philadelphia organizing his collection, but he was keen to return to Inner Asia as the head of his own expedition.[16] He needed financial backing, however, and for this, Schäfer pinned his hopes on his SS connections. From the office of the German consul general in Shanghai, he wrote a five-page letter to August Heissmeyer, an SS-Gruppenführer, describing his success in salvaging the entire trip. "The expedition," he boasted, "is considered to be the most successful—in terms of results and collected zoological expedition material—that has ever been conducted in Inner Asia and should be the model for future expeditions."[17] Schäfer left little room for doubt that he was the right man to lead those expeditions.

When the Nazi party caught wind of Schäfer's success, it quickly called him home.[18] In Philadelphia, Schäfer received an official summons to return to Germany as soon as possible and a telegram from Himmler's office informing him that he had been awarded a commission as an honorary SS-Untersturmführer. All this attention from high quarters was immensely gratifying to the young SS-Mann. "I am so proud and happy that I am not able to express it," he declared in his reply to Himmler. "I hope I will be able to show my gratitude through my actions. All my expectations were exceeded in each and every respect, though the greatest honor for me is to have been promoted."[19]

IN JUNE 1936, the new Untersturmführer was ushered past the sentries at the door of the former art school on Prinz Albrechtstrasse. He was led up the stairwell and down a bright corridor decorated with busts of Hitler and other prominent German figures, and shown into Himmler's office, a plain, impersonal-looking room that displayed little of the gilded magnificence favored by other ministers of the Reich.[20] Himmler was in a talkative mood. He warmed quickly to Schäfer and the conversation soon turned to Schäfer's plans for the future.

The young Untersturmführer had spent seven months mulling over the best way to present his proposal for a new expedition. Like every other member of the SS, he was well aware of the immense importance that Himmler placed on archaeology and the ancient history of the mythical Aryans. As luck would have it, Schäfer's last expedition had stumbled upon some archaeological and anthropological finds that would interest Himmler greatly.

While out hunting near the small town of Batang, along the Chinese-Tibetan border, Schäfer and his companions had come across several ancient graves dug into the fertile terraces of a riverbed. Curious about the contents, they had not hesitated to open the graves, although they had no permission to do so from local authorities. Schäfer and his companions unearthed several skeletons curled in a fetal position.[21] Mourners had sent them to the next world wearing jewels and accompanied by all manner of urns and pots. Schäfer was fascinated when he saw the decorative motifs on these objects. They corresponded in his opinion to "old Aryan symbols."[22]

As the expedition moved on, Schäfer observed what he considered to be other proof of an ancient Aryan presence. During his university studies, he had taken classes in ethnology—the history of human groups and their origins and distribution—and without doubt was familiar with the works of Günther.[23] The entire subject interested him greatly.[24] While trekking along the mountainous northeastern fringes of Tibet, he had encountered several striking individuals. Unlike the local farmers who were "pure Mongol types" or the citified tradespeople who possessed "strong Semitic streaks," these individuals seemed in Schäfer's eyes to possess "pure Aryan facial characteristics, strong hook noses, red-black hair and almost gray-blue eyes."[25] From these brief observations, he concluded that they were "representatives of a racial group that differs from other inhabitants."[26] Such Aryan-looking individuals, he observed, "are spread over the entire area and are strongest in the feudal noble class, as well as among landowners and the warrior-robber group."[27] All this struck the young SS officer as a significant discovery for Germany's *Rassenkunde* experts. "This opens up a whole new area of science in a productive and appreciative field—to establish how far the Aryan race interspersed on the Roof of the World."[28]

Schäfer almost certainly related these racial discoveries to Himmler, who was a very receptive and attentive listener when the occasion suited him.[29] And during the ensuing discussion, Himmler seems to have offered his own thoughts on how Aryan tribes could have ended up in the mountains of Inner Asia—thoughts that clearly blended Wirth's theories on Atlantis with Günther's speculations about Asia. After the war, Schäfer recalled some of this conversation for his American interrogators, who then distilled his recollections into one sentence: "Himmler believed that ancient emigrants from Atlantis had founded a great civilization in Inner Asia, the capital of which was a city called [Obo]."[30] Himmler also seems to have confided to Schäfer his own personal view on the origins of the Nordic race, which was again summarized by the American interrogators. "Himmler mentioned his belief that the Nordic race did not evolve, but came directly down from heaven to settle on the Atlantic continent. He also mentioned Oshima's belief in a similar theory concerning the origin of the noble castes in Japan."[31]

Schäfer thought Himmler's views on the divine origins of the Nordic race were ridiculous. But he seems to have kept these opinions to himself in his first meeting with the SS leader. Certainly, Himmler retained a very favorable impression of the young scientist, so favorable, in fact, that in 1937, he agreed to make Schäfer's great dream come true. He put him in charge of a major new Ahnenerbe expedition that was to travel to Tibet and also to the Gobi Desert in Mongolia, where it would look for the ruins of the mythical city Obo.[32] He then ordered the Ahnenerbe's managing director, Wolfram Sievers, and his own trusted aide Bruno Galke to begin piecing together the financing. He also personally instructed Schäfer to consult with Karl-Maria Wiligut.

Schäfer obliged, presenting himself at the old colonel's villa in Berlin. A pair of SS sentries waved him up the driveway and a young assistant, Gabriele Dechend, led him into a dank, ill-kept tropical conservatory. A deep gloom seemed to envelop the room. As Schäfer waited uneasily, he detected a strange sweet smell that he recognized at once from his travels in the East. It was opium. The old colonel finally shuffled into the room. He embraced Schäfer and kissed him on both cheeks, then took a seat. Slowly, the elderly officer began slipping into some kind of altered state, his eyes glazed and his hands visibly trembling. It reminded Schäfer of the trances he had seen some Tibetan lamas enter.[33]

When Wiligut at last broke the silence, he astonished Schäfer. In a deep, rough voice, he began to speak of the Dalai Lama and rattle off names of Buddhist monasteries and other places in eastern Tibet known only to Schäfer. The young zoologist felt as if Wiligut were literally reading his mind. Schäfer later noted in his memoirs that he was rather shaken by the encounter and left at the first opportunity. But in later years, Wiligut's assistant remembered things differently. The two men, she recalled, had a fascinating conversation about the telepathy of Tibetan lamas, a type of extrasensory perception that Schäfer claimed to have witnessed himself. He told Wiligut that the lamas in Tibet often seemed to be expecting him when he arrived at their monastery.[34]

Nevertheless, this bizarre experience and Himmler's remarks on the divine descent of the Nordic race raised a large red flag in the young zoologist's mind about allying himself with Himmler on matters of science. The question was clearly how to clinch the Ahnenerbe's support without compromising his own sense of integrity. It was a difficult problem, but Schäfer was young and supremely confident in his own abilities. Indeed, he seems to have believed that he could handle Himmler just as deftly as he had managed the brigands and warlords of China. In Inner Asia he had learned that one "could not lose his nerve. The success of the expedition was based on such games."[35] It was a matter of wile.

SCHÄFER WAS ACCUSTOMED to dominating others and bending them to his will. Short, stocky, and barrel-chested, with a thick crop of wavy dark-blond hair, a strong jutting jaw, and a sturdy—as opposed to handsome—face, he radiated confidence. He possessed immense reserves of physical strength that allowed him, day in and day out, to chase big game up into the rarefied altitudes of the Himalayas as if it were no more than child's play. He spoke in such a loud, booming voice that his friends often joked that he had no need of a telephone.[36] He wrote in a bold hand with a thick nib, covering page after page in a dark scrawl of black ink, every second sentence or so punctuated with one or more exclamation marks. He was egotistical and a master of self-promotion. He relished taking risks. He drove like a maniac.

He tended to be blunt and direct to a fault, but he did not hesitate to lie when stuck in a tight spot. A moody man of dark corners and edges, he had a trigger temper, which he lost frequently—sometimes violently so. In his later years, he fell into strange somnambulatory fits. Once while sleepwalking in a train, he nearly strangled a doctor sleeping in a berth nearby. Life with Schäfer, as his second wife later conceded, was never easy.[37] Yet the zoologist rarely bore grudges, and to those he considered peers or friends or family, he was kind, loyal, and generous. He possessed a talent for binding men to him.

His willfulness revealed itself at an early age. Born in Cologne in 1910, the son of a wealthy businessman, Schäfer had little interest as a boy in following in his father's footsteps. To escape the sprawling family home in Waltershausen, he roamed the glens and pine-covered hills of the neighboring Thuringian Forest. He brought home squirrels and small animals and built aquariums and cages, transforming his room into a miniature zoo.[38] He spent hours watching the behavior of birds he had captured. As a teenager, his love of the outdoors was rivaled only by a passion for hunting. He succeeded in shooting his first roebuck at the age of twelve. By his mid-twenties, his prowess with guns was legendary. In Inner Asia, he once shot three fleet-footed antelopes from a distance of two hundred yards, dispatching each with a single bullet. "Consistently he shoots as the average hunter shoots once in a lifetime and bears the memory for the remainder," his friend Brooke Dolan observed later.[39]

When Schäfer reached his late teens, his father insisted that he abandon his feral ways and prepare for a business career, taking suitable courses at university. By then, however, the young naturalist abhorred the thought of spending the rest of his life cooped up in offices and boardrooms. He had read with fascination the books of Sven Hedin, a Swedish explorer who had trekked through Inner Asia, and he dreamed of a similar life in the wilds of the East. He enrolled at the University of Göttingen and told his father he was studying law.[40] In reality, however, he was taking classes in zoology, botany, geology, and ethnology.[41]

It was while studying migrating storks, guillemots, and doves as a student on Helgoland, an island just off the North Sea coast of Germany, that he met a distinguished German zoologist who had traveled through the remote highlands of western China and eastern Tibet. Hugo Weigold was much impressed by Schäfer and his remarkable marksmanship, so when the older zoologist received an invitation to join a new scientific expedition to the headwaters of the Yangtze River in 1930, he recommended Schäfer as the team's hunter. The expedition's American backers were particularly anxious to collect specimens of a giant panda and other large exotic species. "If anyone can kill a giant panda bear," Weigold reportedly told the team leader, "it would be Schäfer."[42]

The young zoologist needed no arm-twisting. He joined the team and departed by train from Berlin in January 1931. In Chung-King, the expedition haggled and bargained for a large caravan, complete with more than a hundred porters as well as horses and pack animals. Thus equipped, it began trekking north and west toward the mountainous headwaters of the Yangtze. Schäfer was agape at the exotic mountain world he found himself in, with its roaring rivers and dripping bamboo forests and its hundreds of species of strange new birds and insects, snakes and mammals, trees and vines, flowers and grasses. He felt reborn, and he set about hunting zoological specimens with an intense single-minded passion.

For months, he stalked the forests in search of the giant panda. He found broken shoots of bamboo—the jagged leftovers from their meals—but failed to catch even a brief glimpse of a panda. One day, however, he noticed bits of bamboo shoots littering the high boughs of a tree. He realized the bears were capable of climbing up into the forest canopy. The next morning he and a Tibetan assistant returned to the spot and silently clambered up a nearby tree. Some four hundred meters in the distance, he noticed branches gently rustling. He took aim and fired at the spot. Sight unseen, he knocked a panda to the ground, killing it for his collection.

A SEASONED VETERAN of Inner Asian research, Schäfer began drawing up plans in the summer of 1937 for the Ahnenerbe trip to Tibet. He set his sights firmly on the high plateau region of northeastern Tibet, bordered by the Amne Machin Mountains. The remote wilderness, he observed in an eight-page letter to Himmler, was "the last completely unexplored realm of Central Asia," an exotic land that would greatly appeal to geographers, botanists, zoologists, and ethnologists.[43] Schäfer was also highly optimistic about the prospects for racial studies in the region. He planned to examine the "mixed Aryan groups of people and their social behavior which I discovered on my last trip."[44] He also intended to excavate the graves of Batang, recovering ancient artifacts that "trace back to the Aryan-Scythian migration."[45] All this he planned to capture on a professional expedition film.

He proposed departing sometime between the fall of 1937 and the end of January 1938. This would give him time to finish two popular books he was writing on his recent adventures in Inner Asia—he clearly hoped to model himself on his hero Sven Hedin—and prepare for the oral examination for his doctoral degree before throwing himself into preparations for the expedition. Himmler perused the plans and approved them. By this time, both men had dropped the notion of an extended trip into Mongolia's Gobi Desert. In all likelihood, Schäfer had convinced the SS leader that such a side trip would needlessly add expense to an already costly itinerary. Schäfer believed that the racial and archaeological studies in Tibet would uncover ample evidence of Aryans.

Over the next ten months, Schäfer slaved away at his studies and his writing, and in his spare time, he wooed a willowy young farmer's daughter, Herta Volz. Soon after their marriage on July 14, 1937, Schäfer took up preparations for the expedition. He estimated that the team's journey to Tibet would cost at least 60,000 reichsmarks, 30,000 of which would be needed before departure.[46] The Ahnenerbe, however, was unable to shoulder such a heavy burden, so Wolfram Sievers and Himmler's trusted aide, Bruno Galke, immediately set to work locating the necessary money. Galke met with representatives from Eher Verlag, the central publishing house of the Nazi party, to arrange a publishing deal for Schäfer.[47] He and Sievers also attempted to sell the rights for the expedition film and obtain free steamship passage to Shanghai for the team members. In addition, they contacted the German Research Foundation and the Advertising Council of the German Economy, the government agency responsible for disseminating business propaganda.[48] In the meantime, Schäfer agreed to search for corporate money through a contact at IG Farben, the world's largest chemical cartel, based in Frankfurt.

Himmler's staff also began passing on names of suitable researchers for the expedition. Schäfer listened politely, but having barely survived the near collapse of one trip after his teammates deserted, he had his own definite ideas on the matter. He intended to pick scientists who were young, strong, physically fit, and capable of handling themselves if push came to shove. He also planned to select researchers of a high scientific caliber. For the expedition geologist, he immediately poached an assistant of Wilhelm Filchner, Germany's most seasoned Tibet explorer. Twenty-four-year-old Karl Wienert had a doctorate degree in geophysics and a passion for fieldwork and adventure. For the technical leader—the person responsible for organizing the expedition in Germany and managing all the provisioning, transport, and communications in Inner Asia—Schäfer selected twenty-four-year-old Edmund Geer, a former regional policeman turned SS officer who possessed strong organizational skills and extensive SS connections. In addition, he recruited thirty-eight-year-old Ernst Krause, a versatile man capable of doubling as the expedition's filmmaker and its entomologist.

One of the most important slots was still vacant, however. The team needed an anthropologist with a solid background in ethnology. A senior official in RuSHA passed on to Schäfer the name of a twenty-six-year-old, up-and-coming *Rassenkunde* expert, Bruno Beger.[49] Beger was a student of Hans F. K. Günther. Sociable and inquisitive and very much a self-made man, he possessed many of the social skills that Schäfer lacked and had mastered the art of networking long before it became fashionable. He was just finishing off his doctoral degree in anthropology. Tall, wiry, and naturally athletic at six foot two, he was the very image of a Nordic man, with his piercing blue eyes, long sharp nose, and neatly parted straight blond hair.

Beger's sense of ambition easily matched that of Schäfer. He was born in 1911 into an affluent old Heidelberg family that had made its money from the region's tanneries.[50] But the family had fallen on hard times. His father died at the front in the First World War, and most of the family's financial assets slipped through his mother's hands during the financial tumult of the 1920s. As luck had it, the mother of a good friend came to his rescue, paying for his university studies. Beger had intended on studying mathematics at the University of Jena. But after attending a lecture by the racial scholar Hans F. K. Günther, he gravitated toward the subjects of anthropology and ethnology.[51] Günther, in turn, seems to have encouraged the budding young racial scholar. Indeed, Beger ended up designing the maps for Günther's book on the Nordic race in Asia.[52]

In 1934, Beger joined the SS. He landed a part-time job in RuSHA and swiftly rose to the position of section head. From its inception, RuSHA was, as one recent historian has pointed out, less a place for opportunists and careerists and "more a place for convinced ideologues."[53] Beger fit the mold extremely well. RuSHA's director, Otto Hofmann, for example, liked the young scholar's approach to racial problem-solving. "Beger," he wrote, "is an inventive and enthusiastic practitioner with an extraordinarily well-founded knowledge, yet he never gets stuck in pure theory."[54] This opinion seems to have been widely shared. When one colleague, Dr. Erich Karl, proposed leading an expedition to Hawaii—a "huge laboratory of racial mixture"—he invited Beger along.[55] Karl intended to study miscegenation, which was of course anathema to the SS, and gather data on its purported perils.[56]

Discussions over funding of the Hawaii trip dragged on, however, and while waiting for final approval, Beger received a postcard, then a phone call, from Schäfer, inviting him to join the Tibet expedition.[57] Beger had read about Schäfer's exploits in Inner Asia in the German newspapers and was flattered that the young zoologist had approached him. Moreover, he was captivated by the prospect of trekking through the Himalayas and conducting racial studies there. Schäfer's new expedition was custom-made for an ambitious man. What better way was there to establish himself as one of Germany's preeminent racial scientists and cultivate favor with Himmler?

So Beger signed on. And with the immensely powerful SS political machine spinning behind him, Schäfer continued sorting through the names of other candidates for the expedition—archaeologists, botanists, zookeepers. He was so close to his dream of returning to Inner Asia that he could almost hear the water dripping in the dense bamboo forests. On November 9, 1937, however, his plans suffered a tragic setback. While weekending at the country estate of a friend, the young zoologist went out duck hunting with his wife Herta and two servants in a rowboat. The weather was gray and rough, and as Schäfer stood up to shoot, a wave suddenly smashed against the boat, knocking him off balance just as he was about to pull the trigger. His shotgun slipped from his hands, broke in two, and discharged accidentally, shooting his bride of four months in the head.[58] By nightfall, she was dead.

SCHÄFER WAS WRACKED with guilt and grief. Immediately after the tragedy, he threatened to kill himself.[59] He soon managed to pull himself together, however, and a mere eight weeks later he plunged back into work, revising his book manuscripts and pushing harder than ever at preparations for the expedition. He desperately wanted to escape to the remote frontier of western China and eastern Tibet, where he could lose himself in the beauty of nature, the pleasure of the hunt, and his duties as an expedition leader.

He drew up page after page of equipment lists—pistols, saddles, a shortwave receiver and transmitter, a collapsible boat, Leica cameras, tents, first-aid kits, animal traps, tools for processing zoological specimens, geophysical gear, crates of food and cigarettes and brandy, and boxes of presents for the Tibetans, including used binoculars, used and rejected pistols, folding knives with decorations, wrapped biscuits and cookies, and cheap watches. The lists were all passed on to Gestapo officials, who ran checks on the suppliers, right down to the producer of the expedition's dried fruit.[60] Each company had to be Aryan-owned. Sievers then composed letters to manufacturers and other potential outfitters, requesting that they donate the necessary gear.[61]

As the preparations proceeded, however, political conditions along the proposed expedition route rapidly deteriorated—to Schäfer's dismay. Fourteen months earlier, in November 1936, Hitler had signed an important anti-Soviet pact with Japan, swallowing his immense disdain of the Japanese as mere "bearers of culture" in order to secure a vital alliance with them.[62] The new allies shared a fierce hatred of Bolshevism. Moreover, many Japanese researchers and politicians also regarded the world through the fractured prism of race. They proudly described themselves as *shidō min-zoku*, or the "leading race" of the world and believed that they traced their origins to the gods.[63] Japan's rulers were immensely keen on carving out a new Japanese empire in Asia and planned on planting colonies of Japanese settlers in the conquered lands.[64] Already they had invaded, conquered, and occupied Manchuria, and they were hungry for more.

So in July 1937, Japan embarked on a full-scale conquest of China. Its divisions soon occupied the major coastal cities, including Shanghai, the commercial capital of China. By November, hundreds of thousands of Japanese troops were advancing on what was then the Chinese capital, Nanking, and waging pitched battles for control of the vital rail lines along the Yangtze River. Schäfer had been counting on steamship travel up the Yangtze as the swiftest and most direct route to Chung-King, where the expedition would then proceed by foot for Tibet. But the Yangtze had now become a major battleground, one that was far too perilous a place for a scientific team from Germany, Japan's new ally.

Determined to find another way to northeastern Tibet, Schäfer began poring over maps of the region.[65] The route that most attracted him wound from the Indian state of Assam into eastern Tibet, but there again he saw major hurdles. The British government feared correctly that a major war with Germany was looming. It would not look kindly on permitting an SS scientific team to traipse through the remote mountain valley of India. If war broke out in Europe, Britain would desperately need Indian troops to man its fronts and Indian factories to churn out munitions. It did not want to have to stamp out an SS-led guerilla movement in the eastern Himalayas, a movement that might ignite all those who advocated Indian independence.

Schäfer refused to concede defeat. He flew to London in early March 1938 with cap in hand. He courted men with influence in the India Office and paid visits to prominent Nazi sympathizers in London who might put in a good word for him.[66] But the British authorities were deeply suspicious of Schäfer's real motives for the trip, and turned him down flat, and the Tibetan government, keen to remain on good terms with the British, followed suit. These obstacles, however, only made Schäfer more determined. He proposed booking the passage to India with a small team—just Wienert, Geer, Krause, and Beger—hoping he could negotiate the necessary permissions on the spot. To his immense relief, Himmler agreed to the plan.

With his egotism and his headstrong ways, however, Schäfer had already antagonized both Wüst and Sievers.[67] Indeed, Sievers had taken the extreme step of writing to Himmler, explaining that the Ahnenerbe was no longer prepared to take responsibility for the expedition.[68] But Himmler did not seem worried by this. As he must have known, Beger, one of RuSHA's best young racial experts, would conduct a thorough search for traces of the supposed master race in the Himalayas. Moreover, Himmler was well aware that Schäfer's maps and Tibetan contacts would prove invaluable if the Reich wanted to engage in a guerilla action later in India.[69] Besides, most of the preparatory work was done. Sievers and Galke had lined up a good deal of the equipment and organized nearly all of the necessary financing for the trip, from the Advertising Council of the German Economy, the German Research Foundation, and Eher Verlag.[70] So Himmler agreed to allow Schäfer and the expedition's technical leader, Edmund Geer, to finish off the remaining preparations on their own.

Even so, Himmler continued to help out with the last-minute glitches. He wrote to Hans F. K. Günther at the University of Berlin, requesting that he move up the date of Beger's doctoral examinations, so that Beger could "arrive abroad as a German scholar with completed exams."[71] He also requested assistance from the German foreign minister, Joachim von Ribbentrop, and the German ambassador in Calcutta: Schäfer needed introductions to the highest members of the Indian government.[72]

By mid-April, almost all of the equipment was packed and all the necessary paperwork was stamped and sealed. The team—known officially as "The German Tibet Expedition/Ernst Schäfer under the Auspices of the Reichsführer-SS and in Connection with the Ahnenerbe Association Berlin"—was ready to depart.[73] Schäfer and Geer took the train to Genoa, arriving on April 20. Beger, Wienert, and Krause arrived the following morning. At one in the afternoon on April 21, 1938, the five men boarded a German steamship, the *Gneisenau*, bound for Ceylon. From there, they planned to travel by freighter to Calcutta.

SCHÄFER, SOMETHING OF a taskmaster, saw to it that the passage out was neither relaxing nor idle. He had purchased first-class tickets for the team so that they could rub shoulders with all the important people aboard ship—diplomats, engineers, businessmen, and officials, men of influence who might in some way prove useful to the expedition. He also insisted on turning the team's spacious cabins into classrooms, pressing team members to brush up on their English and to give mini-lectures to one another on their subject areas. At night, the five men burned the midnight oil, working out problems in the research plans and poring over the maps to plan their routes.[74]

In Ceylon, however, Schäfer received troubling news. The day before they boarded the *Gneisenau*, the Nazi party newspaper, *Völkischer Beobachter*, ran a glowing article on the Tibet expedition, making much of its SS connections, something Schäfer had hoped to keep quiet abroad.[75] This story raised eyebrows on London's Fleet Street. A journalist ran a scoop that an SS expedition had set sail for India, and before long the press in both Britain and India were running headlines about a "Nazi invasion."[76] During the stopover in Ceylon's capital, Colombo, the local press besieged Schäfer. He later raged to the SS headquarters about the "unheard of attention in the press in India, exactly where there are many Jews residing" and expressed his worries about "difficulties to be faced in Calcutta."[77]

The news infuriated Himmler, spurring him into action. He wrote per-

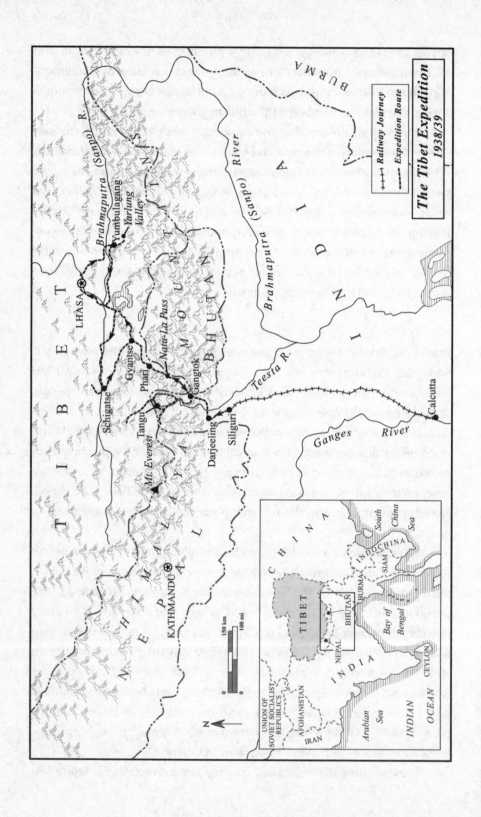

The Tibet Expedition
1938/39

+++++ *Railway Journey*
–··–··– *Expedition Route*

BURMA

Brahmaputra (Sanpo) R.

Yumbulagang
Yarlung Valley

LHASA

Brahmaputra (Sanpo) River

I N D I A

T I B E T

M O U N T A I N S

Gyantse
Natu-La Pass
Phari
Gangtok
BHUTAN

Schigatse
Tangu

Teesta R.

Mt. Everest
Darjeeling
Siliguri

Ganges River

Calcutta

H I M A L A Y A

N E P A L

KATHMANDU

150 km
100 mi
0
0

N

CHINA

UNION OF
SOVIET SOCIALIST
REPUBLICS

AFGHANISTAN

IRAN

Arabian
Sea

TIBET

NEPAL
BHUTAN
BURMA
INDOCHINA
SIAM

South
China
Sea

INDIA

Bay of
Bengal

CEYLON

INDIAN OCEAN

sonally on May 19 to an influential Nazi sympathizer in England, Admiral Sir Barry Domvile, a former head of British naval intelligence. In his letter, Himmler ridiculed the notion that he would be so clumsy as to dispatch a spy in such an open manner and threatened to retaliate against British citizens in Germany.[78] Domvile passed the missive on to Prime Minister Neville Chamberlain, a strong believer in placating the leaders of the Reich.[79] Not long after this, British authorities granted the SS team permission to conduct scientific studies in the small Himalayan kingdom of Sikkim. The kingdom bordered Tibet, but it lay much farther west than Schäfer's planned staging point in Assam. Worse still, other scientific teams had already pored over the flora and fauna of Sikkim. It was a poor bone that the British had thrown Schäfer. They hoped he would turn it down. But he stubbornly accepted.

Calcutta did little to improve his mood. Even the small German community there appeared hostile to the team. Unwilling to dawdle among enemies, Schäfer wrapped up the negotiations for the entry permit to Sikkim as quickly as possible. Then the team packed their gear and boarded a northbound train. The annual monsoon season had arrived a month early, and Schäfer worried that travel would be slow, with frequent mud slides in the Himalayas. But he was ecstatic to be under way again. "And thus," he wrote proudly by hand in a letter to Himmler, "we weigh anchor as an SS expedition!"[80]

In the capital of Sikkim, Gangtok, the team paid their respects to local British officials and set up camp near the British Residency, a lovely stone house set in a sprawling expanse of manicured lawns and well-tended beds of hollyhocks and asters. Gangtok was a thriving hub in the wool trade with Tibet, and as such its citizens knew a great deal about Tibetan ways. After settling in, Schäfer and Geer began hiring porters with strong backs, and interpreters who could speak Tibetan. All the while the British political officer in Gangtok, Sir Basil Gould, watched Schäfer carefully. The young German, he later noted, was "interesting, forceful, volatile, scholarly, vain to the point of childishness, disregardful of social convention or the feelings of others, and first and foremost always a Nazi and a politician."[81] It was obvious to Gould that Schäfer intended to complete the trek to Tibet, with or without official British permission.

On June 21, the team set off north with porters and horses and fifty heavily laden mules, each brightly ornamented with silver bells and scarlet tassels of dyed yak hair. The mule track teetered precipitously above the roaring Teesta River, known locally as "Cleft of the Winds."[82] It was a nightmarish trip. In the heavy monsoon rain, landslides threatened ominously from the sheer, denuded valley walls: in some spots muddy torrents had already roared down out of the morning mists, plastering the trail with a thick, impassable layer of mud. The porters had no choice but to chop down a new pathway in the forest, and the heavy work in subtropical heat and humidity exhausted them. The team's misery was further compounded by a plague of tropical leeches that lay in wait for them in the tall grasses or curled about the tips of twigs. One evening alone, as Schäfer unlaced his boots and threw aside his sodden socks, he counted fifty-three fat leeches clinging to his right foot; another forty-five fed contentedly on his left. He noticed that the ankles of the caravan's porters, who walked barefoot, were coated in a crust of dark black blood.[83]

Despite all the hardship, however, Schäfer discovered a kind of magic along the Teesta. At night he watched tiny lights dancing amid the ferns— "fantastical and surreal: the motion of the shining tropical glowworm army. The sight is eerie."[84] He found a great deal more of fascination as the road wound northward. From the Teesta Valley, the team began climbing upward through forests of fir and pine, which gave way outside the village of Changu to alpine hillsides carpeted with dwarf rhododendrons and azaleas, all blooming furiously in the June heat. Schäfer dispatched men to collect specimens for German botanists to identify, and as the team stopped to set up camp, Schäfer set off into the hills to hunt game.

Slowly, as the team roamed the highland valleys, Schäfer's companions fell into the simple rhythm of fieldwork. In the early morning, Krause and an assistant visited the light traps they had set the previous evening. They gently plucked out the moths and other nocturnal insects captured inside and collected the new species. Wienert, the geophysicist, set off into the surrounding hills with mules laden with his theodolites and other delicate instruments, while Geer, the team's technical leader, climbed into the alpine zone. Schäfer had assigned him the task of collecting all bird species that resided above an elevation of four thousand meters.

Beger, meanwhile, visited the tents of travelers and ventured into the small hamlets, searching for people who would submit to his racial measurements. He soon learned that offering medical assistance was the surest way of winning the confidence of those he encountered.[85] There was always someone who needed help, and Beger was perfectly willing to pull teeth or perform first aid or dispense medicines from one of three large trunks he had packed with pills and salves and tonics from Germany. Only then, after he had finished dosing small babies and frail elders, could he unpack his bag of anthropological instruments—large and small spreading calipers, sliding compasses of various sizes, three steel tape measures, and a somatometer—and begin taking measurements of the local inhabitants for his racial studies.[86]

13. TIBET

FEW THINGS IN LIFE so disturbed German *Rassenkunde* experts as the subject of racial mixing. In their version of history, the human propensity for crossing racial boundaries had reduced an empire of golden-haired gods to a hodgepodge of racial mongrels. It had sown Nordic boldness and creativity and the Nordic capacity for reason into the short, squat bodies of inferior races. It had instilled a penchant for laziness, vacillation, and cowardice in the tall, angular bodies of northern Europe. It had spawned a kind of racial pandemonium in the world, a tumult so bewildering that *Rassenkunde* experts had been forced to develop various intricate anthropological measurements, followed by lengthy mathematical analyses, in their determined effort to trace the ancestry of individual humans. Before leaving Berlin, Bruno Beger had carefully studied these techniques.

In the highlands of Sikkim, he followed a strict research regime. As each of his nervous subjects stood before him, he held up an eye-color table to measure the precise hue of his subject's eyes. He then pulled out other charts to identify the exact shading of the hair and the color of the skin on the forehead, carefully recording the results in special notebooks he had brought for the purpose. He then proceeded to the cranial measurements. He charted the length, breadth, and circumference of his subjects' heads; the height and width of their foreheads; the breadth of their mouths, noses, cheekbones, and lower jaws; the width between their eyes;

the depth of their noses; the position of their ears, noses, and mouths relative to the rest of their face.[1]

Like many other racial scholars of the age, Beger considered certain key cranial features to be reliable indicators of the Nordic race. The most important was the shape of the skull. Anatomists had learned to quantify this shape by using a simple mathematical formula, the cephalic index, which was the ratio of maximum head breadth to maximum head length, multiplied by one hundred. When racial scholars began to use this formula in their early studies, they noted that inhabitants of Scandinavia, Great Britain, and northern Germany—the supposed heartland of the Aryan race—possessed a cephalic index between seventy-seven and seventy-nine, indicating a long and narrow head.[2] This delighted racial scholars. They interpreted the numbers to mean that northern Europeans possessed high foreheads—a sure sign, in their opinion, of high intelligence. And so for many German researchers, the cephalic index became a prized pillar of racial theory. Indeed, in the late 1920s Günther proudly proclaimed that one of the distinguishing features of the Nordic race, along with hair and eye color, was a cephalic index of "round about seventy five."[3]

Günther, of course, had played fast and loose with science, ignoring powerful evidence that ran contrary to his beliefs. In 1908 prominent American anthropologist Franz Boas had begun questioning whether environmental factors could influence the cranial traits regarded as racial markers. So he put the cephalic index to a major test. He and a group of assistants measured 17,821 immigrants and their children in New York City, selecting a wide range of nationalities, from Scots to eastern European Jews, and from Bohemians to Sicilians. The team then divided the children into two groups—those born in the United States and those born in their parents' homeland—and compared the cephalic indices of both to those of their parents. Boas found that environmental factors played a profound role in shaping the human body. The American-born children of round-headed eastern European Jews, for example, did not take after their parents. They developed long heads. And the American-born children of long-headed Italians developed shorter, rounder crania. Boas did not speculate on what specific conditions had brought about these impressive cranial changes, but he left little doubt about the significance of the study. "Not even those characteristics of a race which have been proved to be most permanent in their old home," he concluded, "remain the same under our new surroundings."[4]

The publication of Boas's study in 1910 struck a heavy blow at scientific racism and biological determinism. It convinced leading anthropologists in the United States and Europe to abandon the cephalic index as a scientific indicator of race and to question the whole system of racial classification.[5] But it did not dissuade Günther or other German racial experts from proceeding with their racial studies. And it certainly did not deter Beger from putting Sikkimese men and women through the dehumanizing experience of racial measurement.

AS BEGER CONTINUED to search for Sikkimese subjects, Schäfer hunted for a suitable pretext for entering Tibet. In August 1938, fate smiled on the team when a richly dressed official turned up at the German camp. Schäfer was away on a short hunting trip, so Beger beckoned the man over to his tent, hoping to win the visitor's confidence sufficiently to measure him.[6] He had encountered few members of the nobility—individuals whom, he believed, would most likely possess Nordic blood—and he hated the thought of letting any opportunity slip through his fingers. The expedition's interpreter hastily intervened, however, staving off a major diplomatic blunder.[7] The visitor, he whispered, was a high official of the Rajah Tering, a member of the Sikkimese royal family who was living in Tibet.

When Schäfer returned, Edmund Geer met him on the outskirts of the camp and informed him of their important guest. Schäfer was delighted. He hastened to arrange an appropriately regal reception area in his tent. He asked team members to place all their most impressive-looking scientific gear—altimeters, binoculars, and cameras—around the tent and ordered the servants to make hot tea and to set out biscuits and other delicacies on plates.[8] When all was ready, he took his seat atop an inflated air mattress, hoping this would make him look taller and more imposing. Then he invited his noble guest to enter.

Schäfer's careful management of the occasion succeeded wonderfully. The dazzled official returned to Rajah Tering with several muleloads of gifts—biscuits, chocolate, an eighty-pound bag of potatoes, tinned vegetables, soap that Schäfer and his team were secretly glad to be rid of "because it smells like civilization," woollen socks, rubber boots, and a rubber mattress—as well as a letter from Schäfer expressing his fervent wish to visit Tibet and its legendary monasteries.[9] These tokens of esteem and Schäfer's clever diplomacy sparked months of intense behind-the-scenes talks and deliberations in both Sikkim and Tibet. In December, Schäfer received a letter from Lhasa. The Tibetan council of ministers—an august body accustomed to forbidding foreigners entry to Tibet—invited both him and his colleagues to their capital for a two-week stay.

It was hardly a welcome with open arms. But Schäfer read the letter with intense satisfaction. No German had ever stepped foot inside the holy city, and even the formidable Sven Hedin had been turned away. The young German zoologist had pulled off a major political coup, trumping British officials. Moreover, the journey to Lhasa, with swastika pennants flying boldly from their tents when they encamped, would make striking Nazi propaganda. In exchange for this rare favor, however, Tibet's rulers had laid down one major condition. They forbid the team from killing any birds or mammals inside Tibet, for this "would deeply hurt the religious feelings of the Tibetan people, both clergy and lay."[10] The stipulation ruled out all zoological collecting, restricting the scope of Schäfer's own work. But the expedition leader was determined to reach Tibet—he knew how keenly Himmler awaited racial studies in the mountain kingdom. So he grudgingly accepted the council's demand.

The team returned to Gangtok to stock up on supplies, and in the midst of their reprovisioning they received a message from Himmler. As a reward for their accomplishment, he had promoted the entire team—Schäfer to the rank of SS-Hauptsturmführer and his colleagues to SS-Obersturmführer.[11] Cheered by this, the expedition set off for Lhasa. It was December 20, a bitter time of year to be climbing into the alpine zone and crossing a high Himalayan pass. The next evening, the team held a winter solstice celebration in fine SS style. They piled up wood and built a huge bonfire, and as they huddled about its warmth, they sang an old German military march, "Flame Rise," that was a popular standard among SS men. The next morning, they tramped onward toward the icy mountain pass of Natu La at 14,600 feet, straining for breath in the thin air. At the summit, prayer flags fluttered in the breeze and the team stopped to gaze to their hearts' content at a sight granted few Europeans: the snow-powdered highlands of Tibet.

DESCENDING INTO THE valley below, they passed heavily laden yak caravans and Tibetan wool merchants armed with antique-looking flintlocks and swords sheathed in silver and turquoise scabbards. Along the side of the road, tents of black yak hair revealed groups of Buddhist pilgrims—mothers nursing their infants and men drinking tea. As Beger passed them by, he stared at and sized up the curious faces. In the southern village of Phari, an important caravan stop, he was pleased to spy a high official with what he judged to be a noble warrior mien—the vestige, he concluded later, of some European blood.[12] He hastened to take the man's picture. At the other end of the social scale, he later noted, were beggars of almost indeterminate race. In Beger's eyes, they seem to be walking billboards for the perils of miscegenation. They "are remarkable to me," he noted, "especially because of their irregular and jarring racial-mixture type."[13]

Whenever possible, Beger took a head or partial body cast of his subjects. He had learned a special technique employed by medical professors to make models of diseased human organs, stillborn babies, and on occasion, the faces of convicted criminals.[14] Unlike conventional plaster casting, the technique employed a soft substance called Negocoll that could be heated and applied to the human face. When it set, Negocoll formed a rubberlike mask. A second liquid substance—either Hominit or Celerit—could then be poured into the mask to make a positive. The result looked eerily lifelike by the standards of the day, for as the inventor of the method noted in one of his books, "a beginner is able to make a flawless cast in around ten minutes . . . which shows the finest plastic details of the skin under a magnifying glass."[15]

Such lifelike casts had become valuable commodities in academic circles in the Reich. In the mid-nineteenth century, three German brothers, Hermann, Adolph, and Robert Schlagintweit, had taken plaster casts of many people they encountered during their travels through India and Inner Asia. By the 1880s, copies of the complete Schlagintweit collection, which consisted of 275 heads, sold for 6,000 reichsmarks to museums and universities. Individual heads could be purchased for between 25 and 30 reichsmarks.[16] Moreover, trade in such human replicas had become even more brisk after the Nazi seizure of power. Across the Reich, museum curators sought new and more lifelike casts of exotic *Untermenschen* to liven up otherwise dry museum exhibitions on Nazi racial doctrine.[17] In addition, racial biologists purchased the casts as teaching aids to instruct students in the finer points of racial typing. Head casting had become a profitable line of business.

Beger tried to collect as many casts as possible. He had begun in Gangtok by bribing his personal servant, a Nepalese Sherpa, Passang, who was in poor health. Passang was clearly terrified at the prospect of being replicated, but he tried to keep his composure as Beger first oiled his face, then applied a thick sloplike paste. Before long, some of the paste dripped into one of Passang's nostrils, making breathing difficult. Beger insisted that he remain still, however, allowing the cast to set, but the frightened man could not wait. He leapt out of the chair, clawing at the mask forming on his face "as if he was possessed by the devil."[18] Beger and several others tried pinning the unfortunate man to the ground, but by then he was panic-stricken, flailing his arms wildly and uttering strange roaring sounds.

Beger looked on with growing alarm. "All of us were thinking that if he died, our expedition—which we had embarked upon with so much hope—would be as good as dead. So all of us, with the same thoughts, tried to help him."[19] The German scientists stood watch over Passang as he gradually revived. Later Schäfer threatened the porters who witnessed the scene. If they breathed a word of the incident to anyone outside the expedition, Schäfer would fire the informers on the spot. Despite the strain this created, Beger soon resumed making casts of porters and others he met on the team's travels.

ON THE MORNING of January 19, 1939, Schäfer and his colleagues glimpsed with wonder the steep white and ochre walls of the Potala of Lhasa towering high in the air. Light glinted and gleamed from its golden roof pavilions, a mesmerizing sight. Like other travelers on the road, Schäfer and his team dismounted and "bowed in silence to the stronghold of Lamaism."[20] The Potala was the fabled home of the Dalai Lama, the spiritual and political head of Tibet. The thirteenth Dalai Lama had died six years earlier and Tibetan monks had only succeeded in locating his reincarnation in 1937, in a small village in northeastern Tibet. As a young boy, the new Dalai Lama could not play any part in governing the remote kingdom. Indeed, he had not even arrived yet in Lhasa. In his place a council of ministers and a powerful regent, Reting Rimpoche, ruled the country.

Schäfer had received word earlier that a celebration awaited their arrival in Lhasa. He was expecting a warm reception—an important Tibetan welcoming committee, traditional gifts of white silk scarves known as *khatas*, an invitation to inspect a Tibetan guard of honor, crowds of curious Tibetans, and accommodation in one of the larger mansions of Lhasa. But apart from a swarm of beggars that descended upon them as they reached the western gate of Lhasa, the Tibetans took little notice of their new guests. As Beger recalled later, "a simple official greeted us and showed us to our quarters."[21] This lodging consisted of a small, shabby, foul-smelling government house.

The paltry welcome did not bode at all well for the team's success in Lhasa. Schäfer intended to ask for permission to film and photograph the people of Lhasa and to extend their stay, all of which would require careful diplomacy. So he hurried off to pay his respects to the Tibetan ministers, as well as to Tsarong Dzasa, an important member of the nobility who had once served as commander in chief of the Tibetan army. Schäfer, who had taken lessons in Tibetan manners from one of his interpreters, was gracious and charming and respectful and invariably arrived at these appointments bearing expensive gifts—HMV portable gramophones and records, Philips radio sets, Zeiss binoculars.

To further cultivate favor, Schäfer made frequent mention of what he called a common bond between the Himalayan nation and Germany: the swastika. For centuries, Tibetans had regarded the ancient symbol as a sign of good fortune and permanence. To capitalize on this, Schäfer had brought a supply of Nazi pennants with him and took pleasure in pointing out to his hosts how revered the symbol was in Nazi Germany.[22] The Tibetans were delighted, little knowing that in Europe the swastika had come to symbolize the dark forces of German ultranationalism. Indeed, one influential Tibetan remarked innocently to the team that "it is the first time that the eastern and western swastika could meet under the banner of peace, on the neutral basis of cultural exchange and scholarly activities."[23]

In an official report dispatched to Germany on January 23, just four days after the team's arrival, Schäfer stressed the progress they had made in warming up Tibet's ruling men. The Tibetan government was about to "let their secrets out" to the Germans and "show us life in the capital and in the huge monastery cities."[24] Moreover, the cabinet was prepared to allow the Germans to film Lhasa and its medieval setting and extend their stay much longer than the original fourteen days. "It is a wonderful feeling," Schäfer concluded, "to know that the power of the German Reich is so great that it reaches into the most isolated parts of the Inner Asian continent."[25]

So pleased were the Tibetan authorities with Schäfer's gifts and good manners that they gave the team freedom to explore the exotic streets and temples of Lhasa. As Beger wandered each day past market stalls laden with dark, pungent bricks of dried tea, bales of fine silk and cheap printed cloth, heaps of scarlet chili peppers and fragrant nutmeg, and fine jewelry of amber and coral, he gaped at the exotic wares and the even more exotic press of humanity. There were shaven-headed monks wrapped in dark red robes; burly yak herders muffled in heavy sheepskin robes; delicate-looking high officials dressed in shimmering brocade *chubas,* with an elegant turquoise earring dangling from one ear. Beger wished in vain that he could bring out his calipers and spreading compasses.

When the team finally managed to gain an audience with Regent Reting Rimpoche, Beger stared at the young ruler's face intently "for a long time." He was particularly gratified to see that the frail, spindly young man had "an especially long thin head."[26] The object of this intense scrutiny, however, completely misunderstood the nature of the young German's longing gaze. Reting Rimpoche smiled warmly at the anthropologist and later asked him to join his bodyguard, where the two could presumably get to know each other much better. Afterward, Beger noticed that one of the regent's young male attendants seemed "to serve the inverted desires" of his master.[27]

The team was greatly restricted in the scientific work they could do in Lhasa, but they were resolved to make the best possible use of their time. Schäfer seems to have asked for samples of grains from Lhasa's granaries to add to the collection he had already acquired from Sikkimese and Tibetan farmers en route.[28] The German government was extremely keen to develop new forms of vegetable oils and increase the yield of its cereal crops. It hoped to become agriculturally self-sufficient. Tibet, with its highland fields, seemed a promising place to look for new, hardier, disease-resistant varieties of barley, wheat, and oats.[29] So Schäfer was particularly pleased to discover what he later called Lhasa's "sixty-day grain."[30] It could be harvested just two months after it was sown.

During the colorful New Year's festival in Lhasa, Schäfer and Krause spent days filming the magnificent dances and parades, when crowds of onlookers turned out in their finest clothes. During the quieter periods, Beger set off to visit the most important monasteries in the region, some of which were the size of small cities. As the chief repositories of Tibetan learning, these huge institutions housed important libraries of rare Tibetan books. Beger carried with him a very long wish list.[31] He wanted to collect stories from the ancient Tibetan epic, the *Gesar*; pictures and drawings of the Tibetan gods; copies of the Tibetan astrological tables and calendars; and detailed information on the old holy places of the ancient shamanistic religion of Tibet, known as the Bon, which predated Buddhism. He also wanted to obtain floor plans of each monastery and an exact record of all inscriptions carved on the monastic walls. Last but not least, he wanted to collect copies of the most valuable books—which consisted of loose sheets of paper protected between two separate wooden boards—and interview the scholars who knew them best. All these sources might yield clues to the presence of ancient Aryan lords in Tibet.

To obtain these things, Beger began cultivating the monks and other officials. He filled his notebooks with interesting remarks from their conversations, jotting down Tibetan lore about the swastika and quickly sketching details about the four castes of Tibet. The first caste, he noted approvingly, "hailed from the race of Gods—Second, from non-gods—Third, from the ruling races & Fourth, from subject race[s]."[32] He also managed to collect a priceless copy of a 108-volume encyclopedia of Lamaism. As a rule, Tibetan monks were reluctant to part with such treasures. They had previously bestowed only three copies upon Europeans. All three, however, sat on the shelves of European libraries, gathering dust. Beger intended to see German scholars claim the distinction as the first translators.

After more than two months of visiting the great mansions of Lhasa and sipping yak-butter tea with high officials, Schäfer was growing restless. He felt it was time to push on. By much humoring and wooing of the regent and other key officials, he had wrangled permission to visit what many scholars called the birthplace of Tibetan civilization, the Yarlung Valley. It was a rare coup, one made all the sweeter by the knowledge that such consent had been denied to British officials. So Schäfer and his companions bid their good-byes to their hosts. After all the diplomacy, and the intrigue and undercurrents of tension in Lhasa, Schäfer felt relieved to be on the road again.

The caravan wound its way eastward across a patchwork of farms, toward the Yarlung Valley. It was a place Tibetans treasured. They believed that their first kings were divine beings who slid down from heaven on a sky cord, landing near Lhabab Ri, a mountain bordering the Yarlung Valley. When their reigns had ended, these kings climbed back up the dangling cord to the celestial realm. Eventually, however, an unlucky thing had happened: one of the monarchs accidentally severed his silken escape route while battling a court magician. After that, Tibet's kings died as ordinary mortals did. Tibetans believed that their earliest kings ruled from a great stone fortress—known as Yumbulagang—in the valley. To Schäfer, it sounded like an ideal spot to search for traces of primeval Nordic overlords. He intended to be "the first white scholar to study the secret of Yarlung-Potrang, the ancient capital city of Tibet."[33]

As the team crossed down into the Yarlung Valley, Schäfer reveled in the green peacefulness. He later described the region as "paradisial."[34] But the ruins of the ancient royal stronghold, Yumbulagang, perched upon a high mountainous spur, proved something of a disappointment. Neither Schäfer nor his colleagues could see much sign of a great palace, still less an ancient capital. All that remained were "mighty watchtowers, which are still to this day testimonies to the brave warriors and courageous soldiers who stood at the side of Tibet's ancient kings."[35] The team pitched camp at the foot of the spur and spent two days carefully surveying and mapping these ruins. Then they turned back westward toward Shigatse, Tibet's second largest city. Beger planned on purchasing traditional Tibetan rugs, teapots, *tsampa* bowls, and other goods in the markets there for the team's ethnographic collection.

Relations between the team and British officials in the region steadily deteriorated, however, as the weeks passed. By the midsummer of 1939, family members and friends in Germany were urging them in letters to return home as soon as possible. In March, Hitler had sent armored columns into the old Czech heartland, occupying Prague and declaring the industrially rich provinces of Bohemia and Moravia a protectorate. Now Hitler's eyes were trained greedily on Poland. British Prime Minister Neville Chamberlain had vowed to defend Poland in the event of German military aggression, but no one knew if he would honor his word. Only one thing was clear. If Westminster did declare war, Schäfer and his companions would find themselves in a very difficult spot. British officials would almost certainly arrest them and clap them into an internment camp in India for the duration of the war.

The team had hatched half a dozen adventurous plans for their return trip, even considering at one point driving back from India to Berlin. But Schäfer loathed the thought of capture and imprisonment by the British. He informed the team it was time to wrap things up, and he contacted Himmler's personal staff, who began making arrangements for the team to fly back to Germany. Meanwhile, he, Geer, and Krause carefully packed up the voluminous natural-history collections—animal and bird skins; butterflies, bees, ants, wasps, and other insect specimens; fragile dried plants for the herbarium; packets of seeds containing one thousand six hundred varieties of barley, seven hundred varieties of wheat, and seven hundred varieties of oats, not to mention hundreds of seeds from other potentially useful plants.[36] They also arranged for sturdy crates to be built for the live animals they had captured or acquired for German zoos—three breeds of Tibetan dogs, rare feline species, wolves, badgers, and foxes.[37]

Beger saw to it that his valuable ethnographical collection of nomad tents and lama trumpets was safely stowed in trunks. During his Tibetan travels, he had photographed nearly 2,000 Tibetans, Bhutians, Sherpas, and Nepalis. He had measured 376 individuals and cast the heads and faces of 17, including 2 of the most powerful men in Tibet—Tsarong, a close friend of the thirteenth Dalai Lama and a former commander in chief of the Tibetan army, and Mondo, a Tibetan noble raised and educated in England.

It would take time to thoroughly analyze his measurements, but based on what he had seen, he thought it very likely that the Nordic race had changed the course of Asian history. The first colonists of Tibet seemed to have been ancient Mongolians who settled in the mountainous land after the last Ice Age ended. He believed, however, that racially mixed descendants of ancient Nordic invaders had swept into the Tibetan plateau more recently, giving rise ultimately to "the higher Tibetan classes."[38] The proof, as he saw it, lay in the supposed Nordic characteristics of Tibetan nobles— "tall stature paired with long head," "narrow face," "receding cheekbones," "strongly protruding, straight or slightly bent noses," "smooth hair," and a "sense of themselves as dominant."[39]

THE FIVE SCHOLARS boarded a British Indian Airways flight in Calcutta, and switched in Baghdad to a German flight. For the next leg of the trip, from Vienna to Munich, they traveled in Himmler's own personal aircraft, the *Otto Killenbeth*. And both Himmler and his chief of staff, SS-Gruppenführer Karl Wolff, were on the runway in Munich when the plane touched down at 5:10 P.M. on August 4, 1939. Schäfer was flattered and delighted to see Himmler. It was sunny and warm—a splendid summer day in Bavaria—and the stocky zoologist beamed broadly as he walked side by side with Himmler down the tarmac toward the terminal. His colleagues followed respectfully a few paces behind. Together, the SS chief and the scientists had a cup of coffee in a private room in the terminal.

Schäfer likely explained that he carried with him an official letter from the Tibetan regent, Reting Rimpoche, for Hitler. In addition, the regent had given him three gifts to present to the German leader—a Tibetan mastiff, a gold coin, and the robe of a lama. Schäfer believed that the garment, which was carefully wrapped, had once belonged to the former Dalai Lama, and he hoped to meet with Hitler in person to deliver it.[40] The SS leader listened to the zoologist's account attentively, and together they proceeded to board the plane for a major press conference and official reception in Berlin.

The next morning, all across the country, Germans read sensational news of the expedition. As they sipped their steaming coffees in cafés and unfolded their newspapers on trains, their eyes lit on the headlines: "Hitler's delegation in Tibet," "The First Germans in Lhasa," and "The First White Men in Yarlung-Potrang." Himmler was delighted by the great splash of attention, particularly as it reached all the way to the Reich Chancellery. Hitler, it seems, had finally taken notice of one of his cherished projects. Soon after the Tibet team's return, Schäfer received a rare mark of favor—an invitation to lecture on his Tibetan work to an audience of senior Nazis. Hitler himself would be in attendance.[41]

14. IN SIEVERS'S OFFICE

For many Germans, stories of exotic lamas and romantic Himalayan lands were a welcome relief from the grim invective occupying the pages of most German dailies in late August 1939. For weeks, German newspapers had railed viciously against the Poles, reporting murderous attacks on the German community in Danzig. The increasingly shrill accounts of foreign persecution—part of a massive propaganda campaign orchestrated by the staff of Josef Goebbels—were clearly a prelude to a new German invasion. Poland was next on the list, and many Germans worried whether Hitler wasn't biting off more than he could chew. Would the government of Neville Chamberlain meekly back away from its vow to defend Polish independence? Would Stalin seize the moment and attack? German officials were bracing for the worst. At the end of July, they instructed Berliners to begin practicing air-raid drills.[1] A few weeks later, they began handing out ration cards for meat and such luxuries as jam, sugar, and coffee.

But on the morning of Thursday, August 24, the German government announced a dazzling coup. Hitler, it seemed, had fooled everyone, signing a major non-aggression treaty with Germany's greatest enemy, the Soviet Union. With one stroke of the pen, the German leader had turned one of Nazi Germany's most formidable foes into an ally, eliminating the danger of a bloody war along the eastern borders of the Reich. It was a masterful stroke. The new pact seemed to all but guarantee that the

British would now step aside and allow the German Wehrmacht to begin carving up Poland. Germans were elated. In Berlin, they took to the streets in celebration.[2]

In his elegant office in Dahlem, Wolfram Sievers welcomed the news. He had grown accustomed to the privileged life of a high-ranking Nazi official, riding through Grunewald each morning and attending important receptions in an expensive new dress-suit in the evenings.[3] It was all much preferable to crawling through enemy fire. Sievers, moreover, had important work to do. While Schäfer and his colleagues were roaming Inner Asia, measuring the crumbling glory of ancient Tibetan strongholds, the senior scientists and scholars of the Ahnenerbe had been busy at their desks. Some had planned major new expeditions to remote corners of the world, from Bolivia to the Canary Islands. Others had proposed projects on medieval village life to help organize the future SS settlements. And a few were clinically accumulating racial data for the day when the SS would uproot all Jews and their "mixed-race" descendants from the Reich.

It was up to Sievers to orchestrate all this research, to cut through the flimsy excuses of Nazi officialdom, to slash away the endless rolls of red tape, to prod the unwilling, flatter the important, cajole the incompetent— anything to get the job done. In his spare hours, Sievers was a musician. He played the harpsichord, organ, and piano, and he particularly loved the music of Bach, in all its rich textured complexity. But Sievers reserved his greatest dexterity for his work in the sprawling villa in Dahlem. There he spent his days on the telephone, in meetings, and at his desk arranging financing, foreign currency, steamship company tickets, passports, balloon-mounted cameras, aerial surveillance aircraft—indeed, almost anything that Ahnenerbe researchers needed for their important work. Sievers seemed to be everywhere and anywhere, on top of every file. Without him, everything in the Ahnenerbe would grind to a halt.

BY FAR THE most urgent problem on his desk was the massive expedition of Edmund Kiss. Kiss, an architect and writer by profession, was mounting the Ahnenerbe's largest and most expensive expedition yet. Its destination was the Bolivian Andes. During an earlier trip to the region, Kiss claimed to have located the stone ruins of an ancient Nordic colony in the New World. Bolivians called the site Tiwanaku. Kiss declared that the elaborately carved temples of Tiwanaku dated back more than one million years. This was at least eight hundred thousand years before the evolution of modern humans.[4] And as if all that were not wildly exorbitant enough, Kiss also alleged to have found crucial new geological proof of something known as the World Ice Theory, a crackpot paradigm that many influential National Socialists adored.

Kiss was exactly the kind of man with whom Himmler enjoyed social-izing. At fifty-three, he was a commanding presence, standing six foot three and tipping the scales at 224 pounds.[5] He possessed a broad, sturdy face, ears that splayed out from his head, wire-rimmed spectacles that curled around them, and a determined stamp to his mouth.[6] He spoke bluntly, kept his word faithfully, conducted his affairs with a gentleman's sense of honor, and was, judging from the testimonies of his friends and co-workers after the war, kindness itself toward his subordinates.[7] He also possessed a distinguished military record. He not only survived two gunshot wounds, a serious case of malaria, and four years of mucking in the trenches in the First World War, but also won two Iron Crosses, one of them a first class.[8]

After the war, Kiss took his examinations as a building contractor and settled in Münster, where he began to study the World Ice Theory. The bizarre theory tossed out most conventional scientific ideas about the uni-verse. In their place, it offered a new explanation for just about everything—the origins of the solar system, sunspots, the appearance of the Milky Way, the creation of the human race, the sinking of Atlantis, and some of the more obscure passages of ancient Icelandic creation stories. The theory was the brainchild of an Austrian engineer, Hans Hörbiger, who prided himself on the fact that he never performed calculations and who firmly believed that mathematics was "deceptive."[9] His ideas greatly appealed to right-wing extremists who were always looking for ways of Germanizing sci-ence and junking anything that smacked of "Jewish" science. The Nazis regarded Hörbiger as a genius.[10] They intended to consign Albert Einstein to oblivion.

Kiss was fascinated by Hörbiger, who likened the universe to a giant steam engine filled with hydrogen and water vapor. In the distant past, Hörbiger suggested, small stars thickly clad in ice had collided with steam-ing hot giant stars, spewing stellar material into space. This material then condensed into planets of varying sizes that spiraled around the sun. As the smaller planets edged closer toward the larger ones, they were ensnared by gravity and captured as moons. Hörbiger believed that Earth had known six of these satellites. The serial destruction of the first five, he suggested, had led to vast, almost unimaginable environmental catastrophes on Earth. As each had spiraled downward into the atmosphere, it had revolved faster and faster, creating an immense gravitational pull. This force had then yanked the Earth's waters toward the equator, forming an immense tide resembling a giant spare tire around Earth's girth; beyond the perimeters of this towering wall of water, the land surface froze beneath thick glacial ice. Only in certain mountain refuges—the Bolivian Andes, the Tibetan Himalayas, the Ethiopian highlands—had flora and fauna survived. Each of the plummeting moons had then exploded in turn in the atmosphere, releasing oceans and seas to flow back over the Earth. The last of these celestial explosions, claimed Hörbiger, had taken place more than eleven thousand years ago.[11]

This theory was pure, unadulterated nonsense, condemned in the strongest terms by German astronomers and other serious scientists in the 1930s. The World Ice Theory, noted one prominent mathematician, combined "the tyranny of an Asiatic despot [and] the presumption of a mathematical illiterate who, with childish innocence, strides up to things about which he knows nothing and ventures to substitute a caricature for a scientific picture of the universe."[12] But Kiss and many other Nazis were deaf to such criticism. Hörbiger's talk of giant tides and vast sheets of glacial ice provided a neat explanation for scientists' inability to find any trace of an ancient Aryan civilization in the far north. No less an eminence than Hitler himself had latched onto these outlandish ideas. "I'm quite well inclined to accept the cosmic theories of Hörbiger," he noted one night over dinner, before launching into a muddled description of the engineer's ideas.[13]

What Hörbiger's supporters desperately needed, however, was proof of the primeval cataclysms that Hörbiger had described. Kiss was acutely conscious of this. So in 1927, he began searching about for evidence. He wrote to a silver-haired Austrian expatriate, Arthur Posnansky, in Bolivia. Posnansky had begun a detailed study of ancient stone ruins in the Bolivian Andes—one of the places that the World Ice Theory predicated as a mountain refuge. He had published and lectured extensively on Tiwanaku. Situated just south of Lake Titicaca, the prehistoric capital had once ruled a mighty empire whose power extended all the way from the Bolivian rain forest to the northern coast of Chile and northwestern Argentina. By the twentieth century, however, Tiwanaku lay in scattered pieces. Looters had plundered most of its wealth, leaving only huge inscribed tablets and immense doorways carved with jaguars and strange mythological characters. Some blocks weighed more than four hundred tons.

Most professional archaeologists of the day knew that an indigenous Andean people—forerunners of the Inca—had designed and built Tiwanaku. But Posnansky strongly disagreed. He suggested that a mysterious group of immigrants from the far west designed the great capital and put Andeans to work building it. He also asserted that construction on at least one of Tiwanaku's great temples began seventeen thousand years ago, an erroneous contention based on his own calculations of certain astronomical alignments of the walls.[14] Lastly, Posnansky believed he had discovered an ancient calendar of some sort carved into the stone above a Tiwanaku portal.[15]

Kiss sponged up these ideas, convinced that the mysterious architects were none other than the Aryans. So fascinated was he, indeed, that he journeyed to South America in 1928, subsidized by a 20,000-mark prize he had won in a writing contest. For months, Kiss studied Tiwanaku's ruins, sketching their floor plans and their inscriptions. He was particularly struck by an ancient sculpture of a man's head unearthed from one of its ruins. "It is immediately clear," he noted later, "that this man is not Indian nor does he have Mongolian characteristics, but rather pure Nordic ones."[16] Moreover, he continually jotted down notes about what he thought were European touches in the stone monuments—"the doors are framed as they were in the Baroque period," he observed, and "the construction of the eastern facade shows a series of cross symbols underneath an entablature that is easily identified as Greek."[17]

He brusquely dismissed any suggestion that ancient Andeans had designed the splendid temples. "The works of art and the architectural style of the prehistoric city are certainly not of Indian origin," he wrote later. "Rather they are probably the creations of Nordic men who arrived in the Andean highlands as representatives of a special civilization."[18] The big question in Kiss's mind was when this migration had happened. He peered and squinted at the massive inscribed relief on the ruin known today as Gateway to the Sun. Was it truly a calendar? Kiss thought it very likely and he became convinced he could decipher it.[19] He felt certain he could see symbols for twelve months of the year, each possessing either twenty-four or twenty-five days. He also thought it certain that each of the days had thirty hours, and each hour twenty-two minutes.

Kiss regarded the inscription as compelling proof of Hörbiger's theory. The Tiwanaku inscription, he concluded, was a calendar reflecting the primordial conditions on Earth when an earlier moon rapidly orbited the planet. He had no means of determining when this had happened, but this did not stop him from leaping to a wild conclusion. "One thing we do know—and it would be extremely hard to convince us otherwise—even if the age of Tiwanaku cannot even be guessed, it must be at least millions of years old!"[20]

Back in Germany, Kiss began determinedly popularizing his ideas, first in a series of fantasy novels set in Atlantis and South America, and then in a more scientific-sounding book entitled *The Sun Gate of Tiwanaku*. He illustrated the latter with his own architectural drawings of massive Nazi-style monumental temples and tall, slim inhabitants dressed in a strange futuristic fashion. The editors of Nazi party newspapers and magazines were delighted. Both *SA Mann* and *Die Hitler Jugend*, the official magazine of Hitler Youth, ran popular articles on Kiss's research, illustrated by the architect's drawings. These pieces extolled the beauty of the lordly Nordic colony of Tiwanaku, and described, as if it were now proven scientific fact, how the ancient Andean capital had collapsed during the cataclysms triggered by a falling moon.[21] Himmler was so pleased with the book that he ordered a copy to be expensively bound in leather as a Christmas gift for Hitler.[22]

But Kiss longed to expand his field research. He yearned to return to Bolivia with a large interdisciplinary team of scientists to search for fossil evidence of ancient flooding and to conduct extensive excavations at both Tiwanaku and nearby Siminake. He hoped to unearth compelling new evidence of the ancient master race in the Americas and requested backing from the Ahnenerbe. Himmler and Wüst were both wildly enthusiastic. "One can now quite certainly expect results which might have a revolutionary importance for the history of mankind," opined Wüst.[23]

Kiss drew up meticulous plans and relayed his needs to Sievers. Over the next year and a half, the two men worked together on the project, interrupted only when Himmler dispatched Kiss on a brief research trip to Libya to scour the Mediterranean coast for fossil evidence of the World Ice Theory.[24] By late August of 1939, the mushrooming plans called for a team of twenty—archaeologists, geologists, zoologists, botanists, meteorologists, pilots, and underwater experts—to toil on the project for a year.[25] In addition to the archaeological digs, Kiss planned to explore the deep waters of Lake Titicaca by underwater camera. He also proposed flying across the Andes so that film crews could shoot footage of the famous Inca roads, which Kiss believed were the work of Nordic lords.[26] Last but certainly not least, he also intended to conduct extensive geological fieldwork from Colombia to Peru to find evidence of ancient celestial cataclysms.

Sievers estimated that the salaries of the team members alone would cost 100,000 reichsmarks, or some $520,000 today, taking inflation into account.[27] But Himmler did not flinch at the cost, and by late August 1939, Sievers was deeply immersed in the final arrangements for the trip—booking the team's passage to South America, locating a pilot experienced enough to undertake aerial photography in the high Andes, and organizing payment of all the team members' salaries. It was a mammoth task greatly complicated by the plodding bureaucracy of the Nazi state and the pressing nature of Sievers's other duties.

IN MUNICH, WALTHER Wüst was planning a small expedition to the gray and forbidding mountains of western Iran.[28] The Ahnenerbe's superintendent had never traveled to Asia, but since arriving at the Ahnenerbe he had succumbed to expedition fever. In 1938, he proposed leading a group of researchers to study and record the famous Bisitun inscription in Iran. The inscription preserved the autobiography of Darius I, one of the greatest kings of ancient Persia.[29] In the sixth century B.C., Darius had seized the throne by murdering a rival. He had then successfully extended the borders of his realm as far east as the Indus Valley, before launching an ill-fated invasion of Greece.

Wüst considered Darius to be a great Aryan monarch. Moreover, he thought his story particularly relevant to the Reich of the 1930s.[30] Darius had usurped a throne, ruthlessly extinguished other contenders, stamped out rebellions, and forged a vast empire of diverse peoples. This sounded like the blueprint for Hitler's career. Wüst also believed that empire building was an integral part of the Nordic psyche, a trait that characterized all major Nordic leaders through time. "In all places on earth where the Indo-Germanic peoples turn up," he wrote, "they enter into history because of their creation of states and empires, whether it be the empire of Darius I,

Alexander the Great, the Roman Empire, or the empire of Charlemagne. Each introduced the Germanic form of empire to the Occident."[31] Such claims were patently false, but they greatly appealed to Himmler, suggesting as they did that Hitler was following his destiny as a Nordic leader by founding the Third Reich.

If Wüst was to popularize the story of Darius and curate an important museum show about him, the Ahnenerbe had to acquire an exact copy of the inscription. This would be no easy matter. Darius had ordered his sculptors and scribes not only to inscribe the story of his victories at Bisitun, but to preserve the account for posterity. So his servants had constructed a high stairway up a steep cliff at Bisitun. Perched precariously upon a narrow ledge one hundred feet in the air, they carved Darius's words in three languages and delicately sculpted a large relief of Darius standing majestically in front of a long line of his vanquished foes.[32] When they were finished, they destroyed the wooden stairs up the cliff, placing Darius's words beyond the reach of most vandals. Some twenty-five hundred years later, a young British linguist learned of the mysterious carving. In a feat of great daring and athleticism, Henry Rawlinson scaled the crumbling cliff and copied by hand almost two-thirds of the inscription for scholarly research. His young Kurdish assistant then recorded most of what remained by applying damp pulp to the surface of the inscriptions and pressing it against the indentations to make exact casts.[33] With these copies, Rawlinson and other scholars deciphered three dead languages— Old Persian; Elamite, the administrative tongue of the Persian empire; and Akkadian, the language of Babylon—and opened a lost door to the early civilizations of the Middle East.

Rawlinson had copied the inscription with marvelous accuracy, but he had erred somewhat in his interpretations of badly weathered and nearly illegible passages. Moreover, neither Rawlinson nor his assistant had been able to reach or copy a small portion of the text. So Wüst proposed to journey to Bisitun to create a new scientific recording of the inscription "in accordance with the will of the RF-SS and head of the German police, Heinrich Himmler."[34]

At thirty-seven, Wüst was a bulky, deskbound scholar plagued by knee problems.[35] He was incapable of climbing his way up a sheer cliff, and he could see that building a large scaffold would take too much time and manpower. So Wüst intended to rely upon a new technology devised by an

American archaeologist—suspending a camera from a tethered balloon and floating the device alongside the cliff face at Bisitun.[36] With a long cable-release to activate the shutter, Wüst and his colleagues could snap photographs of the inscription. Already SS archaeologists had enjoyed some success with the technique, employing it to take aerial photographs of an excavation at Tilsit in eastern Prussia.[37]

Wüst planned to travel to Bisitun with his wife and a team of four—a scientific amanuensis; an Iranian student who would be responsible for dealing with the local inhabitants; a photographer to take care of the imaging; and a mountain climber to clamber up and down the cliffs to guide the balloon-mounted camera to the right spot.[38] Wüst was convinced that there was no time to lose. "The inscription itself," he noted in one letter, "is situated on a steep cliff wall, and with each passing year it is more and more in danger of being damaged or even destroyed in the most important parts by a strong torrent."[39]

Wüst had already approached the German Research Foundation for financing.[40] But it was Sievers's job to take care of other logistical matters—from locating a suitable photographer for the team to persuading officials at the Reich Aviation Ministry to provide the Ahnenerbe with the necessary balloons.

AS SIEVERS CONTEMPLATED solutions to these problems, he also labored on the arrangements for two other expeditions. The first was a relatively small affair—a field trip to the Canary Islands led by Dr. Otto Huth. Huth worked in the Ahnenerbe offices in Dahlem as an expert in "religious science," specializing in ancient Aryan spiritual beliefs.[41] He had read nearly everything ever written on the aboriginal inhabitants of the Canary Islands—a small archipelago located off the coast of northwest Africa—and had arrived at a novel conclusion. Huth believed that the original Canarians were in fact members of a pure, undiluted line of the Nordic race, who had preserved ancient Aryan religious practices well into the fifteenth century.

Huth was a former protégé of Herman Wirth. A fine-boned man, with a long face, sharp nose, heavy glasses, and a thin cupid's-bow mouth frequently drawn into a self-satisfied smile, he spoke seven languages—including Hebrew—and was a fervent Nazi. He had become politically active as a teenager, joining an ultranationalist group in the Rhineland that was later absorbed into the Nazi party. At twenty-two, Huth took out membership in the SA, the party's political combat troops, and for a time he worked as a reader for the Official Party Department to Protect Nazi Writing, the agency that dispatched the Gestapo to seize books failing to conform to Nazi doctrine.[42]

Huth seems to have acquired his passionate interest in the Canary Islands from Wirth. The older scholar firmly believed that the islands once formed the southern edge of a vast primeval Aryan homeland, Atlantis, and had somehow escaped devastation.[43] Intrigued by this notion, Huth had sopped up historical accounts of the ancient Canarians, a tribal society of prosperous farmers and herders. The Canarians, he learned, had shaved with stone knives; painted their bodies green, yellow, and red; dressed in dyed goatskins; and mummified the bodies of their leaders. On islands scattered some sixty miles off the coast of northwestern Africa, they had long lived in relative isolation.

During the thirteenth century, however, European navigators began to frequent their ports, carrying news of the islanders back to Europe. Eventually Spanish ships arrived to begin baptizing the Canarians by the sword. The inhabitants fought tooth and nail, but they could not long resist European muskets and European germs.[44] By the early sixteenth century, only a few of the aboriginal Canarians remained. They had little choice but to marry into the families of European settlers. It was a story that appalled Huth, who was no lover of Christianity. "This conquering of the Canary Islands by the Christian Spaniards," he observed in one article, "is a shocking tragedy and one of the most appalling examples of the poisonous effects of Jewish-Christianity on the soul of European people."[45]

Huth's sense of this tragedy was considerably compounded by his particular concept of the Canarians. He noted with delight that some early European chroniclers had observed islanders with golden locks, rosy cheeks, and white skin. He was also fascinated by the accounts of later travelers who found Canarian mummies with blond tresses. But Huth was a very selective reader. He deliberately ignored the warning of a contemporary, the prominent American anthropologist Earnest Hooton, who had written a major book on the ancient Canarians. As Hooton pointed out, the chemical nature of the preservatives and time itself often had "a bleaching effect" upon the mummies' hair.[46] Huth, however, saw only what he wanted to see. "Separated from the disturbances of European world history," he cooed later in print, "the ancient Nordic civilization blossomed undisturbed on the happy islands until it was destroyed."[47]

Huth was anxious to study the religious practices of the ancient Canarians, certain that they would shed valuable new light on the beliefs of the primordial Aryans. First, however, he had to clinch the racial origins of the Canarians.[48] For his expedition, he planned on taking a racial scientist to perform detailed measurements on both the living and the dead.[49] He also intended to take an archaeologist to sift through collections of Canarian pottery shards and stone tools in hopes of detecting similarities to those of ancient Nordic peoples. This, he firmly believed, would give "numerous results."[50]

Huth planned to depart for the Canary Islands in the fall of 1939, and was bubbling with enthusiasm for the venture. "We have a headstart in the source material," he wrote to Wüst, "and now we have to obtain a headstart in the fieldwork, thereby securing the best Canary research for the Ahnenerbe."[51]

IN ADDITION TO completing the arrangements for Huth's trip, Sievers also had to find a way of salvaging an important Ahnenerbe field trip to Iceland. The leader of the trip, Dr. Bruno Schweizer, was an old classmate of Himmler.[52] He was also one of the Ahnenerbe's most senior researchers. An expert on Germany's complex maze of dialects, Schweizer headed the Ahnenerbe's Research Center for Germanic Studies in Detmold. There he supervised a wide range of projects—deciphering rune stones, translating ancient Germanic documents, compiling old Germanic place-names, and offering public tours of Externsteine, a site that many Nazi scholars deemed to be a primeval Germanic shrine. Even so, Schweizer had still found time to plan a major Ahnenerbe expedition to Iceland.

Nazi scholars took a peculiar view of Iceland. Many saw its rocky lands as a kind of racial icebox, a place that preserved some of the purest strains of Nordic blood and the richest legacy of ancient Germanic tradition. In reality, however, Iceland's inhabitants traced their roots largely to Scandinavian ancestors. In A.D. 874, an adventuresome Norse chieftain, Ingólfur Arnason, had crammed his family and his retainers and thralls into large wooden ships and sailed westward to Iceland, which was inhabited at the time by a few Irish hermits. Arnason settled in what is now Reykjavík; the hermits promptly left.

Other Scandinavians soon followed, and along the coast they cut down forests, raised timber halls, slept in warm sod houses, and tended their livestock. Many followed the old pagan religion of Scandinavia, and the most stubborn of their descendants clung to these beliefs long after Iceland formally recognized Christianity in the year 1000. As a result, in the twelfth and thirteenth centuries the old bards of Iceland still sang and wrote of heathen gods, committing their beliefs to paper in the sagas.

Nazi scholars, however, stubbornly insisted on seeing the founders of Iceland as their German forefathers.[53] As one SS researcher noted in a letter to the German Research Foundation in 1935, "Nowhere can the primeval history of our people be recognized in a more thorough and true way than in Iceland, where it has been maintained free from foreign influences on race, customs and language—due to its historical development and geographical position. The new knowledge about old Germanic traits will not be collected from the sagas, because there are already good translations of them, but rather from the study of family records, state records, and traditional customs. . . . Iceland gives an untainted Germanic picture, free of Roman ideas, even in places where people have embraced baptism."[54]

Schweizer strongly subscribed to these beliefs. He had already journeyed three times to Iceland, and had brought back an Icelandic wife for himself—"of old farmer stock"—as well as a fine pair of Icelandic horses.[55] He installed the horses in the SS nature reserve near Externsteine, where guides showed them to visitors "as living continuations of the ancient Germanic horse race."[56] Schweizer believed that this hardy Icelandic breed could be of great value to future SS colonists. They "are raised half wild and are only brought into the stalls in the winter," he observed in a short article for the SS Kalender. "In three days, they can travel around 200 kilometers, only needing grass at the rest stops for their food."[57]

Schweizer was convinced that Iceland held many other treasures of value to the SS. In 1938, he proposed a major Ahnenerbe expedition to the island. He wanted to excavate an ancient farm and a heathen temple to learn more about ancient Germanic agricultural and spiritual practices.[58]

He was also keen on making a detailed inventory of the old assembly places, the *Things*; examining the architecture of ancient Icelandic sod houses; photographing artifacts in Iceland's national museum in Reykjavík; and gathering soil samples for pollen analysis. The latter would supply data on Iceland's paleoclimate. Last but not least, he hoped to record old Icelandic songs in the countryside and in Reykjavík, and film the renowned ballad dances of the Faroe Islands, en route to Iceland.

Himmler lent his full support to the project.[59] Like Schweizer, he regarded Iceland as an invaluable archive of ancient Germanic lore, and he relished the data that Schweizer's research would bring.[60] New details on the *Things*, the architecture of medieval sod houses, old agricultural tools and customs, as well as traditional Icelandic songs and dances would all greatly assist SS planners in arranging and regulating practices in future SS colonies in the East.[61]

Schweizer initially proposed departing around the summer solstice, and picked a team of seven young scholars—from an expert on ancient German buildings to the prominent Ahnenerbe archaeologist Herbert Jankuhn, whose research on bog bodies seems to have inspired Himmler's justification for the arrest, imprisonment, and brutal abuse of gay men. But the entire expedition fell seriously off the rails in late February 1939, when the German embassy in Copenhagen forwarded to Berlin a series of scathing press reports. Scandinavian reporters had learned, much to their amusement, that a German expedition was heading to Iceland.[62] They regarded the entire project as ludicrous, and did not hesitate to point out its faulty logic to their readers:

> Today a private telegram from Berlin arrived at [the newspaper] *Politiken*, announcing that the head of [the secret state police], Heinrich Himmler, wants to outfit a comprehensive 'ancestry-research expedition' to Iceland in order to find his ancestors. A large number of genealogists have received the order to travel along in order to 'excavate the ancestors.' At the same time, they intend to attempt to establish the degree to which the Third Reich can be traced back to the Icelandic Vikings. We showed the telegram to the genealogist Director Th. Hauch-Fausböll, who had the following comment: 'I must assume that there is a misunderstanding because in my opinion this is pure nonsense. No genealogical connection can be made

between Germany and Iceland and the German genealogists who are to be sent will have a hard time with it. It is well known that there aren't any church records dating back to the Vikings, so I cannot understand how they will prove the suspected relationship. Everything that we know about the Vikings regarding families and tribes is taken directly from the Icelandic sagas. Herr Himmler doesn't really have to mount an expedition to get acquainted with this source, as it is readily available in every bookstore, presumably including in Berlin.'[63]

Himmler loathed being the object of ridicule. He was furious that news of the SS expedition had leaked out in such a "careless" way.[64] He forbade any further work on the trip and prohibited all direct contact between the Ahnenerbe and Iceland. SS investigators immediately set to work searching for the informant, but they never found the leak.[65] After Himmler's initial rage abated, he permitted planning for the expedition to proceed and Schweizer and Sievers quietly picked up where they had left off.[66] But a few months later, a second major problem surfaced. Himmler's personal staff was unable to lay hands upon sufficient Icelandic crowns to finance this trip.[67]

There was no immediate solution for it, so once again, Sievers rescheduled the team's departure—this time for the summer of 1940.

THE TOWERING STACKS of paper that crossed Sievers's desk daily would have overwhelmed most other officials, but Sievers sorted through them effortlessly, navigating the labyrinthine channels of the Nazi government with ease. Indeed, the young administrator even found time to oversee other key projects for the SS. At the beginning of 1939, for example, Himmler had instructed the Ahnenerbe to mount a major new research project on Jews.[68] Sievers was only too happy to lend a hand with the arrangements.

Rassenkunde specialists in the Reich had failed to come up with any quick, absolute way of physically identifying men and women of the "Jewish race." Most of these researchers believed Jews to be an elusive blend of many purported races—the Hither Asiatic and the Oriental, the Hamitic and the Inner Asiatic, the Negro and the Nordic—a blend that shifted and changed from group to group, country to country.[69] As a result, they found it nearly impossible to put their finger squarely on the essential physical trait—the biological bar code—that set Jewish men, women, and children infallibly apart from their neighbors.[70] There seemed to be no defining measurement—such as the cephalic index they used for the supposed Nordic race—to neatly separate Jews from others.

For a man such as Himmler, who planned in one way or another to dispose of all Jews, including the elusive *Mischlinge,* or individuals of "mixed Jewish blood," this was a serious problem.[71] He intended to eradicate every trace of Jewish "vermin" from the Reich, so that there would be no chance of introducing Jewish blood into the new SS colonies. So he ordered his own SS research organization, the Ahnenerbe, to look into the matter. Perhaps they could devise some new index of Jewishness.

Sievers and Wüst found a thirty-one-year-old SS researcher, Dr. Walter Greite, to take charge of the project. Greite was a biologist by training.[72] He had studied the pigmentation of bird feathers as a student at the University of Göttingen, but his attraction to Nazi politics greatly influenced the direction of his research. He began delving into *Rassenkunde,* eventually becoming a lecturer on racial matters for a teacher-training school in Frankfurt and a racial researcher for the Reich Health Office. Like many German *Rassenkunde* specialists (and unlike many of their superiors, including Himmler and Hitler himself), Greite looked like a walking billboard for the mythical Nordic race, with his golden hair, blue eyes, and long, narrow face.

Sievers arranged for Greite to conduct measurements on Jewish men, women, and children who flocked each day to the Reich Central Office for Jewish Emigration in Vienna. The office was the purview of another efficient SS official, Adolf Eichmann.[73] It was located in a newly "Aryanized" palace that had until recently belonged to the Baron Louis de Rothschild. Each morning, hundreds of desperate Jews lined up outside its black gates, quietly waiting to enter so that they might apply for the papers they needed to flee Nazi Austria. The SS guards saw to it that this was a terrifying experience, accompanied by much shouting, cursing, and brutality.[74] Only after the applicants had handed over their assets and life savings could they obtain the necessary papers. Greite and his research assistants added considerably to the humiliation, requiring applicants to submit to racial measurements—a cold, dehumanizing experience.

By the summer of 1939, Greite and his team had completed measurements of nearly two thousand Jewish men, women, and children, a sufficient number for the project, and they had begun analyzing the data, making use of photographs and film footage taken at the examinations.[75] Sievers—a fervent anti-Semite who had often heard his father-in-law, a physician, speak on the perils of racial mixing—awaited the results with interest.[76] He knew that Himmler was counting on something of value turning up. Privately, he found this new line of research on Jews full of possibilities. Already Germany's leading racial experts were beginning to court his favor at official receptions, hoping to ally themselves with an increasingly influential research organization.[77]

15. THIEVES

In the last week of October 1939, Himmler and his large entourage of senior SS and Gestapo officers rode in sleek comfort through the Polish countryside, with windows shut and shades drawn. As their private train clattered eastward—past shelled villages and bombed airfields, abandoned cars and houses pocked with bullet holes—the SS chief and his staff pored over a thick stack of dispatches and reports. Himmler had equipped *Sonderzug Heinrich* with everything a mobile SS and Gestapo headquarters needed: three cars fitted with antiaircraft guns, a baggage car, plush parlor, secretarial and office car, Mitropa diner, refrigerator car, six sleepers, and a radio car equipped with telegraph facilities.[1] The din from the office was nearly deafening as secretaries clattered at typewriters and tall, blond-haired men in uniforms dictated orders in loud, rough voices. Himmler, who was tireless when he relished his work, did not want to waste a minute in completing the business at hand—the destruction of Poland.

Hitler had launched his assault on Poland at 4:30 on the morning of September 1. Without any declaration of war, he had unleashed squadrons of lethal Stuka bombers upon the sleeping Polish Air Force and hurled five armies into the heart of Poland. The speed of the assault, the blitzkrieg, was terrifying, and on the morning of September 3, Britain declared war, followed later in the day by France. Hitler took the news poorly. He had gambled that Britain would merely stand by as he methodically dismembered Poland. The German military, in the opinion of many German experts, was simply not ready for a widespread European war.[2] But Hitler was not about to back down from World War Two.

In the weeks that followed, the Western Allies did little to save Poland, and Hitler, who had privately vowed to "annihilate the Polish people," showed no mercy on his new subjects.[3] He directed Himmler to crush any sign of resistance.[4] For this, Himmler dispatched six specially trained *Einsatzgruppen*, or roving killing units, to Poland. Each was some five thousand strong, and made up of men from a wide range of forces under Himmler's command—the SS, the Criminal Police, the regular German Order Police, and the Secret State Police or Gestapo. Following generally on the heels of the armies, the *Einsatzgruppen* searched for any sign of opposition.[5] Equipped with lists of potential enemies, they dragged priests, rabbis, landowners, peasants, doctors, teachers, and lawyers from their homes and executed them in public squares and streets. Often, they freely improvised on the terror. According to one British eyewitness in the small Polish town of Bydgoszcz:

> The first victims of the campaign were a number of Boy Scouts, from twelve to sixteen years of age, who were set up in the market-place against the wall and shot. No reason was given. A devoted priest who rushed to administer the Last Sacrament was shot too. He received five wounds. A Pole said afterwards that the sight of those children lying dead was the most piteous of the horrors he saw. That week the murders continued. Thirty-four of the leading tradespeople and merchants of the town were shot, and many other leading citizens. The square was surrounded by troops with machine guns.[6]

In this way, Himmler's *Einsatzgruppen* slaughtered an estimated sixty thousand Poles in the early weeks of the war.[7]

On September 27, Warsaw surrendered after fierce bombardment. Almost immediately, Hitler began carving Poland up into three separate entities. He brought the western flank, a region where one in every six inhabitants descended from German families, into the Reich.[8] The ethnic Germans would be permitted to stay; everyone else would eventually be deported eastward. Hitler transformed the central region, which included Warsaw, Kraków, and Lublin, as well as the poorest, rockiest, and least fertile lands in the country, into a kind of colony known as the General Government. It would become a vast no-man's-land for future slave laborers—Polish Christians, German Jews, Polish Jews, and Gypsies.[9] That left only Poland's eastern flank outside German control. The Soviet Union had invaded and occupied it in mid-September, but Hitler viewed this as a purely temporary arrangement until he could unleash the Wehrmacht on the Red Army.

No sooner had the first corpses been trucked off Polish streets than many senior Nazis began eyeing the possibilities of plunder. For centuries Polish princes and merchant families had collected fine art, rare books and coins, and ancient manuscripts. The country's cathedrals harbored hundreds of artistic masterpieces, and its numerous museums exhibited important archaeological and historical treasures. All these prizes were now up for grabs among senior Nazis.

JUST FOUR DAYS after the armored columns of the Wehrmacht pounded across the Polish frontier, Sievers wrote to Himmler with an important suggestion. The outbreak of World War Two had brought all expedition planning to an abrupt halt and the offices of the Ahnenerbe had grown strangely quiet. Kiss's expedition to South America, Wüst's journey to Iran, Schweizer's field trip to Iceland, and Huth's trip to the Canary Islands—all had been postponed indefinitely, and it was clear to Sievers that many of the Ahnenerbe scholars would need new projects to occupy their time and justify their salaries. Sievers had come up with a plan. "In the formerly German part of Poland," he noted in a letter September 4, 1939, "there are numerous museums which have irreplaceable finds, documents, and monuments for the study and proof of prehistoric and historic German culture in the eastern area."[10]

Sievers proposed sending an Ahnenerbe scholar to Poland to seize all potentially useful materials—"catalogues, reports of grave excavations, drawings and photographs"—and ship them back to Germany.[11] Such records would greatly assist scholars in fabricating evidence for claims that Germany was merely righting an ancient wrong and seizing land that legitimately belonged to it. While it was certainly true that Poles and Germans had fought over their borderlands for hundreds of years, the Reich now wanted all of Poland, and it intended to present its criminal acts of mass murder and deportation as legitimate policies.

Sievers had already discussed the idea of seizing Polish materials with a young Ahnenerbe scholar and SS-Untersturmführer, Dr. Peter Paulsen. Paulsen was a professor of archaeology at the University of Berlin.[12] He was a Viking expert with an international reputation and a dedicated Nazi who had worked as an archaeologist for RuSHA. At thirty-six, he had published half a dozen monographs, on subjects ranging from medieval gold treasures to the symbolic meaning of weapons, and had taken part in excavations in Poland, Hungary, and the Middle East.[13] Senior German archaeologists praised his abilities highly. According to one prominent scholar, Paulsen was "the best expert on the Vikings among the young German prehistorians. Also abroad he has a good reputation among Swedish scholars."[14]

Paulsen, who had grown up in a little town just south of the Danish border, had worked hard to get where he was. His salesman father had died in a train accident in California when Paulsen was just three years old.[15] After that, the family was forced to scramble to make ends meet. Paulsen was a good student, and he managed to put himself through university thanks to a string of part-time jobs. His professors encouraged him in the study of prehistory and art history. In 1927, at the age of twenty-five, he took out a Nazi party membership.[16] Like many other Nazis of the day, he was drawn to the mystical view of archaeology and was exceedingly fond of the old Icelandic sagas. He named his children Sigurd, Hetha, and Astrid, after Norse heroes and heroines.

In 1938, he and his young family moved to Berlin, where he had landed work at the university and at the Ahnenerbe. His ailing wife suffered from a serious thyroid condition that required surgery, forcing him to borrow money from the SS, and he seemed notably lacking in the kind of high-octane confidence that characterized many of the young scholars there.[17] Indeed his official portraits showed a rather harried-looking man, with deep-set, wary eyes, a thick thatch of dark curly hair, and an air of being rather uncomfortable in his skin.[18]

Even so, Paulsen was eager to begin the plunder of Polish museums, and Sievers was convinced that time was of the essence. As he noted in his letter to Himmler, they would need to move swiftly once the hostilities were over, particularly if they hoped to confiscate not only important records and documents but also Poland's most important prehistoric treasures. Polish authorities would be expecting looters, and "as all wars have shown up to now, the enemy side will make efforts to hide or evacuate the most valuable finds."[19]

While Sievers waited impatiently for Himmler's reply, he asked Paulsen and a colleague, Dr. Ernst Petersen, to draw up lists of the most important Polish museums. Paulsen quickly obliged. In his office at the University of Berlin, he jotted down the names of more than a dozen Polish institutions worthy of looting—from the famous Wavel Castle in Kraków to the Museum of Archaeology in Warsaw.[20] Moreover, since he took a personal interest in the fine arts, he also included a museum of art in Lemberg, as well as the bronze doors of the Gnesen Cathedral and a major sculpture by Veit Stoss in Kraków. Petersen, who was one of Germany's leading experts on the prehistory of Eastern Europe, submitted a second list three days later. This itemized thirty-six major Polish museums and academic institutes and included the names of their directors and curatorial staffs.[21] In addition, it tabulated dozens of smaller archaeological collections in schools and local museums, as well as the names of Polish scientific societies and the locations of the most important archaeological excavations in Poland.

But it was Paulsen's brief mention of German artworks that seems to have galvanized Himmler. On September 21, Himmler approved a plan to send a detachment of scholars to Poland to secure both art and archaeological treasures.[22] He put Paulsen in charge and placed him under the command of the Reich Main Security Administration, a newly created SS umbrella organization that directed all police and security affairs in the Reich, including the operation of the concentration camps. Paulsen reported to Dr. Franz Six, a thirty-year-old SS-Standartenführer with a doctoral degree in political science and a long history as a stormtroop leader.[23] Six outlined the mission to Paulsen. The archaeology professor was to travel to the old royal capital of Kraków with a small team and three furniture trucks to locate, seize, and transport back to Berlin one of Poland's greatest and most beloved art treasures—the Veit Stoss altar.[24]

THE ALTAR OF St. Mary's Church in Kraków is one of the masterpieces of fifteenth-century Gothic art. Its creator, Veit Stoss, or as he is known to Poles, Wit Stwosz, was a man of immense energy and great misfortune. Stoss was born in either 1447 or 1448 in the little town of Horb am Neckar, southwest of Stuttgart.[25] He gravitated as a young man to Nuremberg, where he developed a career as a master carver. In 1477 he accepted a commission in Kraków to carve a massive altar for the Church of St. Mary. He spent the next seventeen years in Poland, laboring over a variety of commissions in stone as well as wood, always breathing life into his carefully observed human figures. When at last Stoss returned to Nuremberg, tragedy awaited. He lost money in a confidence scheme, and to compensate for his losses, he forged a promissory note with the name of the man who'd introduced him to the huckster. The local magistrates condemned him to public branding on both cheeks, turning the carver into a marked man. Emperor Maximilian eventually pardoned Stoss, but his life was ruined and he never recovered from the humiliation.

The altar in Kraków, however, is from the most splendid period of Stoss's career. The artist spent twelve years toiling on it, carving two hundred figures—saints, apostles, magi, and angels—from a single piece of limewood, then delicately painting and gilding each one. The finished altar spans nearly thirty-six feet in width. Its central panel portrays the death of the Virgin Mary and her glorious ascent to heaven; the two adjacent panels depict more than a dozen scenes from the life of the Holy Family. Stoss modeled many of the figures upon his neighbors in Kraków, and it is clear that he had observed their foibles with a loving eye. He depicted one of the three magi, for example, as an exuberant young nobleman, who, as one art critic noted, "strides briskly forward, his drapery swirling because of the swiftness of his approach, exposing to view his trim cuirass of burnished gold, as he ostentatiously doffs his hat, much to the consternation of the old attendant behind."[26]

After his death, Stoss fell into obscurity in Germany, largely because his finest works lay in Poland. In 1933, however, the Germanic Museum in Nuremberg hosted a major Stoss exhibition to commemorate the four hundredth anniversary of the sculptor's death. The display of his exquisitely lifelike figures in wood and stone stirred a newfound pride in his artistry and a deep-seated envy in the breasts of some art collectors, particularly those prominent in the Nazi party. It irked the ultranationalists terribly to think that Stoss's finest work adorned a Polish church, instead of the walls of a German art gallery.

Paulsen departed from Berlin on October 1, 1939, with three furniture wagons and attached containers. Polish officials in Kraków had already taken measures to hide the altar, suspecting that someone like Paulsen would soon be turning up on their doorsteps. They had spent lavish amounts of money to restore the altar six years earlier for their own commemoration of Stoss's death and were now desperate to protect it. As an aid to concealment, they cut up the altar into thirty-two massive pieces, packing them carefully in crates and sending them into hiding places in the Polish countryside.

Paulsen, however, seems to have known exactly where to look. Almost certainly, he had received intelligence from the Reich Main Security Administration, for within days of his arrival, the archaeologist had sniffed out the whereabouts of the boxes. He had located part of the altar, for example, in the fourteenth-century cathedral of Sandomierz, a small town some one hundred fifty miles northeast of Kraków. The four boxes he discovered there weighed seventeen hundred pounds each, and Paulsen and his assistants had to ferry them back to Kraków, across a remote hilly countryside that had yet to be fully subdued by the German army. "Transportation of the Veit Stoss figures," Paulsen complained in a private letter dated October 5, "turns out to be rather difficult. Military movements are a serious hindrance to the ride. On the way from Sandiomierz to Kielce, a car—fortunately without a load—broke down. . . . On account of the bad road conditions, we had to drive without a trailer, and for reasons of security, the drive could only be made during daytime. Today I finally arrived at Krakow with the first load. And tomorrow I am going to pick up the second and last load at Sandomierz."[27]

While in Sandomierz, Paulsen searched for other objects worthy of plunder. He stopped in at the "nice district museum" and ascertained that the building and all of its contents were secure and under the control of German police forces.[28] Then, according to a later report, "I took the valuable card index [the central record system that listed all the artifacts] from the museum in Sandomierz, which I found being hidden by a Jew."[29] Such indexes recorded all pertinent details about the artifacts and would allow the Ahnenerbe staff to pick out at their leisure objects worthy of looting. Paulsen made no further mention of the fate of the brave Jewish curator who attempted to hide the index. With these records in hand, Paulsen departed for Kraków with the second shipment of the Stoss figurines.

In Poland's old royal capital, Paulsen proceeded to organize the transport of the various crates he had located. He realized to his consternation, however, that he would not be able to convey the massive shrine section of the altar back to Berlin without the help of technical workers, so he agreed to store it until they arrived. In his free time, he toured Kraków's museums, seizing their card indexes and registers.[30] When he was at last ready to escort the altar back to Berlin, the city's prince-bishop made a final passionate protest against the theft to Kraków's new German mayor, explaining the deep religious significance of the artwork.[31] "The Veit Stoss altar is as important to Kraków," noted Paulsen in his later report, "as the painting of the Black Madonna is to Czestochowa."[32] But Paulsen was unmoved by the pleas of the prelate. He set off with the figurines, arriving in Berlin on October 14, 1939. He personally delivered the boxes of the altar to a treasury in the new Reichsbank. Only the bank's director, he noted confidently, possessed the key.

News of Paulsen's success traveled swiftly through the Nazi grapevine. A month after his return, the lord mayor of Nuremberg, Willy Liebel, penned a letter urging Hitler to bestow the altar on his city, the place where Stoss had endured his greatest humiliation.[33] Goebbels, in turn, pressed Hitler to give him the artwork for a touring exhibition to mark its triumphant return to Germany, and Himmler seems to have had his own designs on the massive sculpture. Already, he had helped himself to one of its beautifully worked panels, Christian motifs and all.[34] All this politicking astonished Hitler. No one had bothered to seek his permission for the theft.[35] After mulling over the problem, however, Hitler agreed to send the altar—all of it—to Nuremberg, where it was stowed, secure from aerial attacks, in a huge underground vault beneath the city's medieval castle, until it could be safely displayed. It remained there until the summer of 1945.[36]

IMMEDIATELY AFTER DROPPING off the altar at the Reichsbank treasury in Berlin, Paulsen met with Reinhard Heydrich, the head of the Reich Main Security Administration.[37] Heydrich ordered Paulsen to return to Kraków, where he was to begin seizing important museum collections. Sievers saw this as a golden opportunity for the Ahnenerbe. "Paulsen," he wrote after learning the news, "has control over all of the material from various museums and collections in Kraków because he has secured the registers and catalogs. As it is to be assumed . . . that the cultural goods are to be transferred to Germany as completely as possible, it is necessary to view, seize, and transfer to Berlin those parts of the collections important to the work of the Ahnenerbe, especially those which focus on prehistory and early history, valuable collections of house markings, gables, and weapons studies, as well as scientific collections on nature and ethnology, etc."[38]

Sievers wanted the plundering to be carried out with scientific exactitude. This, he realized, would require a team of highly trained specialists. He began hurriedly assembling an SS scholarly command—Dr. Eduard Tratz, the head of one of Austria's most important natural history museums, the Haus der Natur in Salzburg; Dr. Ernst Petersen, the archaeologist who had compiled one of the lists of Polish institutions to be plundered; Dr. Theodor Deisel, an art historian; Dr. Paul Dittel, a historian and geographer who specialized in archival, library, and museum collections; Dr. Wilhelm Mai, a specialist in folktales and legends; and SS-Hauptscharführer Luismann, who would serve as the driver.[39]

Sievers also realized that Paulsen and his colleagues would require considerable administrative support. He sent a proposal to Himmler on October 16 suggesting that he be sent with Paulsen to Poland "to oversee and complete the assignment, which needs to be done as quickly as possible so that Paulsen can continue on from Kraków to Warsaw."[40] Himmler immediately recognized the wisdom of this. A day later, his personal administrative officer, Dr. Rudolf Brandt, informed Sievers that he was to proceed to Poland with Paulsen and his scholars.

If Himmler thought that his unit of experts could simply take what they wanted in Poland, however, he must have been extremely disappointed. In Kraków, Paulsen learned that Göring had dispatched a similar team of experts on an identical mission. Göring's avarice was legendary. The Reich minister of aviation owned eight sumptuous residences—castles, villas, hunting lodges, and mountain chalets—each lavishly decorated with German, Flemish, Dutch, and Italian Old Masters as well as fine Gobelin tapestries and costly Persian rugs. Göring never missed out on an opportunity to add to his collection. To lay hands on the finest Polish pieces, he had chosen SS-Sturmbannführer Kajetan Mühlmann, an art historian with a very shady reputation, to head up a team of nine experts.

Mühlmann was a domineering man with a short temper, a criminal record for petty offenses, and a talent for "ferreting out art treasures."[41]

From the beginning, he seems to have taken complete control of the situation in Poland. He swiftly forged an alliance with Hans Frank, the new leader of the General Government in occupied Poland, and proceeded to outmaneuver Paulsen at nearly every turn. By the end of October, Mühlmann had obtained precisely what Göring most wanted—first crack at all of Poland's rich art treasures. In exchange, Mühlmann agreed to give to Paulsen and his scholars carte blanche on Poland's archaeological, ethnological, and natural history collections.

Paulsen reluctantly accepted these terms, and he and Petersen set to work, assessing the collections of Kraków's scientific institutes and museums. But before they could begin transporting the valuables back to Germany, they received new orders from Franz Six at the Reich Main Security Administration. Six had just returned from a tour of Warsaw, which had suffered heavy shelling and bombing during the siege, and he ordered the unit to begin its pillaging there before others beat them to the punch. So Paulsen and his team swiftly moved their base of operation northward. To assist with the work, Sievers sent three other researchers—Dr. Heinrich Harmjanz, an expert on folklore and ethnology; Dr. Hans Schlief, an architect who had become a prominent classical archaeologist; and Dr. Günther Thaerigen, an archaeologist.[42]

To those familiar with the beautiful parks and gardens of Warsaw before the war, the Polish capital must have seemed a shocking sight. The Royal Castle was in smoldering ruins. The palace of the papal nuncio, the National Theater and Opera House, the City Hall, and most of the city's hospitals and principal railway stations were either destroyed or severely damaged. Entire residential streets in both the suburbs and old town had been flattened, and the sickly smell of decaying human flesh seeped from the ruins. Those who survived walked about in a state of shock. Warsaw's citizens had put up a strong, spirited defense, but the German army, with its superior armaments, had finally overrun their streets. In the days that followed, citizens were forced to turn in their radios to authorities and spend hours each day lining up in the bitter cold to get bread and other rationed food. As they stomped their feet to stay warm, new loudspeakers in the streets blared the latest Nazi propaganda.

If Paulsen took much notice of the misery, however, he made no mention of it, even in passing, in his letters. He had other, more pressing matters on his mind. To inspect the museums and carry off their treasures, the team needed transportation. Cars were scarce in the Polish capital in early November 1939; trucks were almost impossible to requisition. But Paulsen managed to patch together transport for his command, and the Ahnenerbe scholars buckled down to work with a cold orderliness, interpreting their orders liberally.

One of their first targets was the State Archaeological Museum in Lazienki Park, a former royal hunting preserve. The museum was the hub of archaeological investigation in Poland. Fourteen of Poland's leading archaeologists worked there, and it served as a central repository for their collections of artifacts—flint axes, bone points, amphorae, swords, sickles, halberds, scabbards, necklaces, fibulae, bronze collars, bronze cauldrons, face urns, figurines. The museum also housed extensive files and card indexes on all the country's archaeological sites, as well as an important twelve-thousand-volume archaeological library.[43] The Paulsen unit intended to seize all of the most important material and cart it back to Germany, where they would put it to use for the Nazi cause.

They also planned on stamping out, once and for all, the particular brand of research that the museum specialized in. The Polish archaeologists were a patriotic lot who had spent nearly two decades searching for the origins of the Slavic peoples. The quest had led them at times to parts of Europe that Germans had earmarked as their own. This infuriated Paulsen's men, who continually referred to the museum as a "poison kitchen."[44] So they decided to steal the entire research base of their Polish colleagues. As one of Paulsen's subordinates later noted, this would allow German scholars to comb through the Polish data, "to establish where the Poles: 1.) forged the results of discoveries, 2.) suppressed them if they appeared altogether too unfavorable against Poland, 3.) exaggerated Slavonic influences, or 4.) discontinued investigations at the very moment when they met with Germanic remains beneath the Slavonic ones."[45] The terrible irony in all of this seems never to have occurred to the German scholars.

Paulsen delegated Petersen, Schleif, and Thaerigen to take care of the museum, which was under German guard. The trio arrived on their first day in Warsaw and were none too pleased to see a young rising star in Polish archaeology, Dr. Konrad Jazdzewski, and two colleagues in the offices.[46] Jazdzewski was a thirty-year-old native of Upper Silesia, the much-contested borderland that curved along the southwestern edge of Poland.[47] He had studied in both Germany and Poland and spoke German well. Almost certainly, Jazdzewski recognized Petersen from scholarly conferences and meetings he had attended in Eastern Europe, and perhaps he felt a moment of relief seeing a fellow archaeologist turn up at the museum. If so, the sensation must have been fleeting. Petersen despised the young Polish researcher and made no effort to hide it.[48] He thought Jazdzewski belonged to "the worst anti-German agitators."[49]

Together with his SS companions, Petersen asked Jazdzewski to show them the museum collections. He spoke more like a conqueror than a colleague. Jazdzewski knew that the German archaeologists had arrived to case the collection. He gave them the required tour, but as soon as they left he and his colleagues went through the glass display cases, removing the most valuable pieces and hiding them as best they could in the storage area. The following day, Schleif and Thaerigen returned. They were enraged to discover the empty display cases and immediately searched the storage area, locating some of the missing items. Then they threw Jazdzewski and his colleagues out of the museum.[50]

With the Polish researchers gone, Schleif and Thaerigen began crating up the museum's extensive collection of artifacts, its official records and documents, and its library, for shipment back to the Reich. The packing must have taken days, for the museum had extensive holdings, including many delicate ceramics that needed careful wrapping. The German team were reluctant to leave anything significant behind; the material, as Paulsen later observed, would be used to "build the SS research."[51]

As Schleif and Thaerigen wrapped up the key holdings of the archaeological museum, Paulsen combed through Warsaw's other major institutions. He crated up prehistoric treasures from the National Museum and sifted through the display cases and storage areas of Warsaw's Military Museum, seizing the sword of Sandomir and several other splendid ceremonial weapons. At the Crazinski library, he was delighted to find two handsome Viking swords and two ceremonial battle-axes. These he also took and dutifully shipped off to Berlin.

Meanwhile another detachment member, Eduard Tratz, rifled through collections at the State Zoological Museum, examining its collection with a connoisseur's eye.[52] At fifty-one, Tratz was one of the most respected citizens of Salzburg. He had personally founded the city's natural history museum, the Haus der Natur, in 1924, and with assistance from Austria's new Nazi masters, he had rapidly expanded its facilities. Tratz believed museums played an essential role in society, as "the link between science and the people, between humans and nature."[53] To better communicate Nazi party doctrine to the public, he had recently added eight new departments to the Haus der Natur. These specialized in such subjects as racial development, racial hygiene and eugenics, and animal domestication and breeding.

Tratz spent two days at the State Zoological Museum in Warsaw, selecting specimens to send back to the Haus der Natur. He and a colleague chose 147 of the museum's most exotic bird specimens—from the resplendent quetzal of the Central American cloud forest to the crested serpent eagle of Japan—as well as three huge European bison, a massive Nile crocodile, and two wildcats.[54] Tratz also carefully sorted through the museum's collection of skulls and skeletons. The Haus der Natur was planning important new exhibition rooms on human heredity to popularize Nazi ideas of race and prehistory.[55] As part of this exhibit, entitled "The Ancestors," Tratz and his staff intended on displaying head casts of the Nordic and Jewish "races," as well as the remains of ancient humans, such as the Neandertal and the Cro-Magnon. So from the Warsaw collections Tratz selected a variety of human, chimpanzee, and gorilla skeletons; a plaster model of a Neandertal; and casts of the braincases of Pleistocene humans.[56] In addition he crated up a mammoth jaw, the skull of an Ice Age rhinoceros, and dozens of expensive reference books on butterflies, mollusks, protozoa, snails, crabs, paleozoology, bird migrations, the history of philosophy, and anatomy.[57] All this he dispatched to the Haus der Natur.

Paulsen delegated other scholars to tackle Warsaw's libraries. The Reich Main Security Administration, which directed all mass murder in the Third Reich and in the newly annexed territories, was assembling a library to educate its staff on Jews and other enemy groups. Paulsen's commanding officer, Franz Six, believed that it was "necessary for research purposes to carefully study the written works produced by the enemy in order to understand the mental weapons of ideological enemies."[58] Indeed, officers in the Reich Main Security Administration would later use reference volumes on the Jewish diaspora to help trace the origins of ethnically mixed communities in the Soviet Union. Those communities identified as Jewish were then slated for liquidation.[59]

To line the shelves of this new SS library, Paulsen and his colleagues carted off the Sejm Library in Warsaw and approximately forty thousand books from the Judaic library in the Great Synagogue on Tlomacka

Street.[60] In addition, Paulsen crated up the library in the Ukrainian Science Institute and packed away some fifteen hundred books from what was likely the Seminar for Indo-European Linguistics at the University of Warsaw. The latter volumes were almost certainly intended for Walter Wüst, the superindendent of the Ahnenerbe.

Göring's experts had given Paulsen a free rein in all these areas, but the two groups of scholars fought like vultures over the large private libraries of the Polish nobility. These, after all, contained many rare works of art. After much wrangling, for example, Paulsen and Sievers succeeded in laying hands on one of the most important treasures from the Zamoyski library—the Suprasl Codex, an eleventh-century manuscript containing the oldest-known written example of the proto-Slavic language. The two SS officers wrapped it up carefully and sent it to a Reich Main Security Administration storage facility in Berlin. It was a very valuable document. Indeed, Sievers later gleefully estimated its value at between 4 and 5 million reichsmarks, the equivalent of some $20 to $26 million today.[61]

Paulsen also persuaded Mühlmann to release several important Jewish and Freemasonry artifacts from Poland's National Museum. The archaeologist was under orders to send these goods to Wewelsburg, the German castle that Himmler was refurbishing as a senior SS academy.[62] Almost certainly the items were intended for a private exhibition at Wewelsburg, one resembling a Freemasonry "museum" once installed in the SS Security Service headquarters in Berlin. Before the war, SS officers had led groups of SS men and Hitler Youth clubs through the display, which warned in the most dire and lurid terms of the perils of Freemasonry. As one visitor later recalled, "I was shown papers illustrating the work and methods of the Masons, seeking to prove that they used poison to remove the traitors from their own ranks. There were skulls all over the place, a coffin marked with Masonic signs, aprons and insignia—really quite a gruesome display."[63]

As Paulsen and his team of experts stripped Warsaw's museums and libraries bare, Hans Frank, the new leader of the occupied Polish colony known as the General Government, could not shake the feeling that the specialists were robbing him blind. Frank intended to live like a king in Poland, and to do this he needed to put an end to the thievery.[64] He issued a decree prohibiting any further shipments of property to Germany without his government's express approval or without payment from Berlin, a regulation that was to take effect on November 22, 1939. Paulsen, who had yet to plunder all the museums on the Ahnenerbe's list, deeply resented the interference. But he felt powerless against Frank—an old friend of Hitler—and saw little alternative but to bow to his orders. Paulsen's colleagues were infuriated by this timidity, but none more so than Hans Schleif, who had spent days packing up the voluminous collections at the State Archaeological Museum in Lazienki Park.

Schleif was a loose cannon in Ahnenerbe circles. He was arrogant and brutally direct, and he considered many of his co-workers fools.[65] He saw little chance at all of shipping the State Archaeological Museum collection to the Reich before the November 22 deadline that Frank had set. Nonetheless, Schleif traveled twice to Poznan in hopes of wrangling some kind of rail transport for the collection. It was a huge shipment—five freight cars' worth of plunder—and no one was able, or perhaps willing, to help him.

The November 22 deadline came and went, but Schleif refused to give up. After days of haranguing and storming and browbeating others, he and a colleague finally managed to finagle transport of the collection to Poznan on November 30. It meant disobeying an explicit order from Hans Frank, but Schleif was beyond caring. He desperately wanted to strip the archaeological "poison kitchen" bare and cart its collection back to the Reich. He waited for the crates to be loaded on the freight cars, then hurried back to the Reich, where he wrote a letter to Sievers explaining his own criminal actions and complaining about Paulsen's ineptitude. A few weeks later, after opening the crates in Poznan, he gloated over his success. "The Warsaw material is now entirely unpacked and registered. Now for the first time, it is possible to obtain a survey of the truly excellent stock."[66]

Back in Berlin, Paulsen penned a final report to the Reich Main Security Administration, listing his detachment's achievements and taking credit for the transport of the State Archaeological Museum collection to the Reich. He was proud of the successes—so much plunder, it seemed, in so short a time. But he deeply regretted leaving so many valuables behind. Many fine collections still lay untouched in Warsaw, "and in Kraków, everything still needs to be done."[67] But by then it was clear to the senior SS staff that Paulsen lacked the brazen arrogance needed to be a thief among thieves.[68] So Sievers quietly arranged for a reassignment, finding him a teaching job at an SS officer-training school far away from the front.[69]

Working quietly in the background, Himmler searched for some legal way of plundering Poland's riches. Through clever political maneuvering, he took control of a public corporation that Göring founded to confiscate the assets of Jewish and Polish citizens. The corporation had a very forgettable name, the Haupttreuhandstelle Ost, or Main Trust Center East, and served largely as a cover for further piracy in Poland. Most of the profits went straight to Göring, but Himmler arranged to siphon off part of the proceeds for his own SS projects. He placed the Ahnenerbe in charge, named Sievers as the corporation's managing representative, and rewarded Schleif for his earlier audacity with the "poison kitchen" by appointing him trustee for Wartheland, one of the Polish regions incorporated into the Reich.

The new Nazi regime in Wartheland had already begun expelling Jews, and officials there intended to treat Polish Christians much the same. "Everything that is Polish is going to be cleared out of this region," boasted the new Nazi *Gauleiter*.[70] Each deportee would be allowed to take only a small valise, with room enough for a change of shirt perhaps and some underwear and socks; everything else would have to be abandoned. It was a prime opportunity for looting, and the Ahnenerbe staff was delighted at first by the possibilities. "Quite a few works of art and libraries have lost their owner," Schleif pointed out smugly in a letter to Sievers.[71]

Under Sievers's orders, Schleif and the other Ahnenerbe scholars fanned out into the Polish countryside. They inventoried archives, museums, public collections, castles, manors, and other wealthy Polish and Jewish homes, then registered and seized all portable valuables—historic and prehistoric artifacts, old property deeds, books, documents, paintings, sculptures, wood carvings, furniture, silverware, fine carpets, and expensive jewelry.[72] Schleif, a sarcastic and overbearing man, was not much of a team player, and the Ahnenerbe scholars soon wearied of him. They grew to hate the Polish countryside.[73] The local farmers were continually scattering horseshoe nails over the roads, puncturing the tires of their vehicles. The roads were poor and the scholars got stuck in the mud. If they were wearing civilian clothing, the farmers ignored their requests for help, and often when they arrived somewhere promising, they were too late. The Gestapo had already beaten them there, cleaning out all the best booty.[74]

When Schleif lost his enthusiasm for the work, another Ahnenerbe scholar, Ernst Petersen, replaced him. Petersen expanded the efforts. In fifteen months, the scholars of the Main Trust Center East ransacked 500 castles, estates, and private apartments; 102 libraries; 15 museums, 3 art galleries;

and 10 coin collections.[75] They plundered the silverware of Prince Radzi-will; the pearls and gold and silver jewelry of Karl Albrecht von Habsburg-Lothringen; the Dürer drawings at the Lemberg Museum; and important collections from the Museum of Ethnology and Natural Sciences at Plock. At Golochow Castle, they made off with priceless treasures—a rare collection of vases, the oldest of which dated back to the seventh century B.C.; an eleventh-century Italian fountain; a portrait of Copernicus; and dozens of costly paintings, including works by the modern French master Jean François Millet. By March 28, 1941, they had amassed a large storehouse of treasures—some 1,100 paintings, 500 pieces of furniture, 35 boxes of church treasures, and 25 sets of rare metal objects.[76]

The staff of the Main Trust Center East sold some of these valuables immediately to avid buyers. The profits went to Göring. But the trust center officials packed up most of the treasure in crates and sent them with an armed guard to a central collection point in Berlin. Sievers estimated that by the end of 1941 the Main Trust Center East had confiscated goods worth 3 million reichmarks, or some $15.6 million today—a figure that is likely far too low.[77] Göring received the lion's share of the proceeds, but the Ahnenerbe submitted a bill for its services, charging 10 percent of the total. Göring, however, seems never to have paid.

But the pillaging of Poland had given Sievers and many other Ahnenerbe scholars an appetite for piracy. They had looted entire museums and libraries in Poland without a qualm, stealing their greatest treasures. In the elegant Ahnenerbe villa in Berlin–Dahlem, staff members followed the latest reports from the front avidly and watched the advances of the Wehrmacht with new, avaricious eyes.

16. THE TREASURE OF KERCH

On the evening of July 27, 1941, Adolf Hitler lingered over dinner, his eyes glinting with pleasure as he mused upon the future for the benefit of a select audience in his stronghold in the East Prussian forest. Wolfschanze was a dark, dismal, depressing place, more suited to an army of troglodytes or trolls than the triumphant new warlord of Europe. But Hitler liked its Spartan simplicity. He felt invincible there, surrounded as he was by nearly two thousand military personnel, thick windowless concrete walls, barbed-wire fences, sentry posts, and mile upon mile of northern wilderness. His existence at Wolfschanze bore no resemblance to ordinary life, a fact not lost on most of his subordinates. Indeed, one member of the German high command, Alfred Jodl, later described it as "a cross between a cloister and a concentration camp."[1]

Hitler, however, had no desire to leave Wolfschanze, no inclination really to step foot into the bloody apocalypse he and his armies had unleashed upon Eastern Europe. Just five weeks earlier, on June 22, 1941, he had launched a mammoth surprise attack on the Soviet Union, the bulwark of what many committed Nazis called "Jewish-Bolshevism."[2] Emboldened by his earlier military successes—the invasions of Norway, Denmark, the Netherlands, Belgium, northern France, Yugoslavia, and Greece—he had hurled nearly three million German troops against the Soviet military in Operation Barbarossa, opening a two-thousand-kilometer-long front that stretched from the Baltic Sea in the north to the Black Sea in the south.[3] In just two and a half weeks, the Wehrmacht had seized Lithuania, Latvia, and parts of Estonia, Belorussia, and Ukraine, capturing hundreds of thousands of Russian soldiers. German troops were now marching north to Leningrad, a city that Hitler admired, and south to the Crimea, the peninsula that jutted from the north shore of the Black Sea.

Hitler was elated and energized, convinced that his troops would take Moscow in a matter of weeks.[4] He now ruled over more of Europe than any man since Napoleon, and often in the evenings, when the spirit took him, he indulged for hours after dinner in meandering monologues that blended dreamy fantasy with monstrous cruelty. The servants would bring the guests tea and cake—chocolate cake being Hitler's favorite—and as the German leader sipped and nibbled and stared at a large map of the Soviet Union on the far wall, he would begin to talk to the assembled guests: military men, visiting Reichkommissars or Reichministers, members of his inner circle, young secretaries in pretty dresses. Always it was the same. The room would go silent, and a little man sitting discreetly off to the side would bend down over a sheaf of paper and begin taking shorthand notes.[5]

On these occasions, Hitler would casually hold forth in a lengthy stream of consciousness on whatever was uppermost in his mind at the moment—the perfection of the German army, the inferior nature of English music and theater, the natural aptitude of the Swiss for hotel-keeping, the necessity of eradicating insects and dirt in Vienna.[6] But on the evening of July 27, Hitler chose to talk about his future empire in the East. He particularly relished the thought of a new German colony he planned to create in the Crimea. The southern Russia peninsula was blessed with a pleasant Mediterranean air. It possessed forested mountains and splendid seacoasts where dolphins frolicked. It boasted vineyards that produced fine sherry and muscatel, and orchards that yielded apricots and peaches, and it was endowed with the kind of singular beauty that attracted important visitors. The Russian imperial family had built a fine summer palace in the Crimea, and the grand dukes and duchesses of St. Petersburg had followed, putting up lavish dachas. Famous writers and artists took up residence there. In 1903, Anton Chekhov penned one of his most famous plays, *The Cherry Orchard*, while sitting in his Crimean country home.

Hitler, however, envisioned a very different future for the Crimea. He believed that the sunny region possessed special properties of great importance to the Aryan race. "There are few places on Earth," he later observed, "in which a race can better succeed in maintaining its integrity for centuries on end than the Crimea."[7] To illustrate this contention, he pointed to the history of the Goths, wandering herdsmen from northern Europe who settled in the Crimea in the third century A.D. and whose language could still be heard on the peninsula some thirteen hundred years later. Hitler, like many other German ultranationalists, regarded the Goths as ancestral Germans and the Crimea as a kind of southern German homeland.[8] On the strength of this meager claim, he had decided to transform all of the peninsula into an "exclusively German colony."[9] He planned to rid the Crimea of all those he deemed undesirable—Jews, Tatars, Gypsies, Russians, Armenians, Georgians, Ukrainians—and replace them with racially sound German colonists. "My demands are not exorbitant," he explained smugly one night. "I'm only interested, when all is said, in territories where Germans have lived before."[10]

When news of these plans reached Himmler, he was electrified. The Crimea would be a perfect place for dozens of the feudal SS settlements he had been dreaming about for nearly a decade.

ON SEPTEMBER 26, 1941, the German 11th Army, under the command of Lieutenant General Erich von Manstein, began bearing down on the Crimea.[11] Manstein, a West Prussian aristocrat with nearly thirty-five years of experience in the German army, had an excellent reputation as a fighting man, but the Soviet forces put up unexpectedly heavy resistance. All along the eastern front, it was much the same story. Instead of folding and crumbling and buckling under the crushing force of the Wehrmacht, surviving Soviet officers from one battle simply re-formed their units, dragooning reservists and bystanders and arming whoever else they might find quietly tending their crops or walking the streets. Then they threw these makeshift troops back again against Panzer divisions and artillery units, sacrificing ten Russian lives in order to kill one German soldier.[12]

Over the next seven weeks, Manstein's troops bludgeoned their way across the Crimea, but they were unable to take the heavily fortified port of Sevastopol, once a thriving link in the region's grain exports. So on December 17—ten days after the Japanese air force bombed Pearl Harbor, prompting the United States to declare war first against Japan and then enter the war against Germany and Italy—Manstein launched a major attack on the port and its thirty-two thousand Soviet soldiers. The German objective was to cut through three major defensive rings that Soviet troops had constructed around the city.

As Hitler and the German high command waited impatiently for Sevastopol's fall, Himmler directed his forces to begin ethnic-cleansing operations in the Crimea. He ordered Einsatzgruppe D, one of four large roving killing detachments in the Soviet Union, to liquidate all the Jews living in German-occupied Crimea. The first target there was Simferopol. Some twenty thousand Jews had lived in the city before the war, giving it the largest Jewish population in the Crimea. Since then, many Jews had fled for safer quarters, but an estimated eleven thousand remained. So in mid-December, members of three forces—Einsatzgruppe D and the Wehrmacht's Field Police and Secret Field Police—set about methodically massacring the Simferopol Jews.[13] The plan was to liquidate the entire community before Christmas.

Officials informed local Jews that they were to be resettled, and instructed them to gather in a public meeting place. Drivers then conveyed the families to a prearranged kill site some fifteen kilometers outside of Simferopol.[14] There, by the side of the road, officers instructed the frightened families to climb down from the truck, take off their jackets and shoes, and leave behind their suitcases. Armed guards then led the barefooted victims through the snow to an excavated grave, some three hundred meters from the road. Most of the Jewish captives could see at once the fate that lay in store for them. "There were disturbing scenes," recalled one of the German executioners later. "The Jews cried because they were aware of what was happening."[15]

When it came their turn, the victims were lined up opposite their killers, each of whom was armed with a machine pistol, a type of submachine gun. An SS officer gave the order to fire. "Some of the victims immediately fell into the grave," observed one member of the firing squad after the war, "others fell on the edge. These fallen Jews were then thrown into the grave by the waiting Jews."[16] After several rounds of this, the executioners needed no order to fire: they did so automatically. There was no possible escape for the victims—no opportunity to run or flee. The SS Security Service had cordoned off the area and set up guards around the perimeter. "Some Jews who tried to flee," noted one of the executioners later, "were shot down by the unit who ensured that the area was closed off."[17]

The killing squads were very efficient. On December 15, Einsatzgruppe D reported to the SS command that Simferopol was *judenfrei*, "free of Jews."[18] And it immediately set about orchestrating similar massacres in other Crimean cities—Feodosia, Yevpatoria, Kerch, Yalta, and Bakhchysaray.[19] By then, however, some of the squad members had begun to complain about the psychological stress of shooting such large numbers of women, children, and babies in cold blood. Rather than putting an end to the terrible bloodshed, however, Himmler and the SS leadership suggested a more impersonal method of slaughter—mobile gas wagons. As the massacres continued, Einsatzgruppe D obtained three gas wagons—two large ones capable of killing eighty people at a time, and a smaller one that could execute fifty people at once. Squad members used these to kill women and children.[20]

In all, German forces were to shoot and gas to death nearly forty thousand Crimean Jews during their occupation.[21]

THE ASSAULT ON Sevastopol in late December 1941 failed dismally, despite Manstein's brilliance as a tactician. The Soviet navy succeeded in dispatching reinforcements, ammunition, and food to the city on December 20, bolstering the spirits of the defenders considerably; six days later it pulled off a daring amphibious landing of twenty thousand troops on the eastern tip of the Crimea. With this assistance from the sea, Soviet troops fended off the assault on Sevastopol and they retook the cities of Feodosia and Kerch. News of this disaster so infuriated Hitler that he sentenced to death the German officer who ordered German troops to withdraw from the Kerch region, Count Hans von Sponeck.[22]

These and other serious setbacks along the eastern front threatened to undermine the morale of German troops. Himmler, the schoolmaster's son, believed that further political indoctrination was the surest way of rallying the Waffen-SS, the military arm of the SS. "The longer the war draws out," he stated in a later order, "the more we have to educate and convince our officers, junior officers and men about the [Nazi] worldview."[23] In Berlin, SS writers dutifully churned out a flurry of articles casting the invasion of southern Russia as a kind of homecoming, where German forces might once again reclaim their ancient territories in the East. At the center of these fables were the ancient Goths, a favorite propaganda tool of SS writers.

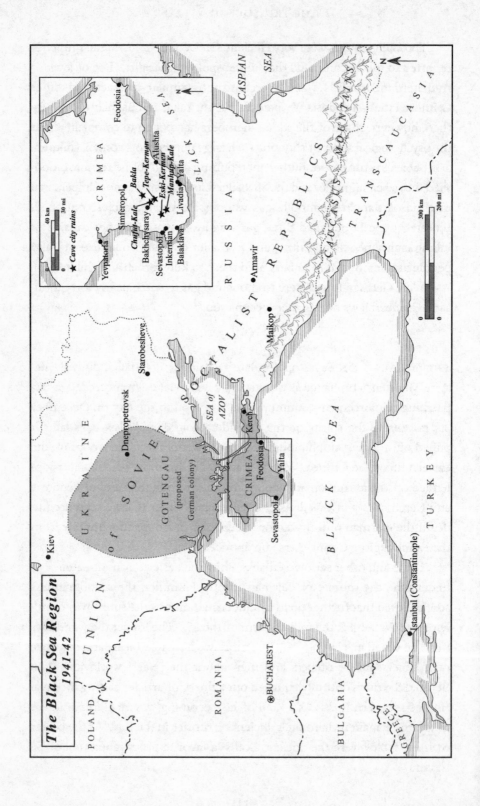

The Black Sea Region
1941-42

POLAND

U N I O N   o f   S O V I E T

UKRAINE

•Kiev

•Dnepropetrovsk

S O C I A L I S T

•Starobesheve

SEA of
AZOV

GOTENGAU
(proposed
German colony)

R E P U B L I C S

•Armavir

•Maikop

R U S S I A

Kerch

CRIMEA

Feodosia

Sevastopol•

•Yalta

B L A C K   S E A

ROMANIA

⊕BUCHAREST

BULGARIA

GREECE

•Istanbul (Constantinople)

T U R K E Y

CAUCASUS MOUNTAINS

TRANSCAUCASIA

CASPIAN
SEA

300 km

200 mi

0

0

N

Inset map (Crimea)

CRIMEA

•Feodosia

Simferopol•

Yevpatoria•

Chufut-Kale
Bakla
Tepe-Kermen
Aliushta
Bakhchysaray
Eski-Kermen
Mangup-Kale
Livadia Yalta
Sevastopol•
Inkerman
Balaklava

B L A C K   S E A

40 km
30 mi
0
★ *Cave city ruins*

N

According to history, the Goths were a tribe of wanderers who originated in a place called Scandza—quite likely in Scandinavia or possibly northern Poland.[24] They spoke one of the Germanic languages, as did just about all the inhabitants of Sweden, Norway, and Denmark.[25] The Goths, however, did not remain in Scandza. Being of a particularly footloose disposition, they wandered south looking for greener pastures. They eventually ended up along the Sea of Azov, a gulflike body of water that adjoins the Black Sea. And in A.D. 238 they made a rather dramatic entrance into the histories of the ancient world.[26] From settlements along the Black Sea coast, they began attacking and pillaging Roman cities to the west. They seized ships moored in Black Sea harbors and became full-fledged pirates, raiding even into the Aegean. They were formidable foes, and according to the old Roman histories, they built a city alternately called Doros or Dorys or even Doras, somewhere in the Crimea. They also converted to Christianity, and one of their bishops, Ulfilas, whose name means "little wolf," invented the Gothic alphabet so he could translate the Bible into his native tongue.

The fiery arrival of the Huns, nomadic horsemen from the Central Asian steppes, convinced many Gothic families that it was time to search for safer homes. They fled across the Danube in A.D. 370 and headed north and west on a lengthy odyssey that eventually ended for some in Spain. But a small group remained behind, clinging to their homes along the Black Sea. Travelers to the region took note of them, remarking in the thirteenth century upon inhabitants who spoke a German-sounding language and who lived side by side with their Tatar neighbors.[27] In 1475, the Turks invaded the region, bringing Islam and a new way of life to the Crimea. As the centuries passed, the speakers of Gothic converted to Islam and dressed as the Turks did. They forgot the old ways, and by the middle of the sixteenth century, the Gothic tongue had all but vanished from the Crimea.

German ultranationalists made a great deal of this slender history. They brazenly claimed the Goths as their own ancestors, although there was not a single shred of evidence to support this contention.[28] Moreover, they grandly portrayed the Goths as the founders of a mighty German empire in the East that once stretched all the way from the Black Sea to the Baltic, and from the Carpathian mountains of Slovakia to the Urals of Russia.[29] To tease out the truth about the Goths, Soviet archaeologists in the 1920s and 1930s began studying Crimean ruins.[30] They combed sun-bleached coasts and thyme-scented plateaus and pored over several ancient mountain fortresses and cave cities, places clearly designed to repel invaders and help a badly frightened people sleep a little easier at night. They surveyed remote cave cities where members of the Gothic tribes had lived, and ancient churches where the Goths prayed, but they uncovered not a trace of a mighty German empire in the East.[31] The story was a fantasy, pure and simple.[32]

Nevertheless, tales of the Goths and their magnificent empire in the East continued to circulate in Nazi circles. SS writers wasted little time in capitalizing on them. SS magazines sported cover photos of sparkling Gothic diadems and ran colorful articles with pseudo-scholarly titles such as "Germanic Empire on the Black Sea" or "Gothic Art—Proof of Culture."[33] They recounted tales of Gothic empire-builders and remarked upon the instinctive German need for Lebensraum. All these stories, they insisted, were fully rooted in facts, and they left little doubt of the importance that the SS placed on this history. "The arrival of the Goths," concluded one article, "marked the first time in history that an organizing power of the highest kind appeared in the still undeveloped and unshaped east of Europe."[34]

This doctored history was clearly intended to inspire the soldiers of the Waffen-SS to new and greater heights in the Crimea.

IN THE EARLY spring of 1942, German troops prepared for a massive new attack on Sevastopol, assembling an enormous siege train of 670 artillery guns, including one behemoth that required the work of two thousand men over a period of six weeks to prepare for its firing.[35] As Himmler waited for the campaign to begin, he carefully examined a detailed proposal for the future German colonization of the Soviet Union. Hitler had named him Reichskommissar für die Festigung des deutschen Volkstums, or the Reich Commissioner for the Strengthening of the German Race, placing him in charge of resettling ethnic Germans from outside the Reich in the new Eastern lands. It was work that Himmler had embraced enthusiastically, and in late January 1942, he had begun working closely with a senior planner and agricultural scientist, Konrad Meyer, to draw up a proposal for the future of the Soviet Union.[36] The Crimea figured prominently in these proposals, which Himmler called the Master Plan East, and which he intended to present to Hitler at an opportune moment.

Himmler hoped to found three large colonies in desirable parts of the East, each of which would undergo what he and Meyer euphemistically termed "Germanization." One of the colonies would encompass Leningrad and the lands directly south. The second would straddle northern Poland, Lithuania, and southeastern Latvia. And the third would embrace the Crimea and the rich fields of southeastern Ukraine.[37] Himmler intended to call this southernmost colony Gotengau, a name that roughly translates as "Goth region."[38] He also intended to rechristen Simferopol as Gotenburg.

Himmler estimated that it would take twenty years to completely "Germanize" all of Gotengau.[39] As a first step, he planned to round up the region's inhabitants. Examiners from RuSHA would perform anthropological measurements on those who appeared to be racially valuable to the Nazis, and men, women, and children thought to possess Nordic blood would be permitted to stay in Gotengau. Himmler's various security forces would then forcibly expel the Slavs and other "racially unwanted" groups from their homes in the Crimea. They would kill most, and enslave the remainder as "helots."[40] When this was done, the undesirables would be replaced with ethnic German settlers and with SS settlers who would inhabit defensive villages along the borders of Gotengau. Such settlements would be the preserves of *Wehrbauern*, or "soldier-farmers"—blond, blue-eyed men of the SS—and their wives and children.

The defensive villages of the proposed German colonies clearly reflected all Himmler's fervent dreams for the SS. He proceeded to draw up detailed blueprints for a prototypical farmer-soldier village in the East and showed them with immense pride to his personal physician, Felix Kersten, in the summer of 1942. "Such a village," he explained to Kersten, "will embrace between thirty and forty farms. Each farmer receives up to 300 acres of land, more or less according to the quality of the soil. In any case a class of financially powerful and independent farmers will develop. Slaves won't till this soil; rather, a farming aristocracy will come into being, such as you still find on the Westphalian estates."[41]

His plans called for settlements closely resembling those that the SS had already built in Germany. Dominating each would be a "manor house" occupied by an SS or Nazi party leader.[42] In addition, each settlement would feature a local party headquarters that Himmler envisioned as a "center for general intellectual training and instruction"; a *Thingplatz*, where inhabitants could hold outdoor celebrations for the summer solstice and other important Nazi holidays; and a special graveyard, where families could honor their ancestors.[43]

Himmler was not content, however, with simply Germanizing the Crimean population. He also planned to turn the landscape of Gotengau into his vision of a Teutonic homeland.[44] "Germanic man," he explained to Kersten, "can only live in a climate suited to his needs and in a country adapted to his character, where he will feel at home and not be tormented by homesickness."[45] To soothe the new settlers and supply better cover for their defense, Himmler intended to plant hundreds of thousands of oak and beech trees to reproduce the ancient forests of northern Germany. "We'll create a countryside something like that of Schleswig-Holstein," he boasted.[46]

Himmler also planned to develop hardy new varieties of crops in order to boost the agricultural yields of colonies across the Eastern territories. [47] He ordered the Ahnenerbe to found a teaching and research institute in plant genetics, assigning the task to Dr. Ernst Schäfer, the headstrong young German zoologist who had led the Tibet expedition.[48] Schäfer set to work with characteristic vigor. He obtained a staff of seven research scientists, including a British prisoner of war, and set up an experimental station at Lannach, near the town of Graz in Austria. There the new institute set to work, experimenting with samples of grains that Schäfer had acquired from the granaries of the Tibetan nobility.

ON JUNE 2, 1942, after struggling for more than eight months to capture the entire Crimean peninsula, Manstein ordered a massive artillery attack on Sevastopol, determined this time to take the Crimean port. At his command, a deafening bombardment of five-ton high-explosive shells shattered the Soviet fortifications with the force of a high-magnitude earthquake. A prolonged aerial attack by German dive-bombers followed, flattening the city. Manstein's troops then began their final assault. Overwhelming the Soviet gunners in their heavily fortified hill positions in weeks of heavy fighting, they stole the guns of the dead and began fighting their way to Sevastopol's outskirts. By July 2, they had captured the city's airfields. The Soviet casualties were staggering. "I have never seen such a battlefield in all of my life," reported one veteran SS officer. "Thousands of totally destroyed vehicles lie in the area. Heavy weaponry of all kinds, guns and ammunition—in short, everything that an army requires to fight—are simply strewn haphazardly on the ground. The earth is all churned up and shell craters cover the ground between the enemy field positions. Tens of thousands of dead Russians, and uncounted horse cadavers contaminate the air."[49] Manstein's forces finally captured the city on July 4.

Two and a half weeks later in Berlin, Sievers organized the necessary paperwork to send a small Ahnenerbe scientific team to the region.[50] An eminently practical man, Sievers regarded the battle of Sevastopol not as a human tragedy, but as a prime opportunity for new research and plunder. Despite the terrible devastation wreaked by the war, the Ahnenerbe's most senior archaeologist, Dr. Herbert Jankuhn, was anxious to travel to south Russia in order to secure for the Reich the great Gothic treasures of the Crimea and to locate Gothic sites for excavation. For years, SS archaeologists and scholars had enthused over the beauty of the famous "Gothic crown of the Crimea," a small garnet-encrusted diadem discovered in an ancient grave near the city of Kerch and exhibited in one of Berlin's most famous museums.[51] Jankuhn hoped very much to find more of the Kerch treasure. He also yearned to find proof of what he called the "Gothic empire in southern Russia."[52] Such evidence would help build a case for Germany's claim to the future colony of Gotengau.

Jankuhn was one of the most respected archaeologists in Germany. He was a short, muscular barrel of a man whose sturdy physique was strangely at odds with a fine-boned, almost delicate face. Raised in East Prussia, not far from the border of Lithuania, he believed implicitly in Greater Germany and the ultranationalist cause. His own schoolteacher father had played an active part in local politics, publishing a small book entitled *Is There a Prussian Lithuania?* and Jankuhn had inherited his father's conservative views.[53] At university, he had become fascinated by the history of the Knights of the Teutonic Order, a religious group that colonized Prussia in the thirteenth century, founding German towns and marketplaces throughout the region.[54] And these studies led Jankuhn directly into the field of historical archaeology, a discipline he excelled at.

At twenty-six, Jankuhn became the director of one of the most important excavations in Germany—Haithabu, a Viking trading post located just south of the Danish border. It was there he first met Himmler, who toured the dig in March 1937.[55] The SS chief took a keen interest in the site, offering to heavily subsidize the excavation. He confirmed Jankuhn as the leader.[56] A few months later, Jankuhn joined both the SS and the Ahnenerbe.[57] Himmler came to prize his careful scientific approach and his extensive knowledge of the ancient world. The senior scientific staff at the Ahnenerbe also welcomed Jankuhn into their midst. Bruno Schweizer, Himmler's childhood friend, picked him as the leading archaeologist for the ill-fated Iceland expedition.

In 1940, Himmler appointed Jankuhn head of the Ahnenerbe's prehistory and excavations department, making him, as one scholar recently observed, "the most powerful archaeologist in the Third Reich."[58] From this august position, Jankuhn supervised German scientific research at major archaeological sites throughout the Reich. And as Hitler's empire expanded, so, too, did Jankuhn's field of activity. Just weeks after German forces invaded Norway in 1940, Jankuhn traveled to Oslo to inspect the nation's rich Viking sites and assist the SS Security Service in its futile attempt to win over the Norwegian population.[59] Soon after the evacuation of the British Expeditionary Force from Dunkirk, and the fall of France in June 1940, he toured the new occupied zone, examining the major archaeological sites and gathering information for the Security Service on the degree to which French peasants accepted German political ideas.[60] Like many Germans, Jankuhn believed that Britain was on the verge of capitulation. The end of the war, he concluded, was imminent, so he confidently proposed postponing the "stock-taking" of French museums and private collections until after the armistice, when he would have more workers and greater financial resources at his disposal.[61]

But Britain did not capitulate. The Luftwaffe's devastating blitz on London had failed to break the British spirit as Hitler had hoped. Moreover, the Japanese bombing of Pearl Harbor had brought the United States into the war, greatly adding to the strength of the Allies. German casualty lists grew longer by the day, and in recognition of this fact, Himmler had recently urged his senior SS officers to send all able-bodied men on their staffs to military duty at the front.[62] But he did not apply this injunction to the senior scientists of the Ahnenerbe, who were spared military service. In the summer of 1942, Himmler dispatched Jankuhn and two colleagues—Dr. Karl Kersten, an expert on the northern European Bronze Age, and Baron Wolf von Seefeld, a young ethnic German archaeologist from Latvia who spoke some Russian—to the Black Sea region to search for the treasures of the Goths.

Himmler considered this work to be of prime importance. Indeed, he asked that Jankuhn's reports be forwarded directly to him, although he was mired in work—supplying Waffen-SS and police divisions to the great summer offensive in the East, eliminating political opposition within the Reich, administering his vast empire of concentration camps and lucrative SS business enterprises, and proceeding as quickly as he could with the extermination of the Jews.

JANKUHN INFORMED HIS two colleagues that they would be traveling light—with knapsacks rather than suitcases, and with steel helmets rather than SS hats.[63] The three archaeologists departed on July 21, 1942, for the field headquarters of the SS-Panzerdivision Viking. Jankuhn had learned that the major Crimean museums had crated up their most important collections and shipped them to the northern Caucasus before the arrival of the German army, hoping to protect them from theft.[64] Jankuhn was stubbornly determined to find and seize them, however, even if this meant traveling to the front itself, for he believed the artifacts to be "of great scientific worth."[65] He hoped to catch a ride with Viking Division, which was advancing across the northern Caucasus toward the rich oil fields of Maikop.

The three archaeologists endured a hot and dusty train trip to the eastern front, reaching the Viking command post at Starobesheve in the Ukraine on August 1. To Jankuhn's disappointment, however, the division commander, SS-Gruppenführer Felix Steiner, had just departed on a mission to the front, forcing the archaeologists to cool their heels for five days at Starobesheve. While marking time at the camp, Jankuhn, an immensely intelligent and observant man, must have discovered that the division was traveling with Einsatzkommando 11—one of the roving killing units in Einsatzgruppe D—as well as a gas wagon to facilitate the slaughter of Jews.[66] If Jankuhn found these traveling companions repulsive, he gave no indication of it in his surviving letters to the Ahnenerbe. Indeed, Jankuhn seems to have befriended some of the senior officers of the murder squad. The new head of Einsatzgruppe D, for example, made a point of passing on information and advice to Jankuhn concerning the holdings of museums in the region.[67]

Jankuhn yearned to get to work tracking down the treasures of the Crimea. Tired of marking time at Starobesheve, he ventured off to find Steiner at the front. When the two finally met, the commander explained that the Caucasus campaign had reached a critical stage. He did not want any distractions from the work at hand, but he reluctantly agreed to cooperate with Jankuhn. He advised the archaeologist to be cautious.[68] The military situation in the region, he noted, remained volatile and required "clarification." Jankuhn, however, was not deterred by the prospect of danger. He, Kersten, and Seefeld prepared to head south immediately with the Viking division and its accompanying *Einsatzkommando*.

THE JOURNEY TO Maikop must have been memorable for Jankuhn. Viking Division took few prisoners, generally executing captives and suspected partisans on the spot.[69] Its tank drivers tended merely to run over refugees and others on the roads, instead of stopping or going around them.[70] And if the senior officers of the accompanying Einsatzkommando behaved like those of better-known killing groups, they made little secret of their work, even casually posting notices of their murderous assignments on bulletin boards in their quarters for anyone to see.[71] Traveling with the division did not appear to disturb Jankuhn, however. As the tanks fought their way toward Maikop, Jankuhn and his two colleagues searched the passing countryside for ancient grave mounds, secured local museum collections, and kept their ears open for rumors about the Kerch treasure.

On August 9, German forces captured Maikop as Hitler had directed, but it was a Pyrrhic victory. Before retreating, the Soviets had sabotaged the oil refineries, shutting down their production. As Jankuhn waited to enter the city, he received a radiogram from Sievers, relaying urgent orders from Himmler. Himmler had recently obtained from Ludolf von Alvensleben, the SS and Police Leader of Taurien, a description of an ancient Crimean site known as Manhup-Kale.[72] Alvensleben had toured the Crimean mountain fortress with two companions, a historical novelist and a physician, and the trio had become convinced that Manhup-Kale was once the residence of Gothic princes.[73] Himmler wanted an immediate investigation. Jankuhn, however, was loath to abandon his search for the Goth treasures that the Soviet army had spirited away from Kerch. So he ordered Kersten to depart immediately for the Crimea to start archaeological surveys of Manhup-Kale and other possible Gothic sites.

On August 26, Jankuhn obtained a truck from the division and set off into Maikop. Already, the *Einsatzkommando* had set up a killing facility in the city.[74] As eyewitnesses later recalled in court depositions, the SS and police forces had plastered notices on lampposts and storefronts, ordering Jews to gather in the courtyard of a building formerly belonging to the Soviet state security service, a place of ominous reputation. They were told to pack one suitcase in preparation for resettlement. In the courtyard, an *Einsatzkommando* officer greeted the crowd in a friendly way, patting one of the Jewish girls on the shoulder. This helped break the tension in the air. Someone then asked the assembled families to enter the building. Inside, men in uniforms ordered the crowd to strip and submit to an inspection to ensure they concealed no valuables. When this terrible indignity was over, the troops herded the frightened families into a gas wagon hidden away in a smaller courtyard.[75] By such assembly-line methods, the *Einsatzkommando* methodically murdered the city's Jewish men, women, and children.

Jankuhn and Seefeld made their way across Maikop to the museum, where they proceeded to conduct a leisurely inspection. Jankuhn was very pleased. The Red Army had failed to ship off to safety some of the most important valuables, and while the building had sustained some damage, much of its collection escaped unscathed. The display cases still gleamed with the splendid grave goods of an ancient Scythian noble—a bronze helmet, basin, and cauldron—which delighted Jankuhn, for he considered the Scythians, like the Goths, to be ancestors of the Germans.[76] He and Seefeld also spotted dozens of other desirable antiquities, including a Greek bronze helmet, decorated bronze mirrors, bronze equestrian gear, two war axes, two iron swords, bronze figurines, and Paleolithic stone tools.[77] To Jankuhn's disappointment, however, he could see nothing made by Goth craftsmen. Nevertheless, after sizing up the value of the collection, he decided to ship off the most important antiquities to Berlin.

He discussed the problem of transporting these valuables with Dr. Werner Braune, the commander of Einsatzkommando 11b and the man who had supervised the massacre of Jews at Simferopol six months earlier.[78] Braune took an avid amateur interest in archaeology and had even worked with the Ahnenerbe at one time on educational reforms in Germany.[79] He had often talked to his troops about finding the "Gothic treasure of Kerch" and was clearly delighted that Jankuhn had turned up something valuable. He ordered his men to assist Jankuhn. They found a large crate, and Jankuhn proceeded to pack up the objects that were "scientifically and artistically the most important."[80] Jankuhn was immensely grateful for Braune's help, praising in his final report the "total support" of the *Einsatzkommando*.[81] When Sievers learned of this assistance, he sent Braune a photo of the bronze helmet that Jankuhn had seized at Maikop. This memento, wrote Sievers in an accompanying note, was "supposed to serve as a nice reminder of this part of his work in the mission."[82]

Jankuhn was still anxious, however, to locate the prize museum collections from Kerch. He and Seefeld kept their eyes and ears open, hoping that they might track down the hiding place, but increasingly Jankuhn worried that the treasure had been shipped off beyond reach. On August 28 in Armavir, an important railway junction in the northern Caucasus, Seefeld received a key piece of intelligence. A medical warehouse in the city had received a transport of seventy-two wooden crates. They were reputedly filled with museum treasures from Simferopol, Sevastopol, and Kerch.[83]

Seefeld hurried to track down the crates, greatly excited by the thought of the treasure he was about to find. When he arrived at the warehouse, however, his heart fell. The depot had been reduced to a smoldering ruin. He got out of his car and took a look around. Out in the courtyard, he discovered twenty sealed crates and several others that had been pried open and plundered.[84] He notified Jankuhn of his discovery, and when the senior archaeologist arrived, they proceeded to pore over the contents of the sealed crates, artifact by artifact. They unwrapped ancient Greek vases, Greek terra-cotta statuettes, pearl necklaces, important Stone Age artifacts, ancient coins, a marble relief, valuable geography books on south Russia, Tatar mother-of-pearl chests, and carved marble reliefs.[85]

But there was not a single Gothic artifact to be found amid the rubble there. Jankuhn gazed up with immense disappointment and frustration. The precious boxes concealing the treasure of Kerch had eluded him. Nevertheless, he and Seefeld packed up fourteen crates of the most valuable antiquities and dispatched them back to the Ahnenerbe offices in Berlin.[86]

17. LORDS OF THE MANOR

In the last week of August 1942, members of the SS high command were much struck by the exceptionally high spirits of their leader. As Himmler strode purposefully between the airfields, meeting rooms, banquet halls, and elegant homes at Hegewald—his secret field headquarters in northeastern Ukraine—he greeted his senior officers with a smile and spoke warmly, even jovially at times, to his aides. On one memorable afternoon, he took time out from the briefings and telephone calls to play a popular European game, fistball, with some of his staff at the *Sportplatz* in the compound.[1] He believed, as Hitler did, that competitive sports honed and strengthened the human body, making it more fit for warfare.[2] Later, he spent an hour in Hegewald's shooting gallery, practicing his marksmanship with a pistol.

Part of Himmler's happiness stemmed from the satisfactory way that he had finally managed to arrange his personal life. His marriage to Marga had long been a source of frustration and acrimony. His fifty-year-old wife had given him a blond-haired daughter, Gudrun, whom he adored, and the couple had adopted a son. Often when Himmler returned to Germany, he visited his family in their chalet near Gmund. But his relationship with Marga had long ago dissolved, and it was exceedingly unlikely, given her age, that she would produce any more children. This greatly troubled Himmler. So he chose a willowy young blond secretary on his personal staff, Hedwig Potthast, twelve years his junior, to become his mistress.

He had installed Potthast, whom he affectionately referred to as Häschen, or "Little Bunny," in considerable luxury in Haus Schneewinkel-lehen, near Berchtesgaden, not far from Hitler's mansion. The two were happy together, for they saw eye to eye on a great many things. Potthast, for example, saw little amiss with her lover's inhuman treatment of Jews. On one occasion, when her friend Gerda Bormann and her children dropped by for a visit, she offered to show them something "very interest-ing."[3] She led her guests up to the attic and ushered them into a small room. Inside, she pointed to some furniture—a chair made from the pol-ished bones of a human pelvic girdle, and another made from human legs and human feet. Then she picked up a copy of Mein Kampf, explaining "clinically and medically" that its cover was made from human skin.[4] The Bormann children shrunk back. They were horrified by the ghoulish display.

In February 1942, Potthast gave birth to her first child by Himmler, a boy named Helge.[5] Himmler was delighted. Three years earlier, just shortly after the war began, he had issued an order to all SS men on the subject of fathering children. In a directive set in an old-style German type-face, he ordered all members of the SS to produce as many children as possible—within marriage or outside of it, it made no matter. Only in this way, he declared, could the superior bloodlines of the SS survive the mis-fortunes of war. "The old wisdom that only he who has sons and children can die peacefully must in this war again become reality for the Schutzstaffel [the SS]," he asserted.[6] It was advice that many of his senior officers in Germany—including Wolfram Sievers and several other senior Ahnenerbe staff—gratefully took to heart, acquiring lanky, blond-haired mistresses and fathering second families.[7]

The birth of a son contributed considerably to Himmler's buoyant mood in the summer of 1942. But beyond his personal situation, Himmler believed in mid-1942 that his cherished dream—the creation of an SS landed nobility in the East—was at last within reach. Ever since he had em-braced the goals of the agriculturally oriented Artamanen society in his twenties, he had dreamed of founding settlements of perfect young Nordic men and women who would defend the Reich from its enemies to the East and who would return to the pure ways of their ancestors. In these feudal settlements of Wehrbauern, or "soldier-farmers," Nordic families would till the earth, sow ancient grains, tend antique cattle breeds, live in medieval-style houses, heal the sick with traditional plant remedies and age-old magical incantations, play time-honored musical instruments such as the *lur*, practice the old Germanic religion, and generally follow the traditions of their ancestors—as revealed by the scholars of the Ahnenerbe.

Himmler believed that SS colonization of the Crimea and other select regions of the Soviet Union was not far off, and in his field headquarters in the Ukraine he marveled at the speed with which his most cherished dream was unfolding. "Who would have dreamed ten years ago that we would be holding an SS meeting near the Jewish-Russian city of Zhytomyr?" he gloated before a gathering of his SS officers in September. "This Germanic East extending as far as the Urals must be cultivated like a hothouse of German blood. . . . The next generations of Germans and history will not remember how it was done, but rather the goal."[8]

HIMMLER WAS, OF course, not the only one contemplating the future. For some time, Hitler had been mulling over the disposition of prize territories in the East. In the summer of 1942, he had read a paper on the Crimea that interested him greatly.[9] The writer, Alfred Frauenfeld, advocated resettling the region with ethnic German families from an ancient borderland between Austria and Italy. The region, known as the South Tirol, was a restive place. Long part of the Austrian Empire, South Tirol had been given to Italy at the end of the First World War. A few years later, Mussolini had tried to impose the Italian language on all of its residents.[10] He failed miserably. To calm growing dissent, he and Hitler agreed to allow the South Tirolese to choose their own fate in a vote in 1939.

Most of the region's German-speaking residents opted to migrate to the Reich—an outcome that delighted many high-ranking Nazis. Legend had it that the South Tirolese descended from the wandering Goths. Indeed, one German writer dubbed them "Goths conserved in glacial ice."[11] But many questions remained about the origins of the South Tirolese. After the vote, scholars from the Ahnenerbe conducted detailed studies of their folk customs, music, house markings, clan symbols, architectural styles, folk art, and prehistory. They concluded that the ethnic Germans in the region were a valuable racial stock of ancient *Wehrbauern*.[12]

Hitler had little intention, however, of settling tens of thousands of immigrants from South Tirol on valuable farms in the German heartland.

He much preferred to plant them elsewhere. Frauenfeld's plan to transport South Tiroleans more than a thousand miles east to the Crimea—a foreign land situated on a dangerous border—appealed to him greatly. It never occurred to Hitler that the South Tiroleans might have something to say about such a plan. "Their transfer to the Crimea presents neither physical nor psychological difficulty," Hitler blithely informed his guests one night at Wolfschanze. "All they have to do is sail down one German waterway, the Danube, and there they are."[13]

All this talk at Wolfschanze of new Eastern settlements seems to have delighted Himmler. He had been waiting for just such a moment. For many months, he and his staff had labored over the Master Plan East, channeling all his dreams for the future onto paper. They had drawn up fanciful maps of Eastern Europe and the Soviet Union dotted with proposed villages and *Wehrbauern* settlements, future forest sites and industrial areas, all linked by a vast new system of *Autobahnen,* or "expressways." Himmler was convinced that Germany's future lay in such settlements, and he presented his ideas as forcefully as possible to Hitler.[14] To his astonishment, the German leader listened attentively, barely interrupting the presentation— a rarity in his dealings with others.

In mid-July 1942, Hitler approved these settlement plans.[15] Himmler was euphoric. Writing after the war, his personal physician Dr. Felix Kersten recalled Himmler's mood when he arrived for a therapeutic massage not long after he had spoken to Hitler. It was, Himmler explained, "the happiest day" of his life:

> "Everything I have been considering and planning on a small scale can now be realized. I shall set to at once on a large scale—and with all the vigor I can muster. You know me: once I start anything I see it through to the end, no matter how great the difficulties may be."
>
> I asked Himmler to lie down so that I could begin the treatment. He did not even listen to me, but continued: "The Germans were once a farming people and must essentially become one again. The East will help to strengthen the agricultural side of the German nation—it will become the everlasting fountain of youth for the lifeblood of Germany, from which it will in turn be constantly renewed. These phrases opened my remarks to the Führer and I

linked them with the idea of defending Europe's living space, which I knew lay very close to the Führer's heart. Villages inhabited by an armed peasantry will form the basis of the settlement in the East— and will simultaneously be its defense: they will be the kernel of Europe's great defensive wall, which the Führer is to build at the victorious conclusion of the war. Germanic villages inhabited by a military peasantry and filling a belt several hundred miles wide— just imagine, Herr Kersten, what a sublime idea!

"It's the greatest piece of colonization which the world will ever have seen, linked too with a most noble and essential task, the protection of the western world against an irruption from Asia. When he has accomplished that, the name of Adolf Hitler will be the greatest in Germanic history—and he has commissioned me to carry out the task."[16]

As Himmler was well aware, such sweeping plans for Eastern Europe and the Soviet Union, involving the relocation of millions of people, could not possibly be carried out during a world war. They would have to wait for victory. In the meantime, however, Himmler proposed founding a small German colony around his own field headquarters at Hegewald, not far from the Ukrainian capital of Kiev.[17] Such a colony, he argued, would serve a sound military purpose, for the settlers could grow grain and tend livestock to feed local SS and police forces.

Himmler proceeded with his customary blend of brutality and efficiency. On October 10, 1942, his troops began forcibly rounding up 10,623 Ukrainian men, women, and children from family homes in the region, cramming them into boxcars destined for labor camps in the south.[18] By the middle of the month, most of their houses stood eerily empty. Their dishes still sat on the kitchen table and their linen lay folded in the cupboards. Their livestock ran loose in the fields. When they were gone, trains began disgorging thousands of new settlers—ethnic German families forcibly removed from villages and towns in the northern Ukraine. The bewildered newcomers stretched their legs and stared at the unfamiliar surroundings.

Local SS officers left no doubt, however, as to who controlled the new colony. SS agricultural specialists doled out parcels of land to the new arrivals, giving most families small blocks of fourteen acres as well as a promise to add to these modest allotments when the settlers proved their abilities as farmers.[19] The specialists also notified each family of the SS quotas of milk and produce that they would be required to meet, and informed the settlers that they could expect to have their crops confiscated when the SS had need of them. In addition, the specialists took note of all unarable blocks of land and set them aside for SS factories, where Ukrainian slave laborers would soon be put to work.[20]

It was not quite the SS settlement that Himmler had originally dreamed of, but he intended to set matters straight as soon as Germany won the war, bestowing large parcels of land in the East on his SS men and officers. Members of his staff were well aware of this plan and frequently argued over the size of their own future country holdings. As Himmler's physician, Felix Kersten, recalled after the war:

> They all dreamed of the grand estates in the East that had been promised to them as the first fruits of victory. They waxed hot and eloquent on the subject. There were even quarrels, occasionally, over the exact dimensions of the farms that should be allotted to them, the comparative wealth of the reward according to the years of their service! One man would say he expected to receive a gift of a thousand hectares, at least. Another would pipe up with, "But I've been in the Party a year longer than you. If you get a thousand hectares, what about me? I ought to receive in all justice, two thousand hectares!" And a third: "What about me? I got this wound in my arm during the Putsch!" And a fourth: "Well, I was photographed twice with the Fuehrer. And for a deed of bravery, I was awarded this Party emblem in gold. I should have, by the way you all rate yourselves, at least five thousand hectares!"[21]

Few lands would better suit the new SS gentry than the Crimea, with its languid shores and pretty vineyards and sunny fruit orchards. The new Nazi governor of the Crimea, Alfred Frauenfeld, had described it as a paradise that rivaled the Alps, the French Riviera, and Sicily.[22] Well aware of its many attractions, Himmler was anxious to begin staking the SS claim to this Garden of Eden.

ON SEPTEMBER 17, 1942, Dr. Karl Kersten received a letter from the commander of the Security Police and the Security Service in the Crimean city of Simferopol. Kersten had just arrived from the Caucasus, where he had accompanied Herbert Jankuhn on the search for the ancient treasures of the Goths. The archaeologist was expecting to begin a detailed survey of the Crimean cave cities in preparation for major Ahnenerbe excavations, but the letter in his hands contained new orders. Himmler, he learned, was planning to visit the Crimea in the fall and "intended to take this opportunity to visit the Gothic cities of the Crimea."[23] The SS command ordered Kersten to draw up a suitable itinerary.

In civilian life, Kersten was an expert on the northern European Bronze Age and a curator at the prehistory museum of Kiel.[24] He was a clever, reserved man with hooded, deep-set eyes and a long, sharp nose and the look of someone not easily deceived. He had met Jankuhn as a student in 1929, and the two men became fast friends. To pursue his studies, Kersten had traveled to Eastern Europe and conducted research in Russia. In 1937, he had taken out a Nazi party membership, but he had defended ultranationalist policies for years. He was a strong anti-Semite, and his political attitude, vouched Jankuhn in one official report, was "irreproachable."[25]

A diligent man, Kersten recognized that he knew very little about the complex prehistory and history of the Crimea, with its thousands of years of war and diplomacy, piracy, and trade by a rich blend of cultures—the Goths and Cimmerians, Scythians and Sarmatians, Greeks and Alans, and Khazars, Huns, and Turks. So after receiving his new orders, he spent six days in the Simferopol museum, poring over photographs of ruins and skimming the available scientific literature on the most important sites. On September 23, he set off with a driver and a translator to hunt for the imperial cities of the ancient Goths.[26]

As beautiful as the Crimean countryside was, with its grassy plateaus and steep mesas, it was not a relaxing place to travel in the early fall of 1942. Although Manstein's army had dealt a decisive defeat to the Soviet forces at Sevastopol, many of the local inhabitants had joined partisan groups who had begun carrying on a bloody war of attrition against the Germans. Some hid from time to time in the old cave cities of the region that perched high atop the local mountains. Kersten had no particular desire to surprise them while drawing up an itinerary for Himmler. He called in at local police stations to check on partisan activity before heading into the countryside.

Kersten chose the old Tatar town of Bakhchysaray as his base and arose each morning to the unfamiliar call of the muezzins from the local mosques. Notebook in hand, he headed out in a private car after breakfast, rummaging for traces, any traces, of an ancient Gothic empire. He trekked up to the old ruins of Bakla, wandering among the wild roses and the hawthorns, and roamed about the old cave city of Chufut-Kale, puzzling over its small cliffside dwellings and taking note of an old Tatar courtroom and prison "where torturing and killing took place."[27] But sprawling as it was, Chufut-Kale could not be the old Gothic capital, Doros. Its earliest known construction, he noted in his report, was not Gothic at all, but a Byzantine chapel dating to the sixth century.[28]

He spent a morning in the prehistory museum tucked away in the old khan's palace in Bakhchysaray, examining carved stone reliefs taken from several of the old cave settlements of the Goths, and for a day and a half, he roved the ancient fortress of Tepe-Kermen, rubbing dirt from the pottery shards he found and hunching over graves in a Gothic cemetery. He recommended that the Ahnenerbe conduct excavations at the site. "There should be no doubt as to the Gothic origins of the city," he observed in his report, "because of the situation of the caves and the form of the graves."[29]

But he had still not seen anything that resembled the capital of a great Gothic empire, so he resumed his survey. On September 30, he set off to see the famous ruins of Eski-Kermen. The old Crimean cave city was a formidable fortress, perched along a steep mesa top: below ran an old military road and trade route to the ancient ports of Inkerman and Balaklava. Russian archaeologists had spent five field seasons at Eski-Kermen in the late 1920s and early 1930s. They had surveyed its walls, excavated one of its gates, and studied its graves. They concluded that it was quite likely the old city of Doros mentioned in the histories of the classical world, but they made no mention of a vast Gothic empire.[30] Kersten was keen to explore the shadowy hollows of Eski-Kermen. He slowly tramped up the steep, serpentine path to its southern gate, gazing up at the high cliffs and the chambers there where sentries once guarded the approaches.

He spent nine hours rambling through the old city, or rather what remained of it. He marveled at some of the "356 caves, casements and defensive towers, which have been built directly into the cliff and which have been erected exactly at those points where footpaths lead through small gaps in the cliffs and into the inner city."[31] He traipsed through stables and granaries and found the spot where the city's basilica once stood and roamed through its graveyard. But what he most hoped to find was some trace of imperial splendor, the remnant of some mighty Gothic palace. In this he was disappointed. The old catacomb city bore no resemblance at all to the glories of imperial Rome, with its Colosseum and its Pantheon and Hadrian's villa. If Himmler hoped to gaze upon the Gothic counterpart of ancient Rome, then Eski-Kermen would be a disappointment. But Kersten refused to admit what other scholars had long known—that the cave cities and fortresses were merely part of a Gothic province that was subordinate to the great Byzantine Empire, whose capital lay in Constantinople.[32] Kersten firmly adhered to the Nazi party line. At Eski-Kermen, he concluded, "the Goths founded the main city in the Crimean Gothic Empire in the fifth century."[33]

The archaeologist continued his survey. He was particularly keen to see the old cave city of Manhup-Kale, the one that Himmler's former aide, Ludolf von Alvensleben, and two associates, had visited a few months earlier. To Kersten's disappointment, however, Manhup-Kale had become too dangerous a place to visit. Someone in a nearby village had spotted partisans taking shelter in the ruins. He resumed his journey, continually stopping in to inspect potentially promising sites as he headed west. He took careful notes of all that he saw, and in early October he dutifully dictated short, plodding reports on all the major sites for Jankuhn and the local SS authorities. Then, before heading north to survey sites along the Dnieper River, he carefully drew up an itinerary for Himmler's forthcoming visit to the Crimea.[34]

THE SS LEADER flew into Simferopol on October 27, 1942. He had arrived to see for himself the future colony of Gotengau and to ensure the successful conclusion of a major SS and police operation against the Crimea's partisan forces.[35] From bases tucked away in the mountains, the partisans had succeeded in miring German army and police forces in an ugly little war, preventing the military from handing over responsibility for the region to a civil administration. Their operations had greatly perturbed Hitler. Over dinner one night, Hitler insisted that Germany would not be denied its possession of the Crimea. "The struggle we are waging there against the Partisans resembles very much the struggle in North America against the Red Indians," he explained. "Victory will go to the strong, and strength is on our side. At all costs we will establish law and order there."[36]

To impose that law and order—a necessary prerequisite to the resettlement of the Crimea—local SS and police forces had drawn up plans for something they called Operation Leatherstocking.[37] Informants in the region had told them about a small airfield in the mountains near Alushta where Soviet pilots were secretly landing supplies for the partisans. Based on this intelligence, senior German officers in the region had drawn up plans for a counteroffensive. It called for capturing the airstrip, destroying the supplies hidden nearby, and surprising a large number of the partisans who had gathered in the area.

While waiting for his forces to deal a crushing blow to this local resistance, Himmler had decided to tour the Crimean countryside and inspect the ancient Gothic sites that Kersten had described in his official reports. So on the morning of October 28, he set out from Simferopol on a grand tour of his future German colony. He journeyed first to Bakhchysaray, and visited the old khan's palace, with its pretty grounds and its harem rooms and its museum with the carved relief stones from Manhup-Kale and Eski-Kermen.[38] Then in the bright sunlight, so different from the damp and cold and gray of a Berlin fall, he drove to Sevastopol to see the devastated battlefield, and stopped to view the old Gothic port of Inkerman, one of the cave cities that Kersten had described briefly in his report.[39]

Almost certainly Himmler expected to hear good news from his men that evening about the resistance fighters. But there was no word of a decisive defeat, not that day, or the next, or the next. As it soon became clear, Operation Leatherstocking was a failure. The SS and police forces managed to capture the airfield, as well as a broadcast station and the storage facilities, but they failed to surround the partisans themselves.[40] Indeed, the local fighters eluded them entirely, melting back into the mountain country that they knew so well. For all of Himmler's careful planning, and all the long hours he spent contemplating the maps of the Master Plan East and reveling in the future of Gotengau and its villages of SS farmers, the Crimea remained unconquered.

Angered and troubled by the turn of events, Himmler made no further attempt to visit the other Gothic cave cities on Kersten's itinerary. The golden moment of triumph he craved—the contemplation of which had constituted "the happiest day" of Himmler's life—had been snatched from his hands.

18. SEARCHING FOR THE STAR OF DAVID

SITTING IN THEIR ORDERLY offices in Berlin, SS racial experts were greatly troubled by the extraordinary cultural richness of the Soviet Union. Surrounded by neatly arranged card indexes and carefully alphabetized file folders, they had never grasped before the unruly complexity of the world. They had never understood that a nation such as the Soviet Union could be so vast, so complicated, so chaotic, or that human beings could be so diverse, so exotic, so difficult to pigeonhole. More than eighty different ethnic groups resided in the country—from the Belorussians to the Moldavians, the Ossetians to the Chuvash, the Kazakhs to the Mongols, the Tungus to the Dargins, the Chechens to the Kabardas, the Mordvins to the Mansi, the Nenets to the Koryaks.[1] And this posed serious questions for the racial specialists of RuSHA. Who among all of these peoples was Aryan? And exactly who was Jewish? Each day seemed to bring new doubts.

German racial scholars, after all, had still not devised a way of identifying members of the supposed Jewish race. In their scientific papers, they struggled in vain to define the physical characteristics of Jews. More often than not, they had fallen back on old anti-Semitic stereotypes. They talked about the short stature of Jews; about their flat breasts and rounded backs and weak muscles; their large, fleshy ears and hooked noses and yellowish skin; about the way they shuffled when they walked and the way they mumbled when they talked; and their great susceptibility to schizophrenia, manic depression, and morphine addiction.[2] But in reality, German racial experts could not separate the fictional Jewish race from its fictional Aryan counterpart. Indeed, one famous anthropological study conducted among German schoolchildren had revealed that 11.17 percent of Jewish children possessed fair skin, blond hair, and blue eyes.[3] Nazi researchers lacked a biological equivalent of the yellow Star of David.[4] Nothing offered itself, and this failure deeply troubled racial scholars in the SS.

Already, the uncertainty was stirring up confusion where Himmler least wanted to see it—in the minds of the SS killing squads. Just six months after the Russian campaign began, one of the *Einsatzgruppen* leaders, Otto Ohlendorf, had thrown up his hands in the Crimea, unable to determine what to do with two local groups—the Krimchaks and the Karaites.[5] Reputedly, the Krimchaks descended from Jews who fled the Spanish Inquisition, but they closely resembled their Muslim neighbors, the Tatars. They spoke a variant of Tatar, lived in Tatar-style houses, married into Tatar families, and followed many Muslim customs. Their women, for example, wore veils in public.[6] The second group, the Karaites, equally confounded the SS. They were a Turkish people who spoke a Turkish language, but they practiced Judaism devoutly.[7] Was either truly Jewish?

This conundrum seems to have sparked great anxiety in the SS headquarters in Berlin. Hitler had frequently likened the Jews to a dangerous microbe that threatened the rest of humanity, and he had become increasingly incensed by the Jews of the Soviet Union. "Russia," he had observed in one venomous conversation in July 1941, "has become a plague-centre [*Pestherd*] for mankind. . . . For if only one state tolerates a Jewish family among it, this would provide the core bacillus [*Bazillenherd*] for a new decomposition."[8] Well aware of these views, Himmler intended to wipe out every last Jew in the German-occupied territories of the Soviet Union. He solicited the opinions of various self-proclaimed SS authorities on Judaism, for their views on the Krimchaks and Karaites.[9] He then resolved to eradicate those belonging to the fictive Jewish race in the Crimea, and turn a blind eye to those who followed the Jewish faith. He ordered Ohlendorf and his men to liquidate the Krimchaks, who followed all manner of Muslim customs, and to spare the Karaites, who were devout Jews.[10]

Perhaps Himmler hoped he had seen the last of such troublesome problems. But the confusion had only just begun. As the murder squads moved eastward into the Caucasus—a borderland between East and West,

Europe and Asia—the lines between ethnic groups and tribes became more and more blurred. SS troops stumbled upon villages of the Christian Ossetes, who physically resembled their Jewish neighbors, lived in villages with Jewish names, married their sons and daughters off in Jewish-style marriage ceremonies, and buried their dead in Jewish-style funerals.[11] And they met Mountain Jews who rode their horses superbly, bred fine cattle, and seldom stirred anywhere without strapping on their daggers and guns—all qualities greatly admired by SS men.[12]

Who was who in this great ethnic bedlam? Who was Jewish and who wasn't? All the old certainties were slipping away. At times like this, Himmler counted upon science and scholarship to show the way.

ON DECEMBER 10, 1941, Wolfram Sievers welcomed SS racial specialist Bruno Beger to his office in the Ahnenerbe headquarters in Berlin. Although most young German men had been called up to active service—patrolling the waters of the North Atlantic in U-boats, battling British tanks in the bleak desert of North Africa, or navigating across the frozen forests of the Soviet Union in Stuka dive-bombers and Messerschmitts—Sievers had managed to obtain exemptions for many of the Ahnenerbe's young scientists and scholars. This he did largely by transforming the research organization itself. While the Ahnenerbe had long supplied scientific camouflage for Nazi racial policies and furnished blueprints for future SS farm settlements, it had, since the onset of the war, taken an increasingly active part in the wartime crimes of the SS.

These illegal activities had begun with the plundering of foreign museums, galleries, and private homes, but in recent months they had taken a far more deadly turn. With Himmler's encouragement, a senior Ahnenerbe researcher, Dr. Sigmund Rascher, was preparing to investigate the far limits of human endurance at extremely high altitudes.[13] British fighter planes had pushed German aircraft to higher and higher elevations, and the Luftwaffe command feared for the safety of its aircrews. Rascher, a medical doctor, proposed testing the problem by replicating the effects of extreme high altitude on human beings placed in a vacuum chamber. Knowing that the experiments would inflict great suffering and kill some of the subjects, he had requested the use of concentration-camp prisoners—a suggestion that Himmler readily approved. While Sievers waited for Rascher's high-

altitude trials to begin at Dachau, he began planning one of the most notorious mass murders of the Second World War.

Like many other senior SS officers, Sievers knew all about the difficulties German racial scientists were having in defining the racial characteristics of Jews. Indeed, he had long hoped that the Ahnenerbe could play a
key role in solving this problem. Since 1939, one of the Ahnenerbe's department heads, biologist Walter Greite, had been studying racial measurements that he and a team of assistants had conducted in Vienna in 1939 on
some two thousand Jews anxious to emigrate from the Reich. But the project was an embarrassing fiasco.[14] Greite had assigned the important and tedious task of statistically analyzing the measurements to his secretary, and
had failed abjectly in coming up with anything new.[15]

If the Ahnenerbe intended to make any headway with this problem, it
clearly needed someone more reliable and energetic to handle the Jewish file.
Beger, the ambitious young *Rassenkunde* scholar from the Tibet expedition,
must have seemed a logical choice to Sievers. In their December 10, 1941,
meeting, the two men discussed Jewish research.[16] Beger seems to have
taken an interest in the subject. He had, after all, worked for four years for
RuSHA, a deep mire of anti-Semitism where researchers had helped draft
the Nuremberg race laws and continued to churn out studies to expand the
Holocaust.[17] Moreover, he had risen to the position of division head at
RuSHA before he set off to Tibet. Since his return, he had continued working for the department in an honorary capacity.[18]

In the meeting with Sievers, Beger seems to have discussed going
about the Jewish research in a completely different way. To study Jews
thoroughly and to search for that elusive feature that would define and
label Jewishness—the shape of the ears, perhaps, or the arch of the
cheekbone—he would need a good reference collection of Jewish skulls,
one that contained as wide a variety of "Jewishness" as possible.[19] If the
Ahnenerbe wanted to ensure that its findings would apply to the disparate
Jewish communities of the Soviet Union, the collection would have to include a representative sample of Jewish skulls from across the nation, from
the remote mountain villages of the Caucasus to the bustling streets of
Murmansk in the north.

Acquiring such a collection would be difficult. Few if any universities
or museums in the Reich possessed large numbers of Jewish skulls. Devout
Jews had long regarded tampering with the dead as a terrible sacrilege.

Uniformed members of Romania's Iron Guard stand at attention during a political rally in Bucharest. While touring the Romanian countryside, Altheim and Trautmann gathered intelligence on prominent supporters of the Iron Guard. *Getty Images*

Donning borrowed Arab robes, Altheim and Trautmann pose with a Bedouin sheikh on their trip to Hatra in Iraq. The Bedouin, they later reported to Himmler, would be staunch German allies in a war: "They speak the names 'Hitler' and 'Mussolini' as if they are holy." *Rädda Barnen*

An ambitious schemer, Assien Bohmers joined the Ahnenerbe so that he might one day rise to power in his native Netherlands as a Nazi leader. *Bundesarchiv, Ahnenerbe (ehem. BDC): Bohmers, Assien*

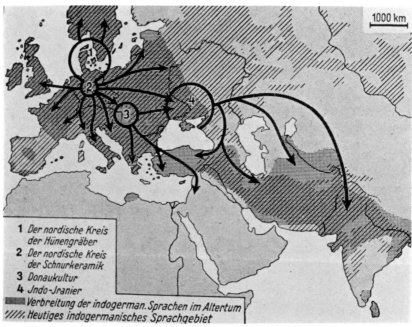

1 *Der nordische Kreis der Hünengräber*
2 *Der nordische Kreis der Schnurkeramik*
3 *Donaukultur*
4 *Jndo-Jranier*
▬▬ *Verbreitung der indogerman. Sprachen im Altertum*
///// *Heutiges indogermanisches Sprachgebiet*

Racial researchers claimed that Germany's ancestors conquered Asia in the far distant past, producing a powerful blond-haired ruling class. Both Himmler and the Ahnenerbe's president, Walther Wüst, yearned to find proof. *Map from* Nordische Urzeit

Hitler, Himmler (third from right), and other prominent Nazis tour a museum in Rome in 1938. To finance the Ahnenerbe, Himmler ordered his senior staff to sign for massive bank loans. The SS leader had no intention of repaying these debts. *Fratelli Alinari*

In 1938, the Ahnenerbe moved into a sprawling mansion in wealthy Dahlem in Berlin. Formerly owned by a Jewish family, the mansion boasted a conservatory, library, laboratory, microscopy room, photographic studio, darkroom, and nearly two dozen offices.

Dressed Tibetan style, Ernest Schäfer (third from the left), Bruno Beger (second on the right), and other SS team members and friends assemble for a photo in Shigatse, Tibet. Organized by the Ahnenerbe and the SS, the Tibet expedition set off in 1938. *ullstein–ullstein bild*

To conduct his racial measurements, Beger carried special tools, which included these calipers. *From* Taschenbuch der rassenkundlichen Meßtechnik

Tibetan women gaze warily at the camera's eye. Racial scholar Bruno Beger analyzed nearly 2,000 Tibetans for supposed Nordic traits. *NARA, Still Pictures Unit, RG-319*

Horses graze among tents in the Schäfer expedition camp. Reaching Lhasa at last, Schäfer wrote exultantly to Himmler, "It is a wonderful feeling to know that the power of the Reich is so great it reaches into the most isolated parts of Inner Asia." *NARA, Still Pictures Unit, RG-319.*

Himmler greets Schäfer and other members of his returning team on the runway in Munich in August 1939. So fascinated was Hitler by reports of the team's adventures and findings that his staff invited Schäfer to lecture on his Tibetan work. *ullstein–ullstein bild*

In the summer of 1939, Edmund Kiss worked feverishly to mount the Ahnenerbe's largest and most expensive expedition—to Bolivia. This is the only known surviving official photo of Kiss. *Bundesarchiv, SSO Kiss, Edmund*

Kiss believed the old Andean capital of Tiwanaku was the creation of Nordic colonists who sailed to Bolivia more than a million years ago. He published this wildly imaginative drawing of Tiwanaku in a book Himmler later gave to Hitler. *From* Das Sonnentor von Tihuanaku.

A photograph by Arthur Posnansky shows the Gateway to the Sun at Tiwanaku as it looked in 1920s, when Kiss first saw it. Modern archaeological studies demonstrate that the forerunners of the Incas built the old Andean capital some 1,700 years ago. *Courtesy of Javier Nuñez de Arco*

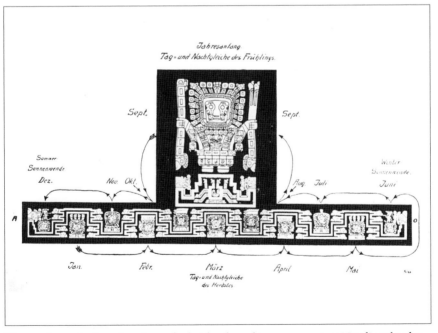

Kiss made meticulous drawings of what he thought was an ancient Nordic calendar carved on the Gateway to the Sun. Armed with such research, he convinced Himmler to send him and a twenty-man expedition to Bolivia, a plan that faltered only when the war began. *From* Das Sonnetor von Tihuanaku

Himmler wanted SS men to shun the moral decay of cities and return to simpler, more traditional lives in the country. *USHMM, courtesy of James Blevins*

Under Himmler's command, SS planners designed a model SS farmhouse. The house at right was built in an SS settlement at Mehrow in 1937. *Brandenburgisches Landesamt für Denkmalpflege*

The spacious house design chosen for the SS settlement at Mehrow dated back to at least the Roman era and combined family home and barn under one roof. Himmler later planned to build similar rustic settlements throughout Eastern Europe for SS families. *Brandenburgisches Landesamt für Denkmalpflege*

They looked with horror upon autopsies—which they believed mutilated the deceased—and cherished cemeteries as holy ground. As a result, few European anthropologists had dared to collect scientific specimens of Jewish skulls. The world famous Museum of Natural History in Vienna, for example, possessed only twenty-two of these crania, and had searched far and wide for others to display in a planned exhibit on "the mental and racial characteristics of the Jewish people."[20] To expand this collection, curator Josef Wastl had proposed digging in Vienna's old Jewish cemeteries in 1939, but Austrian officials had denied permission for the dig—not out of any sense of decency, it seems, but because property developers were hungrily eyeing the land.[21]

So intense was Wastl's desire to study the Jews, however, that he eventually ended up purchasing Jewish skulls by mail order from the anatomy institute of the Reich University of Posen, a new creation of the Nazis in occupied Poland.[22] The director of this anatomy institute, Dr. Hermann Voss, had made a bargain with the local Gestapo. In exchange for the use of his institute's incinerator, he received bodies of some prisoners who had been guillotined or hung. These cadavers he rendered into anatomical specimens that could be sold to interested parties. The porter at the institute later recalled how the system worked. "The heads of the transported victims were thrown into a basket like turnips, and brought in the elevator to the third floor for maceration," he explained. "Here they were prepared and later used in our institute of anatomy, where some can still be found, or sent to various universities in Germany, or sold to the students."[23]

Voss made a tidy profit from the business. He sold individual Jewish skulls for 25 reichsmarks—or $130 in today's currency—and agreed to furnish the victim's date and place of birth, information considered vital for many scientific studies. He also offered a small range of other similar wares. "Together with these Jew-skulls," Voss wrote to one potential customer, "I am able to supply plaster death-masks of the individuals concerned at RM 15,—of especially typical Ostjuden [Eastern Jews]. I can also prepare for you plaster busts, so that one can see the shape of the head (before dissection) and the frequently rather unique ears. The price of these busts would be 30–35 RM but because of a scarcity of time and plaster I could not supply very many."[24]

For statistical reasons, Beger needed a minimum of 120 Jewish skulls to produce significant results.[25] To obtain such a large number could be costly, and there was no guarantee that he could find a broad diversity of Jews. So Beger and Sievers conferred, contemplating ways of laying hands on a large and diverse collection, and at some point in the conversation, Sievers seems to have mentioned the name of Dr. August Hirt, the German director of the anatomical institute at the new Reich University of Strassburg.[26] Just two and a half weeks earlier, Sievers had dined with Hirt at the university's official opening ceremonies, and it seems to have occurred to him that Hirt could help obtain a collection of Jewish skulls.

As fate would have it, Beger knew the anatomist well. He had met Hirt while serving in the SS in Heidelberg in 1934 and both men had worked for the RuSHA in 1937, becoming good friends.[27] Beger welcomed the anatomist as a collaborator.

HIRT WAS THIRTEEN years older than Beger and was a man that many found impossible to forget. As a young teenaged soldier during the First World War, he had received a severe gunshot wound to his upper and lower jaw.[28] When the injury finally mended, his face took on a fierce, scarred, rather cavernous look that tended to unsettle people. Hirt tried to compensate for this with a jocular, friendly, outgoing demeanor.[29] Some people found this bluff manner charming and were greatly drawn to him.[30] But others were unable to get beyond his scarred appearance.[31]

Despite his terrible injury—or perhaps because of it—Hirt studied medicine at university, becoming a talented anatomist. He specialized in the human nervous system, and together with a Jewish colleague, he pioneered an early form of medical imaging that permitted researchers to inject dyes into the organs of living animals and study their function under fluorescent light.[32] But the long hours he sunk into his research did not stop him from getting involved in extremist politics. While he was a member of the medical faculty at the University of Heidelberg, he joined the SS, swiftly becoming its campus leader. And on the strength of his research and the strong connections he was beginning to forge to the SS, he rose to prominence in the German medical establishment.

In 1939, shortly before the German invasion of Poland, Hirt joined one of the army's Panzer divisions as a military doctor and spent the next two years tending the wounded in a series of field hospitals. After Germany's annexation of the long-disputed French territory of Alsace-Lorraine in 1940, he received an important new position in Strassburg. The Reich Ministry of Education had transformed the city's four-hundred-year-old university into a new type of educational institute, the Reich University of Strassburg. Staffed with German scientists and scholars, the university was intended to be a showcase of Nazi research and pedagogy.[33] It hired Hirt as the director of its anatomical institute. From the start, he paraded his authority as an SS officer, turning up at classes dressed in an SS uniform complete with a revolver slung in his holster.[34]

In Strassburg, Hirt began searching for war-related research projects. He believed that a dye he used in his medical-imaging research—trypaflavine— might help heal the terrible burns suffered by soldiers caught in a mustard-gas attack.[35] The German army had employed mustard gas in the First World War, and many Germans feared that the Allies would one day turn the tables. Hirt wanted to test the dye treatment, claiming that he had enjoyed some success with it while assisting a pharmacist accidentally ex-posed to mustard gas. But the treatment had no scientific merit. Trypaflavine is in itself a toxic substance—so much so, in fact, that researchers handling the chemical today in laboratories are warned to wear "a long-sleeved labo-ratory coat or gown, rubber gloves, safety goggles and a face mask as a min-imum standard" of safety.[36]

Nevertheless, Hirt proceeded.[37] He exposed laboratory rats, as well as dogs and pigs, to mustard gas, then attempted to treat them. The experi-ments ended badly. Hirt was so careless with the poison gas that he devel-oped serious lung lesions himself, landing in a hospital in Strassburg. Despite this failure, however, he wanted to move on to human trials. He recognized that volunteers for such research would be scarce, so he began casting around for other options. The growing SS network of concentra-tion camps seemed to him an obvious source of expendable human beings. So while dining with Sievers in Strassburg in November 1941, he broached the matter.[38]

Sievers relayed the substance of this conversation to Himmler on his return to Berlin. The SS leader was greatly interested in the idea of finding an antidote for mustard-gas burns, for Hitler himself had fallen victim to a gas attack during the First World War, becoming temporarily blinded.[39] So in late December 1941, Himmler agreed to furnish Hirt with "prisoners and professional criminals, who would not be given their freedom anyhow, as well as people who are scheduled to be executed."[40] And it may have been at this time that Sievers brought up the problem of the Jewish skull collection. Certainly Hirt readily agreed to assist. In all likelihood, the physician saw the endeavor as a way of building a new anatomical collection for his institute at Strassburg. Perhaps he even harbored thoughts of getting into the skull mail-order business himself. The only question that remained was where to obtain the necessary variety of Jewish skulls from the Soviet Union.

None of the three collaborators—Sievers, Beger, or Hirt—knew of any such collection in the Reich. But one of them secretly came up with a grisly alternative.[41] Under a directive known as the Commissar Order, the German military was expected to execute without trial any Soviet "commissars" that it captured.[42] As was so often the case in the Third Reich, the language of the order was euphemistic. By "commissars," the army actually meant "Jews." Nazi propagandists had skillfully portrayed Soviet political officers and officials as Jews for years, and so deeply engrained was this notion in the minds of many SS and Wehrmacht officers that they simply accepted it as fact.[43] While some army officers refused to carry out the infamous order, others began executing Jewish civilians as they advanced across the Soviet Union.

So in February 1942, Hirt or Beger—or possibly both men together—wrote a proposal for a new research project. Hirt then seems to have forwarded it to Sievers:

> Subject: Securing skulls of Jewish-Bolshevik Commissars for the purpose of scientific research at the Strassburg Reich University.
>
> There exist extensive collections of skulls of almost all races and peoples (Völkern). Of the Jewish race, however, only so very few specimens of skulls stand at the disposal of science that a study of them does not permit precise conclusions. The war in the East now presents us with the opportunity to remedy this shortage. By procuring the skulls of the Jewish Bolshevik Commissars, who personify a repulsive, yet characteristic subhumanity, we have the opportunity of obtaining tangible, scientific evidence.
>
> The actual obtaining and collecting of these skulls without difficulty could be best accomplished by a directive issued to the Wehrmacht in the future to immediately turn over alive all Jewish

SEARCHING FOR THE STAR OF DAVID

Bolshevik Commissars to the field M.P. [Feldpolizei]. The field M.P. [Feldpolizei] in turn is to be issued special directives to continually inform a certain office of the number and place of detention of these captured Jews and to guard them well until the arrival of a special deputy. This special deputy, commissioned with the collection of the material (a junior physician assigned to the Wehrmacht or even the Field M.P., or a medical student equipped with a car and driver), is to take a prescribed series of photographs and anthropological measurements, and is to ascertain, in so far as is possible, the origin, date of birth, and other personal data of the prisoner. Following the subsequently induced death of the Jew, whose head must not be damaged, he will separate the head from the torso and will forward it to its point of destination in a preservative fluid within a well-sealed tin container especially made for this purpose. On this basis of the photos, measurements and other data on the head and finally, the skull itself, the comparative anatomical research, research on race membership [*Rassenzugehörigkeit*], the pathological features of the skull form, the form and size of the brain and many other things can begin.

In accordance with its scope and tasks the new Strassburg Reich University would be the most appropriate place for the collection of and research upon these skulls thus acquired.[44]

Himmler read this proposal with immense interest. A month earlier, in a large villa overlooking Grosser Wannsee, the SS had sought and obtained official government approval for a policy that it had already secretly adopted and embarked upon.[45] This was the Final Solution—the seizing and murdering of all Jews in the territories under German control. During the meeting at Wannsee, officials had debated at some length the problem of the *Mischlinge*, or "part-Jews," and the measures to be taken against them. Himmler was keen to take action. He wanted RuSHA to racially evaluate all children of mixed marriages and their progeny for three or four generations, just as agriculturalists did when attempting to breed superior varieties of plants and animals.[46] Descendants who exhibited Jewish traits could then be at least sterilized, if not murdered. For this, the SS needed a much clearer picture of the Jewish race.

Beyond all these official reasons, however, Himmler was intrigued by the idea of a Jewish skull collection. He believed that a man's character and criminal nature could be clearly read in the assemblage of his bones and he sometimes gave little lectures on this theme to his SS entourage. While touring Poland aboard his private train in September 1939, for example, he had instructed his men to bring forth some of the "criminal specimens" from among the local Jews.[47] With a stick in his hand, he would then point out certain facial features and skeletal characteristics of old men who were visibly quaking with fear. "These people," he concluded, "were vermin."[48]

So at the end of February 1942, he instructed his personal administrative officer, Dr. Rudolf Brandt, to inform Hirt that he would "place at his disposal everything he needs."[49]

TWO MONTHS LATER, over the Easter holidays, Sievers attended an evening meeting with Himmler and a certain SS-Sturmbannführer Petrau.[50] Over a leisurely dinner, Sievers discussed with Himmler the possibility of archaeological research in Bulgaria, and the current state of Rascher's high-altitude experiments at Dachau. The young SS physician had embarked with great relish on the tests—killing several of his subjects—and this news and the recent proposal from Hirt had given Himmler a new idea.[51] The Nazi leadership had long regarded modern medicine as a degenerate science due to the great influence of Jewish physicians.[52] In its place, many prominent Nazis had embraced all manner of alternative medical treatments and drugs. During the war, Hitler had criticized German physicians for not doing enough to save the lives of soldiers at the front.[53] So Himmler decided to take matters into his own hands. He instructed Sievers to found a new research organization within the Ahnenerbe to oversee medical experiments performed on concentration camp prisoners.[54]

Three months later, the Institute for Military Scientific Research was born. Walther Wüst, the cautious superintendent of the Ahnenerbe, assumed direct responsibility for the new institute.[55] But Wüst seems to have distanced himself from its day-to-day activities. Instead, Himmler appointed Sievers to the position of director and approved the creation of two divisions—one headed by Rascher, the other by Hirt. Financing was to come directly from the Waffen-SS.

As the new division head, Hirt began rethinking plans for the Jewish skull collection. Transporting human heads all the way from the Soviet Union would be extremely troublesome. A more practical solution was to find subjects in the extensive network of German concentration camps. In this way, Beger could personally select the victims and perform a first set of racial measurements while the individuals were still alive. When this was done, camp guards could murder the subjects in a tidy manner, making sure that they did not damage any bones. Hirt then could dispatch an assistant to pick up the remains and transport them to his lab in Strassburg. There his staff could proceed to deflesh the bodies and produce skeletons suitable for a reference collection.

Although Beger insisted vociferously after the war that he knew nothing of this plan until it was too late to save the victims, he may possibly have been aware of it from the start.[56] There was, after all, no obvious reason to keep secrets from him. The honorary RuSHA staff member seems to have long agreed with SS plans to eliminate the Jews. Indeed, he later advocated conducting research on the characteristics of the Jewish spirit so that even this ephemeral influence could be rubbed out of German life. As he explained in a letter to Himmler's personal assistant, "I take the view that the complete extermination of the Jews in Europe, and beyond that, in the whole world if possible, will not mean that the spiritual elements of Jewry, which we encounter at every turn, are fully eradicated. The important role of research on racial souls stems from this fact."[57]

Before Beger and Hirt could get down to work, Sievers had to tackle a number of complex logistical problems. He quickly discovered that Auschwitz would be the best place to send Beger, for the sprawling prison served as a major death camp for prisoners from the East. But Auschwitz was located in southern Poland—a long, unrefrigerated train trip away from Strassburg—and Hirt refused to work with half-decayed bodies. So Sievers set about obtaining permission to ship the selected prisoners alive from Auschwitz to a camp much nearer to Strassburg—Natzweiler. This new plan solved many logistical problems, but it created a new one. Natzweiler was one of the smallest camps in the German system, with a large cohort of prisoners who were members of the French resistance.[58] It did not possess a gas chamber. So the SS leadership seems to have ordered the construction of one to accommodate Hirt's work.[59]

Hirt needed additional equipment and facilities in Strassburg, too. He

requested a special elevator for corpses to be built at the anatomical institute.[60] He also ordered custom-made equipment for rendering in a sanitary way entire human corpses, with their hair, nails, tendons, cartilage,
muscles, and other soft tissues, down to pristine skeletons.[61] Museum
preparators had devised several methods over the years for defleshing animal cadavers for scientific and educational collections.[62] To strip soft tissue
from large mammals, they sometimes buried cadavers in the ground to allow soil bacteria and chemicals to eat away the flesh, but the process could
easily take more than a year and the resulting skeleton often smelled horri-
bly, making it totally unsuitable for display indoors. It was also possible to
remove flesh from cadavers by placing them into containers inhabited by
colonies of dermestid beetles. But this procedure required large crawling
masses of beetles, something rather at odds with the image of an antisep-
tic medical institute.

So Hirt chose a third and more sanitary method—that of maceration.[63] The bodies would first be immersed, one by one, in a large tank
filled with a substance such as lime chloride.[64] This would dissolve all the
soft parts. Then they would be placed in a second solution such as gasoline
to remove all traces of fat. A corpse treated in this way would be flensed
within a matter of weeks.[65] The result would be a bleached skeleton that
still contained much of its cartilage, but emitted no foul odor—an important consideration in an anatomical laboratory.

Chemical maceration required a large steel tank and a heat source, as
well as special equipment for bone-degreasing, all of which had to be
custom-manufactured. This would be no easy matter in wartime Germany,
where most factories were dedicated to churning out munitions of one sort
or another. Hirt had already ordered the equipment from the German manufacturer Bergmann und Altmann, but the company had made little
progress in delivering.[66] To speed matters up, Sievers handed the file over to
his personal assistant Wolf-Dietrich Wolff, who began doggedly following
up on the order. So deeply did Wolff fall into the spirit of the project that he
soon began referring to the future victims of the project as "objects."[67]

With Sievers's immense talent for organization, the necessary preparations were rapidly forging ahead by the end of September 1942. But on October 3, Beger learned of something worrying. A typhus epidemic had
broken out at Auschwitz. He immediately discussed the problem with Hirt
to see what should be done, then wrote to Sievers. "It is, of course, very

important to establish if this is true before the ordered racial examinations and recordings are done," he declared, "because otherwise there is a risk that typhus will be brought back into the Reich. Prof. Hirt specifically pointed this out to me."[68]

Beger's information proved correct. In an effort to stamp out the plague, the Auschwitz authorities began marching off all infected prisoners to the gas chambers.[69] In addition, they prohibited prisoners from traveling outside the camp boundaries, even when requisitioned as slave labor for various work projects.[70] It would be impossible under this ban for Hirt and Beger to convey their subjects from Auschwitz to Natzweiler. So the two men were forced to put their plans for the Jewish Skeleton Collection on hold for the next eight months.

THE DELAY MUST have been a particular frustration for Beger. In August 1942, as the Wehrmacht began advancing toward the mountains of the Caucasus, Himmler had issued orders for two detailed scientific studies of the Middle East.[71] He had directed Walther Wüst to prepare a team of eight researchers, including a racial anthropologist, to conduct archaeological, racial, and other studies in Iran.[72] And he had commanded biologist Ernst Schäfer, the leader of the Tibet expedition, to head a military and scientific mission to the Caucasus known as Special Command K.[73] Schäfer in turn had promptly named his old colleague, Bruno Beger, as the deputy leader of the mission and placed him in charge of the "racial exploration of the Caucasian tribes."[74]

Schäfer had spent most of the war at Himmler's beck and call. Apart from a brief and nearly fatal stint as a soldier on the Finnish front in 1941, he had spent the war in offices and hotel rooms—advising the SS leader on the design of winter uniforms for German troops serving in Poland, testing new varieties of grain for the Eastern settlements, and lecturing in occupied Europe as a kind of official poster boy for science in the Third Reich.[75] He had felt trapped behind a desk, but in early 1942 he saw an opportunity to escape.[76] As the German army advanced toward the oil fields of Maikop, he proposed leading a scientific survey to the Caucasus so that the region's natural resources could be suitably exploited after the war. In the late summer of 1942, Himmler ordered him to organize a military and scientific mission to the area.

Schäfer grandly sketched out his requirements. He wanted a team consisting of dozens of scientists—from geophysicists, geologists, geographers, and paleontologists to plant geneticists, livestock experts, zoologists, entomologists, parasitologists, and herpetologists. He also requested page after page of equipment—tropical uniforms, mountain troops uniforms, leather jackets, Lederhosen, klepper jackets, fur vests, pullovers, bathing suits, hiking shoes, Africa boots, helmets, seventy two-man tents, two fifty-man tents, skis, Hindenburg lamps, snow glasses, sunglasses, ice axes, five gramophones with records, ten travel typewriters, and mosquito nets.[77] On and on it went. He had, it seems, quite forgotten there was a war going on and that the German military was stretched to the limit, battling Stalin's massive forces in the Soviet Union; occupying much of Europe and a vast swath of the Soviet republics; defending Europe from an Allied invasion; combating local partisans and resistance groups; waging war against British forces in North Africa; and searching out and destroying enemy ships in the Atlantic and in European waters.

Himmler's personal administrative officer, Rudolf Brandt, soon brought Schäfer crashing back down to earth. "I have already let him know, through SS-Obersturmführer Meine," explained Brandt in a letter to Sievers, "that at the moment his plan for an expedition in the Caucasus area, as he imagines it, is out of the question. Dr. Schäfer is to be ready only for a military assignment in the realm of this current operation. Here and there he will surely be able to conduct some scientific work in his area of interest while completing the assignment, but in no case is this his main goal."[78] In aid of this plan, the SS leadership agreed to supply Schäfer with a small group of researchers and ninety-seven SS men armed with pistols, machine guns, and grenades.[79]

Schäfer told American interrogators in 1946 that the purpose of this special command was to "win over to the German cause the tribes in the Caucasus Mountains."[80] But little evidence of such an assignment exists in the surviving documents. What can be said without a doubt, however, is that racial research lay close to the heart of the operation.[81] Beger and a handpicked team of Rassenkunde experts planned to conduct extensive studies of the native tribes of the Caucasus in order "to facilitate a racial diagnosis of the population."[82] They seemed particularly anxious to "diagnose" the Mountain Jews, one of the tribal groups likely to confuse SS killing squads.[83] During questioning in 1970, Schäfer suggested that the fate of this group—their liquidation or survival—would depend on Beger's team and on their conclusions. "At the time," Schäfer confessed, "it was known that the Jewish people were to be annihilated."[84]

The Mountain Jews, or Dag Chufut, lived mainly in the northern and eastern Caucasus. They had by all accounts resided in the rugged region for twenty-five hundred years and were descendants of several waves of Jewish immigrants—including captives whom Nebuchadrezzar bestowed upon the tribal leaders of the Caucasus, and refugees who fled from the destruction of the First and Second Temples in Jerusalem.[85] They spoke a Persian dialect, which they had once written in Hebrew characters, and they kept their ancient Judaic beliefs alive. While the rulers of many other lands forbade Jews from owning land, the potentates of the Caucasus were more tolerant. They permitted the Mountain Jews to farm. In the nineteenth century, before Soviet collectivization, these Jewish families reaped wheat, corn, and rice crops from mountain fields others thought unarable, and tended vineyards that were famous. They observed nature closely, rode well, and cared for their guns lovingly. By 1942, they numbered about thirty thousand people.[86]

But the lines between the Mountain Jews and many of their neighbors were very fuzzy. The young men often purchased wives in marriage from surrounding Muslim families and insisted that their new spouses dress in modest Islamic fashion. Young and old wore talismans and amulets, believed in demons and black magic, and celebrated agricultural festivals—all customs borrowed from others. The local Muslims returned the favor. Neighboring tribes proudly boasted of Jewish ancestors and took pains to preserve ancient Hebrew bibles in their families. Several tribes in the region—the Tats, Kumyks, Avars, and Tabassarans, to name a few—clearly descended from mixed Jewish and Muslim ancestry.[87]

BEGER'S ASSIGNMENT WAS to neatly pigeonhole all these tribes in a way the SS leadership could understand.[88] To do this, he began assembling a team of racial specialists.[89] He immediately chose two veteran colleagues from RuSHA—Dr. Hans Fleischhacker and Dr. Heinrich Rübel. Fleischhacker had taken a keen interest in Jewish peoples, and was writing a thesis on Jewish skin color.[90] His comrade, Rübel, had studied *Rassenkunde* at the University of Cologne.[91] After the invasion of Poland, the SS sent both men to Litzmannstadt.[92] As *Eignungsprüfer,* or "aptitude testers," there, they performed racial measurements on ethnic German residents, assessing whether they were "racially valuable"—and therefore worthy to be sent as colonists to the German territories in the East—or whether they should be relegated to starvation, slavery, and extermination.[93] So adept did both men prove at this work that their superiors recruited them to train and develop educational programs for other aptitude testers.[94]

In addition to the RuSHA specialists, Beger managed to obtain two other qualified scientists for his team. Dr. Rudolf Trojan had studied under the prominent German anthropologist Dr. Theodor Mollison, whose most famous pupil was Dr. Josef Mengele, the physician who conducted the infamous twins research at Auschwitz and who came to be known as the "Angel of Death."[95] Under Mollison's guidance, Trojan had taken up racial blood studies and conducted racial research on ancient skeletons. Rounding out Beger's core team was Dr. Hans Endres.[96] A scholar with many interests, Endres had studied philosophy, psychology, education, and psychiatry in addition to anthropology. Beger had recruited him to study the racial psychology of the Caucasus tribes.[97]

Throughout the fall of 1942, as the team waited for its orders to depart, Beger drew up a detailed research plan. He proposed taking the team on an inspection trip through the Caucasus soon after their arrival. As they journeyed from one village to the next, the racial specialists could take the lay of the land and calculate the minimum number of men and women they would need to measure in each ethnic group.[98] Then the team would get down to work in a mobile examination facility—a large field tent that could be divided into four separate sections, including a room for disrobing.[99]

Almost certainly, the team planned on using trickery and deception to obtain the cooperation of their subjects. RuSHA racial examiners were accustomed to donning white laboratory coats and masquerading as physicians conducting medical examinations.[100] They had discovered over the years that subjects were far more willing to undress—and less inclined to make a scene—when they thought they were receiving medical attention. Moreover, Beger had already practiced a similar form of duplicity in Tibet.[101]

Beger's racial specialists proposed conducting a wide battery of tests and measurements. As a matter of form, they intended on describing the exact hue of their subjects' hair, skin, and eyes; performing an assortment of racial measurements; and snipping hair samples for later study. They

[handwritten annotation: ✓ MEL BROOKS & MADELINE KAHN: AIRPORT SCENE "HIGH ANXIETY"]

also planned to photograph and film particularly interesting individuals, for German racial experts claimed that Jews moved differently from others, ✗ dragging their feet along the ground.[102] The team's sculptor would "produce casts of the head or the whole body of representative types or entire families of each ethnic group and each race."[103] In addition, Endres would conduct a series of "racial intelligence examinations," employing games, colored glass beads, crayons, watercolors, and a variety of testing machines.[104]

Beger also planned to put the Special Command K doctor to work, conducting studies on racial anatomy, racial physiology, and racial hygiene.[105] To carry out his duties, the physician requested three tattooing needles on his equipment list.[106] His reason for this request is unclear, but shortly after the invasion of the Soviet Union, one RuSHA scientist, SS-Obersturmbannführer Walter Scholtz, proposed sending racial examiners into Russian prisoner-of-war camps in order to test the inmates, divide them into racial categories, and tattoo each prisoner's ear with a letter—rather as ranchers do with cattle—as a permanent visible record of the classification.[107] An "E," for example, would indicate a prisoner who was supposedly of Nordic blood and who would be sent to one of the new German colonies. An "R" would mark someone who purportedly exhibited a balanced mixture of European races and who was therefore suitable to join the workforce in European Russia. An "A" would identify a prisoner who was Asian or a mixture of Asian and Middle Eastern ancestry. Scholtz proposed that these individuals be "extinguished."[108]

Beger labored over Special Command K for months in Munich, consulting with his fellow racial specialists and preparing as best he could for all the unknowns that the Caucasus would present. But the order to depart did not arrive. Indeed, Oktoberfest came and went and the great chestnut trees in the Munich streets lost their leaves. Families opened their Advent calendars and lit candles on their Christmas trees, but Beger and Schäfer and their teams were still waiting. The order from Himmler that they were all expecting, indeed anxiously anticipating, did not arrive. Nor did it appear as the new year approached: 1943.

The reasons were becoming increasingly obvious. Something had gone badly wrong on the eastern front. Hitler's plan to drive south and east through the Caucasus to capture the rich oil fields of Baku and safeguard the pipeline to Batumi had stalled, due to a stunning military catastrophe at Stalingrad. Hitler had vowed to destroy the city, ordering his troops to slaughter every male resident and deport every female.[109] But his staggering indecisiveness and his refusal to recognize the realities of the campaign had left some 200,000 German troops badly undersupplied. On November 22, 1942, Soviet forces had completely surrounded Germany's Sixth Army, cutting off its remaining supply lines. In the dreaded cold of the Russian winter, German soldiers starved and perished in great numbers. "We're completely alone," wrote one of the desperate troops, "without help from the outside. Hitler has left us in the lurch."[110] As the temperatures plunged lower and lower and the winds howled ever louder, Hitler sat grim-faced over the dinner table at Wolfschanze. The little man who once sat discreetly in the corner, jotting notes, had vanished. Hitler wanted no record preserved of the gloom settling over many of his military guests.

On February 2, 1943, Germany's Sixth Army surrendered. Nearly 100,000 German and Romanian soldiers lay dead in the gray streets of Stalingrad and the white fields of the countryside.[111] Another 113,000 were captured by the Soviets. Three days later, Himmler wrote to Schäfer, postponing Special Command K. It would be totally impossible, he explained, to start the mission within the next few months due to the military situation.[112]

19. THE SKELETON COLLECTION

On a bright June day in 1943, Sophie Boroschek stood in front of an Auschwitz barrack, waiting for someone important to appear. Like the other prisoners assembled there, she did not know who the person was or why he was important. Boroschek was thirty-three years old.[1] In a previous life, she had lived in the villa of a prominent cigarette manufacturer and worked as a nurse in the Hospital of the Jewish Community in Berlin, a little-known haven for Jews.[2] But in mid-May 1943, the authorities had forced her parents, Abraham and Lieschen, onto a transport bound for Auschwitz. Five days later, Boroschek boarded a freight car headed for the infamous camp. On her arrival at Auschwitz, a doctor had looked her up and down with a bored glance, then selected her for work in the camp. Since then, Boroschek had learned a great deal about survival.

That morning, she and some one hundred and fifty other prisoners had been excused from their work details and daily routines and dispatched to an area outside Block 28, a red-brick barrack officially described as an infirmary.[3] It was an ominous corner of the camp. As many of the prisoners were well aware, Block 28 was an infirmary in name only. Far from caring for the seriously ill, its staff specialized in conducting gruesome medical experiments on the healthy.[4]

Boroschek and her fellow prisoners waited uneasily. Finally, a tall, blond, athletic-looking SS-Hauptsturmführer arrived and instructed the prisoners to undress.[5] The officer was Bruno Beger. He unpacked several shiny steel instruments—various kinds of calipers and compasses.[6] Some of those standing beside Boroschek blanched at the sight of the metal instruments.[7] They did not know what would happen next. Beger beckoned first one prisoner, then another, forward. He stared at them intently for a minute or two, studying their faces, and ran his eyes down the length of their bodies. Silently, he arrived at some decision. Some of the prisoners he immediately dismissed, as if to say they were not worthy of his attention. Others, however, he began to measure, sliding his calipers across their heads. Then he called out the numbers tattooed upon their arms to a prisoner assistant, who carefully recorded them in pencil on a form.[8] Those prisoners whose survival instincts had been stropped to razor sharpness by long months of imprisonment at Auschwitz suspected no good could come of it.

Finally, Boroschek heard her own name called. She walked over to the blond-haired man and stood in front of him. Beger stared hard at her, sizing her up. Then he reached for his calipers. Satisfied, he called out Boroschek's number.

BEGER HAD ARRIVED in the town of Oswiecim on the morning of June 7.[9] His civilian assistant, Wilhelm Gabel, had already spent a night there.[10] Sturdily built at thirty-nine, Gabel worked for the Ahnenerbe as a sculptor, creating museum dioramas from the casts Beger had taken of Tibetans. He had agreed to assist Beger making casts of the Jewish prisoners at Auschwitz. Beger was also expecting another colleague, a hard-looking man of thirty with a long face, metal-rimmed spectacles, and a pair of heavy eyebrows that merged over the bridge of his nose.[11] Dr. Hans Fleischhacker, an SS-Obersturmführer, was one of the officers Beger had chosen for the Caucasus research. His research specialty was Jewish skin color.

Before heading off to the camp, Beger looked for a hotel room in Oswiecim. SS officers of the day were particularly fond of the Haus der Waffen-SS, located near the railway station.[12] The facility had a lovely garden for sunbathing, and its comfortable restaurant served excellent dinners of roast pork and chicken, fresh vegetables, and, according to one SS officer, "a magnificent vanilla ice cream"—luxuries unheard of in most other parts of the Reich.[13] Beger checked in there.[14] After stowing his bag, he set off to the camp to introduce himself to the commandant and his staff.[15]

SS officers generally regarded Auschwitz as a choice tour of duty. The camp was a good deal safer than a posting to the eastern front, which many saw as a death sentence, and Auschwitz offered daily opportunities to steal jewelry, watches, and other treasures from the condemned. Moreover, the rank-and-file guards shared this sense of enthusiasm for the posting. Indeed, they often competed to take part in the "actions," the mass murders that featured so prominently in Auschwitz life.[16] For this, they received a bonus—one-fifth of a liter of vodka, five cigarettes, and one hundred grams of sausage and bread.[17]

Beger presented his orders, and soon after, camp officials sent him off to get a vaccination against typhus, for the threat of epidemic was still present at Auschwitz.[18] Then they escorted him inside the electrical fences to the men's camp. Some of the buildings standing there were part of an old factory formerly owned by the Polish Tobacco Monopoly, while others had once served as barracks for the Polish army.[19] And even in the summer of 1943, Auschwitz retained a few ironic vestiges of a happier past. Along the grounds, tidy red-brick barracks stood neatly in long rows intersected by paved streets that went nowhere, each with a pleasant sounding name, such as Cherry Street.[20] But its bland appearance was an illusion. Each morning, prisoners were rousted from their beds at 4:30 and forced to don uniforms caked with dirt and sweat.[21] They were ordered to stand for hours outdoors in meaningless roll calls, no matter what the weather, and then put to work for twelve hours each day. Everywhere they went, they carried a red bowl and tin spoon in the vain hope that they would receive something more substantial than the thin, watery soup dispensed from the camp kitchen. They lived in a state of constant fear. At any moment, an SS guard could haul them from a line and beat them brutally or drag them off to the gas chambers, the gallows, or one of the camp's other execution grounds. Throughout the summer of 1943, flames literally leapt from the tall chimneys of Auschwitz's crematoria. The stench of burning hair and scorched human flesh hung over the camp.[22]

Amid all this horror, Beger quickly settled down to work, selecting prisoners for the skeleton collection from among those assembled in front of Block 28. He looked for relatively healthy, robust young people who had not yet lost too much soft tissue from starvation.[23] This meant selecting people in the prime of life who had arrived at Auschwitz within the last few months. He also wanted to find "as many varieties of Jewishness as possible."[24] So he chose Jewish prisoners—men, women, and children alike—from across Europe: Greece, Germany, Poland, Belgium, the Netherlands, France, and Norway. The majority came from the northern Greek city of Salonika, where the SS had only begun deporting Jews on March 15, 1943.

In addition to these prisoners, Beger also selected two Polish Christians and four Asians—two Uzbeks, one person of mixed Uzbek-Tadzhik ancestry, and one Chuvash. The Asians, he observed later in a letter to Schäfer, were selected "just on the side," for the benefit of the Ahnenerbe's new department of Inner Asian research and expeditions, headed by Schäfer.[25] Beger was particularly pleased to have found them at Auschwitz. "One Uzbek, a big healthy guy, could have been Tibetan," he exclaimed in a letter to Schäfer. "His way of speaking, his movements and his manner were simply delightful, Inner Asian."[26] In all, Beger picked 115 individuals from the prisoners assembled in front of Block 28. At least five of them were teenagers.[27] The selection took just three-quarters of an hour.[28]

According to witnesses, camp authorities housed the selected women in Block 10, a two-story barrack whose windows were kept permanently shuttered or boarded to ensure that no one could see inside.[29] The building hid laboratories, X-ray facilities, and rows of crowded wooden bunks for the subjects of Auschwitz's grisly medical experiments. It also provided bunks for the female prisoners forced to work in a camp brothel. As one prisoner who spent a year working in Block 10 recalled later, it was a "horror place" that rivaled Dante's *Inferno*.[30] Many of the selected male prisoners were similarly housed. They were sent to Block 21 and Block 28.[31] Both served as camp infirmaries, and the latter confined men forced to participate in experiments that damaged their livers or in tests of toxic chemicals that provided data on possible techniques malingerers might use to elude military service.[32]

Over the next few days, Beger proceeded with his studies, performing detailed racial measurements in a small room in one of these blocks.[33] As the prisoners filed in, the sculptor Gabel stared at their faces. "When I found a Jew who was especially interesting or remarkable," he later recalled, "Dr. Beger agreed that I should also make a cast of them."[34] In all likelihood, Gabel chose individuals who possessed one of the supposed Jewish traits—short head, fleshy lips, large ears.[35] As Gabel and Beger must have realized, plaster likenesses of these individuals would be highly marketable as teaching aids in Nazi universities and as exhibits on *Rassenkunde*.[36] Gabel took twenty casts at Auschwitz.[37]

Focused as Beger was on completing his work, he could not help but notice the cruel, inhuman conditions in the camp and the terrible toll they were taking on the prisoners. At one point during his stay, he happened upon SS guards dragging out a large pile of corpses from a barracks.[38] The scene disturbed Beger—not, it seems, because so many Jews were dying daily at Auschwitz, but because the SS officers in the camp, the future lords of Europe, were dirtying their hands with this foul business. "I spoke with the adjutant about this," Beger observed in a later court statement, "and mentioned I was concerned that it was specifically the SS who were involved with this. He told me that they would have had so many [prisoner] transports that they would not have been able to cope with them all."[39]

After finishing with each of his subjects, Beger passed the individual on to Fleischhacker, who had finally arrived in Auschwitz and who performed the measurements again to ensure their accuracy. Beger wanted no errors. The two men seem to have worked together amicably, but on June 15, Beger departed abruptly, placing the remaining subjects in Fleischhacker's care and returning to Munich.[40] He explained after the war that he was horrified by the barbaric conditions at Auschwitz and had left in disgust at the earliest opportunity. But correspondence from the time paints a different picture of his motives. Shortly after his hasty departure, Beger informed Sievers that he had finished up early "because of the existing risk of an epidemic."[41] Moreover, his subsequent correspondence shows little trace of compassion for Auschwitz's inmates. Indeed, when he received news a few weeks later about the death of several of the selected prisoners, he offered no comment about their plight. "Herr Gabel," he wrote to Sievers, "is once again in Munich almost fourteen days after the examinations in Auschwitz. He cast the heads of twenty prisoners I had chosen and examined. He told me that the number of examined people has once again declined because of deaths. The head casts that he completed turned out excellently."[42]

Beger's worries about a possible epidemic at Auschwitz were well founded. By early July, typhus was spreading rapidly through the camp, greatly worrying authorities. To combat the outbreak, SS physicians began killing prisoners who turned up in the infirmary admissions area with symptoms. They injected phenol, a common disinfectant also known as carbolic acid, directly into the hearts of the ailing inmates.[43] To ensure that the prisoners Beger had selected would pose no threat to the outside world when they were shipped to Natzweiler, camp authorities placed them in quarantine for two to three weeks and ordered blood tests to determine whether any were infected.[44] In addition, Sievers wrote to request that the individuals be issued clean prisoner clothing.[45]

Only when the quarantine had ended in late July did guards at Auschwitz begin loading Beger's subjects onto a train heading west toward Natzweiler.[46] In Berlin, Sievers's assistant dispatched a deliberately cryptic telegram to Beger, notifying him that the wheels were once again in motion. "Transport from Auschwitz 30.7. Get into contact with Hirt concerning the beginning of the work. Arrival of transport Natzweiler presumably 2.8."[47]

NATZWEILER CONCENTRATION camp was a place of dark, ominous reputation. It sat on a high, forested slope of the Vosges Mountains, thirty-one miles southwest of Strassburg, the capital of Alsace.[48] Before the war, skiers and hikers had flocked to the region, reveling in its solitude and pristine beauty. But in 1940, after France capitulated and Germany annexed the old border province of Alsace, the SS had arrived in the Vosges Mountains on a very different mission: to make money.[49] The local cliffs gleamed with a rare red granite that greatly appealed to Nazi architects. Himmler immediately recognized the potential of the stone. He had founded an SS enterprise, the German Earth and Stone Works Ltd., which used slave laborers from concentration camps to perform the backbreaking work of quarrying rock. So Himmler ordered the construction of a concentration camp in the Vosges Mountains to house the slave laborers needed to mine the rock.

Natzweiler soon became a grim hell for prisoners. Many were forced to haul large slabs of granite up the slopes until they died from exhaustion and starvation. Others were subjected to harrowing and often lethal medical experiments. Himmler had constructed and outfitted a special laboratory at Natzweiler for August Hirt, and the anatomist had begun his monstrous mustard-gas experiments.[50] He and his assistants administered drops of a liquid form of mustard gas to the arms of prisoners, or forced his victims to inhale or gulp down a venomous brew. Fifty men—one of every three subjects in his experiments—perished in agony, suffering severe burns on their skin or fatal internal injuries. Those who survived often went blind or suffered other debilitating conditions, and in the days before Natzweiler acquired its own gas chamber, officials disposed of them by shipping them to other camps, earmarked for death.

It was an appalling situation. But Himmler and Sievers greatly admired the doctor responsible for all this human suffering. Indeed, they came to see Hirt as a martyr for science, placing his own health in peril for sake of the dangerous experiments, and in the spring of 1943, when the anatomist developed intestinal problems, they arranged for him to holiday at a luxury resort at St. Lambrecht in the Austrian Alps.[51] "Hirt," observed Sievers in a letter to Himmler's personal administrative officer, "has not spared his own health and can even be considered a victim of his own science because his research work is in such a new field that dangerous risks cannot be avoided."[52]

By the beginning of August 1943, however, Hirt was sufficiently recovered to carry out the orders for the Jewish Skeleton Collection. Nearly everything was ready. The forced laborers at Natzweiler had completed construction of the gas chamber. Sievers's assistant, Wolf-Dietrich Wolff, had obtained a bottle of a hydrogen-cyanide salts for use in murdering the selected Jewish prisoners in the chamber. This he had passed on to Hirt.[53] Wolff also provided the physician with vouchers for fifty liters of gasoline so that the camp could deliver the corpses to Strassburg. All that was lacking, it seemed, was the maceration machine. Through much dexterous paper shuffling, Wolff had obtained fifteen kilograms of steel—an immensely precious commodity at the height of the war—for the manufacture of the machine. He also managed to register the equipment "as an urgent army commission."[54] The machine was now months overdue, but Hirt was hopeful that it would soon arrive.[55]

On Monday, August 2, a train carrying the prisoners from Auschwitz rolled into Natzweiler. A guard unlocked the door to a freight car, and the passengers climbed down stiff-legged and cramped from four days of travel. Hirt was anxious to get down to work. The original plan had called for Fleischhacker and a second Caucasus team member, Dr. Heinrich Rübel, to travel to the camp to take blood samples as well as skull X-rays for additional osteological measurements.[56] But neither man could obtain a release from his other duties. So Beger, who was visiting his wife and children in a hunting lodge in Rüthnick in northeastern Germany, was forced to take their place. He departed for Strassburg on the morning of Saturday, August 7, and presented himself at the camp a day or two later.[57]

In a rough wooden barrack, Beger proceeded with his final studies, taking two X-rays of each of the prisoners' skulls.[58] By then, Hirt had decided to extract additional medical data from the group. As a senior SS physician, he had learned of Himmler's keen interest in finding cheap new techniques of human sterilization that could be applied in future to many Germans of "Jewish mixed blood," as well as to the large numbers of Jewish, Russian, and Polish forced laborers in the East.[59] So medical researchers at Auschwitz and other concentration camps had already begun testing a wide range of potential sterilization procedures on human subjects, from injecting caustic substances into women's cervixes to exposing men's penises and scrota to dangerous levels of X-rays.[60] These tests inflicted immense human suffering. So intense was the radiation, for example, that many of the victims developed severe burns along the groin and buttocks, and some died soon after. Moreover, the researchers made little effort to spare their subjects additional humiliation. To test the effectiveness of the technique, assistants collected sperm from the irradiated men by rubbing their prostate glands with pieces of wood inserted rectally. Later the men were subjected to orchiectomies, in which the researchers removed one or both testes.

Hirt had devised a new variation on this theme. He wanted to inject a foreign chemical—quite possibly the toxic dye trypaflavine—into the men's testicles in hopes that this would lead to infertility.[61] So he had male prisoners brought to him, one by one, and proceeded to give them the injection.[62] His own ghoulish appearance added to the horror of the experiment, and the result was both humiliating and debilitating. As two French pathologists testified after the war, "assuming the most favorable hypothesis, that is to say that the injections were given under anesthesia, the secondary reaction of congestion and edema must have been very painful."[63]

After administering the injections, Hirt wanted to wait at least eight days before testing the effect of the chemical on the men's ability to produce sperm.[64] So he permitted the male prisoners to live awhile longer. But neither he nor Beger seems to have had any further reason to keep the women alive after the racial measurements were completed. So on August 11, the killing began.[65]

THE COMMANDANT OF the Natzweiler camp, Josef Kramer, took personal charge of the slaughter. He was a notoriously brutal man, whom newspaper reporters later dubbed the "Beast of Belsen" after his barbarous behavior at the Bergen-Belsen camp. The son of an accountant, Kramer had joined the Nazi party in 1931 and had spent nine years working his way up through the concentration-camp system, learning to inflict misery on old Jewish men and split the skulls of Jewish women with one expert blow from his truncheon.[66] At the age of thirty-six, he was a hardened killer. His deep-set eyes glowered from a stern face shaded by a heavy five o'clock shadow; his mouth pressed into a tight, hard line of determination.

Kramer waited until around nine in the evening on August 11 to begin carrying out his orders to kill the female prisoners. He and a few of his SS men gathered together some fifteen of the women Beger had selected and forced them into a small van.[67] "I told these women that they had to go to the disinfection chamber and I did not tell them they were going to be asphyxiated," Kramer recalled after his capture at the end of the war.[68] They drove a short distance outside the camp fences and pulled up by the new gas chamber. The SS officers pushed the women roughly inside and ordered them to take off their clothes. By then, at least some of the women knew what was coming next, but Kramer and his men showed no pity. "Helped by a few SS," he later remembered, "I undressed them completely and I pushed them in the gas chamber when they were completely naked."[69] Then he closed and locked the gas-chamber door. The women, he noted, "started to howl."[70]

He retrieved the bottle of hydrogen-cyanide salts that Hirt had given him. He opened it up and poured its contents into a funnel located above and to the right of the chamber's observation window, then closed the funnel. This sent the salts and a supply of water flowing down toward an opening in the chamber. As the salts mixed with the water, they formed a deadly gas that began to seep into the chamber below. The women inside pounded the door frantically, screaming and pleading to be let out. Kramer watched through the glass, taking it all in. "I lit the inside of the chamber with a switch plate near the funnel and I observed through the observation window what was going on inside of the chamber. I have seen that these women continued to breathe about half a minute and then they fell on the floor."[71]

Two days later, Kramer executed the remainder of the women by the same method. He then ordered the bodies from both executions to be loaded into a small van and driven to the anatomical institute in Strassburg. Two years later, he recalled for investigators his own personal reaction to these mass murders. "I have not felt any emotion in doing these acts because I had received the order to execute these eighty inmates (sic) according to the way I have spoken to you."[72]

THE VAN CARRYING the women's bodies arrived at the anatomical institute in Strassburg at seven the next morning. Located near the old southern wall of the city, the institute was dark and quiet. Hirt had already instructed his staff to clean out the laboratory tanks they regularly used for preserving cadavers and to fill them with a solution of 50 percent alcohol. He informed them—erroneously as it turned out—that they would be receiving 120 cadavers.[73] The laboratory assistant, Henri Henrypierre, a forced laborer, did not think there was anything particularly odd about this order. Strassburg's anatomy institute regularly received shipments of dead bodies. Anatomists were in the habit of dissecting cadavers for medical research, preserving diseased or deformed human organs for teaching specimens. And students at the institute regularly dissected the dead as part of their medical training. For these purposes, Hirt had previously purchased through his SS connections the emaciated cadavers of Russian prisoners of war, who had perished of starvation and natural causes. The cost was 10 reichsmarks per cadaver.[74]

The new shipment from Natzweiler was not like the others, however. It consisted entirely of women, one of whom still modestly wore a brassiere.[75] Henrypierre was shocked to see the condition of the women: most were young, healthy individuals under the age of thirty-two, "all of a commanding appearance," as he told investigators after the war.[76] Moreover, some of the bodies were still warm to the touch and their eyes shiny. Henrypierre guessed that they had perished no more than three hours earlier and he was certain they had not met a natural end. Some had clearly been beaten and abused. All had bled from the mouth and nose, which made Henrypierre think that they had been gassed or poisoned.

When he tried to discuss the matter with Hirt, the gaunt anatomist fixed him with a chilling look and warned that if he did not keep his mouth shut he would end up just like the women in the containers. Henrypierre was stunned. "I knew then that these people were killed for that purpose."[77] But the fact that Hirt had arranged the murder of twenty-nine females for the purposes of research did not disturb Hirt's other assistant, Otto Bong, a German preparator that the anatomist had brought from Frankfurt. As the two men cleaned and prepared the women's bodies, Bong told Henrypierre not to fret; the women were, he said, "only Jews."[78] Henrypierre could not stop thinking about their criminal deaths, however. He noticed that each woman had a long number tattooed on her left arm. When no one was watching, he wrote down a list of the tattooed numbers.[79]

Four days later, the driver from Natzweiler appeared again at the institute, bringing the corpses of a large group of men. Soon, the driver returned with a third and last shipment. Henrypierre examined the men's bodies, once again secretly jotting down their tattooed numbers. Like the women, they were generally young and healthy, and they had been murdered in the same manner. But there was one important difference. Before the French preparator was allowed to immerse the bodies in alcohol, Bong insisted on taking tissue samples. With a scalpel in hand, he severed a testicle from each of the male cadavers.[80]

He placed the organs in a container and sent them on directly to Hirt's personal laboratory.

20. REFUGE

THERE WAS NO MISTAKING the tense atmosphere in Himmler's field headquarters in the late summer of 1943. A string of recent military and political disasters had sent the German high command reeling, and it was impossible to keep all the bad news from the German public. Allied troops had crushed the Wehrmacht in North Africa in May and then promptly turned their attention to Italy. Under the cloak of foul weather, the British army had landed tanks and heavy artillery along the southeastern coast of Sicily on July 10, while American forces had succeeded, against stiff resistance, fighting their way onto the island's southwestern beaches. Less than two weeks later, the Allies had captured the Sicilian capital and were plotting their route west and north.

The news had badly rattled the confidence of the Italians and precipitated a coup d'état in Rome. The Italian king had ordered Benito Mussolini to step down as prime minister on July 25, curtly informing him that one of his most prominent critics, Marshal Pietro Badoglio, would be replacing him. Then the king arranged for Mussolini to be put under house arrest on the Mediterranean island of Ponza. The abrupt shift in leadership had sown much uncertainty in Germany on the eve of the Allied invasion of Italy. Moreover, the crisis had raised serious doubts in the minds of ordinary Germans about the stability of the Nazi regime. As one secret Security Service report noted, the idea that a similar coup d'état could take place in Germany "can be heard constantly."[1]

All this was cause enough for concern to Himmler, but a further disaster had followed. In late July, the British Bomber Command had launched a major offensive on Hamburg. Dubbed Operation Gomorrah, the attack was intended to lay waste to Hamburg, residential districts and all, crushing the German will to resist. The British had discovered a technique for confusing German radar—releasing strips of aluminum from their planes that acted as decoys—and their bombers had begun slipping more easily past German defenses. So over a period of seven days beginning July 24, 1943, British planes brought a terrible apocalypse to Hamburg, unlike anything German civilians had previously seen. On July 28, for example, British high-explosive and incendiary bombs ignited a mammoth firestorm in the city. The flames towered more than a mile in the air and spread across eight square miles, whipping up hurricane-force winds that dismantled roofs, uprooted trees, and flung human beings through the air. Operation Gomorrah killed an estimated thirty thousand people. It left another one million homeless, fleeing in horror into the German countryside.[2]

A day after the Hamburg firestorm, on July 29, 1943, Himmler ordered the immediate evacuation of the Ahnenerbe—with its extensive library, archives, museum casts, and trove of documents—from Berlin.[3] Sievers had already picked out a modest refuge in the German countryside for the staff.[4] It was a small group, for many researchers had been called up to military service, while others had opted to go their own way. The Institute for Inner Asian Studies and Expeditions, headed by Ernst Schäfer, for example, had already begun moving from Munich to a castle in the Austrian countryside.[5] And Walter Wüst, the head of the Ahnenerbe, intended to stay in Munich; he had been appointed rector of the university there. This had left Sievers with just thirty or so people to worry about.

After careful deliberation, he dispatched the Ahnenerbe's prized library to safekeeping in a castle in Ulm. Then he moved the Ahnenerbe's much-depleted staff and documents into a derelict seventeenth-century building known as the Steinhaus or Stone House, in the tiny village of Waischenfeld.[6] And it was there in the quiet German countryside that he secretly signed the papers for and orchestrated some of the Ahnenerbe's most terrible war crimes.

WAISCHENFELD IS SITUATED in the northern Bavarian countryside, just a short drive away from the city of Bayreuth, home of the famous Wagner Festival. But in the summer of 1943, the village was a remote backwater that lacked both a rail connection and good roads to the outside world.[7] It possessed few cultural amenities—no theaters, libraries, symphony orchestras, museums—and little in the way of cafés or newsstands, and as such it came as something of a shock to the Ahnenerbe's staff. Their sense of isolation was further compounded, moreover, by the wary attitude of Waischenfeld's seven hundred residents, who had little use for the SS. During the mid-1930s, SS men stationed in the Steinhaus had bullied the villagers, attacked the local Catholic church, and beaten one local man so badly in a street fight that he later died of his injuries. After that, relations between Waischenfeld's inhabitants and the SS were strained.

In their first few weeks in Waischenfeld, Sievers and his staff worked to convert the dusty Steinhaus into a functioning government office. The cramped, primitive conditions were a far cry from the luxurious villa in Dahlem, but Sievers managed to obtain approval to construct a barracks to house staff members. To pass the long evenings, some of the staff formed a choir and staged occasional evening entertainments of music and dance. But many of Sievers's colleagues had trouble settling in. The Steinhaus, they complained, was cold and drafty. It was infested with mice. It lacked dependable telephone connections and was plagued with electrical problems. As Sievers's own personal assistant confided in one letter to a friend, "It is not so easy to 1.) bring everything together under one umbrella and 2.) to up and move overnight to the country an entity that is accustomed to the city. Even the lack of a barber in town is a big stumbling-block, to say nothing of the lack of functioning, running water; a decent oven; etc."[8] It was becoming painfully obvious that few Ahnenerbe employees were really cut out for the country life that Himmler fondly envisioned for the SS.

Sievers, however, seemed to welcome the isolation. He moved his wife and children into rooms in the Steinhaus, and continued to proudly wear his SS uniform about the streets of Waischenfeld. He took pains to keep the villagers in the dark about the nature of the Ahnenerbe's work. He refused to allow local people to enter the kitchen of the Steinhaus to deliver groceries: the food had to be left inside the guardroom. He also forbade his children to bring their friends from the village home after school. There was a reason for this mania for privacy, however. With most of the Ahnenerbe's senior research staff serving in the thick of battle, Sievers had begun devoting more and more of his time in the small bucolic village to arranging and overseeing the Ahnenerbe's top-secret medical experiments.

With Sievers's gift for administration and with Wüst's support behind the scenes, the Ahnenerbe had greatly expanded its medical research program at Dachau and Natzweiler, inflicting terrible human suffering. At Dachau, Dr. Sigmund Rascher had completed his high-altitude tests and had moved on to a new round of experiments. In search of data on how long German aviators could survive in the North Sea after parachuting from a downed aircraft, he inserted electrodes into the rectums of prisoners and then immersed them for as much as three hours in a tank filled with ice and water. As the men shuddered uncontrollably, lost consciousness, and succumbed to hypothermia, he looked on with clinical indifference, charting their rectal temperatures and failing pulses.[9]

Rascher then attempted to rewarm his subjects by a variety of methods—with mixed success. Several of his subjects died. In some cases, he placed the frozen men in hot baths or heated sleeping bags, or tried wrapping them in covers, but it was often a case of too little, much too late. For others, however, he obtained four women prisoners from Ravensbrück, adding intense humiliation to the pain of freezing.[10] He brought a "spacious bed" into his laboratory, and laid the body of a frozen man in between two of the naked women, instructing them to nestle up as closely as possible and engage the man in sexual intercourse if they could.[11] Rascher closely observed the behavior of the three, later calling these crimes of sadistic voyeurism "Experiments for rewarming of intensely chilled human beings by animal warmth."[12] He then rounded out his data by placing naked prisoners outdoors in the dead of winter for up to fourteen hours.[13] He paid no heed to their screams of pain. In all, as many as 108 of Rascher's 360 human subjects died in the various freezing experiments.[14]

Sievers also obtained approval for Rascher to conduct experiments on a possible styptic for staunching the gunshot wounds of soldiers. The substance in question was called Polygal. Made from beets and apple pectin, it was a gelatinous substance generally used in the manufacture of marmalade.[15] A prisoner at Dachau had proposed making Polygal in tablet form, and Rascher was keen to see whether an oral dose would help coagulate the blood flowing from open wounds. To test this premise, he proposed shooting prisoners at close range. In one documented case, he asked an SS guard to climb atop a chair and shoot a "Russian Commissar" standing directly below.[16] The bullet entered the unfortunate man's right shoulder and exited near his spleen. Over the next twenty minutes, the victim "twitched convulsively," then slumped into a chair and died.[17] Rascher then had the body carried to his autopsy table, where he searched for evidence of any ruptured organs that might be "tamponed by hard blood clots."[18]

Rascher's experiments were monstrous, and he was by no means the only Ahnenerbe physician meting out great suffering. August Hirt, the physician who had helped plan the deaths of the Jewish prisoners for the skeleton collection, continued to expose camp inmates to poisonous mustard gas in order to test potential new treatments. And under Hirt's direction, two other German physicians had joined the Ahnenerbe's medical program. Dr. Niels Eugen Haagen, one of Germany's leading experts on viral diseases, performed medical tests on a group of one hundred healthy prisoners at Natzweiler.[19] He inoculated these victims with an experimental typhus vaccine and tested its effectiveness by exposing them to the potentially lethal disease. Meanwhile, a second physician, Dr. Werner Bickenbach, ran tests with phosgene gas, a substance used in chemical warfare. He hoped to develop an antidote.[20]

Bickenbach used the small gas chamber at Natzweiler for the experiments, and one survivor later recalled how they worked.[21] Bickenbach, he said, injected some of the men with a mystery substance and gave others medicine to drink. He blithely assured them that no one would die and that all would receive first-rate medical treatment. Then he led them, four at a time, into the gas chamber. After instructing them where to stand, he walked toward the door, threw two small capsules onto the ground, and quickly exited. Before long, colorless phosgene gas filled the air and the men began coughing and choking. One man died immediately in the chamber. The other three were retrieved and sent to a barracks near the crematorium. Three days later, a second man from the group perished in agony. "He coughed up pink-red blood," recalled a survivor, "and the longer it took, the more bits of lung tissue came out of his mouth. He was aware of everything the whole time, until the end. We were not allowed to drink anything—the water faucet was turned off. Eckstein died in my arms."[22] In all Bickenbach subjected 150 people to these murderous experiments. An estimated 35 to 40 died.[23]

Sievers made certain that the experiments ran smoothly and that researchers had everything they required. From time to time, he and Himmler paid visits to the laboratory at Dachau. These were more than formal inspections. Both men seem to have relished watching the experiments.

AS THE INHABITANTS of Waischenfeld whispered about Sievers and his secretive work, people in the small Austrian town of Mittersill worried about another group of Ahnenerbe officers lodged in the nearby castle.[24] The castle of Mittersill had once been a fashionable resort for Europe's titled and moneyed set.[25] Perched on a mountain slope overlooking the scenic Pinzgau Valley, some seventy kilometers from Salzburg, the sixteenth-century castle commanded a stunning view of the Austrian countryside. It lay in some of the best hunting and fishing country in Europe and offered luxury accommodations. When Princess Juliana of the Netherlands married in 1937, she and her husband planned to spend ten days of their honeymoon at Mittersill; they ended up lingering for six weeks.

But the *Anschluss* had cast a heavy shadow on Mittersill. The castle's owner had fled to America, and a mysterious fire had swept through the premises, leaving behind a good deal of smoking rubble. Nazi officials had stolen the fine furniture, rugs, and porcelain.[26] Still, parts of the castle were salvageable, and in the summer of 1943, as the tide of war began turning strongly against Germany, the Ahnenerbe's largest research department obtained a lease for the palatial dwelling and moved in. Led by Ernst Schäfer, the department consisted of nearly a dozen scientists, and was known to outsiders by the name of the Sven Hedin Reich Institute for Inner Asian Studies, after the famous Swedish explorer and Nazi sympathizer.[27]

Himmler had supplied the institute with everything it needed: money, plenty of gasoline vouchers, vehicles, and a dozen or so workers—concentration-camp prisoners and forced laborers from Russia. These were ethnic Germans who had been uprooted against their will and put to work in the Reich.[28] While this labor force cleaned and repaired the castle, the senior researchers set about furnishing the drafty rooms. Much to Beger's annoyance, they raided the Tibetan ethnographic collection, stirring trays of chemicals in the institute's darkroom with small arrows that

Tibetans had used for warding off ghosts, and drying their feet on rare Tibetan carpets that doubled as bathmats.[29]

In his office at Mittersill, Beger toiled away on the Jewish skeleton project. The maceration machine for rendering the cadavers of the murdered Jews into tidy skeletons had failed to arrive in Strassburg—quite possibly because Allied bombers destroyed the factory where it was supposed to be manufactured.[30] But this setback did not stop Beger.[31] He had managed to lay hands on some Jewish skulls, perhaps from a museum or a university department anxious to stow a precious collection in a safe place, or possibly from the Strassburg preparator Otto Bong, who may have begun defleshing a few of the murdered Jews by other maceration methods.[32] Either way, Beger had obtained a collection of Jewish skulls, and he seems to have been working on them. His assistant Wilhelm Gabel was finishing off the head casts of the murdered prisoners.[33]

And although little good news made its way from the eastern front, Beger's mind still buzzed with racial research projects. He wanted to send one of his Mittersill colleagues, Rudolf Trojan, to measure and examine Russian prisoners of war.[34] He was also keen to examine the behavior of different races on the battlefield—research he believed would prove of immense importance to the Wehrmacht.[35] He could not stop churning out these ideas, and day after day, he dreamed of new projects for parsing and classifying humanity.

In the mid-spring of 1944, Himmler himself chose to pay a visit to Mittersill. He arrived quite suddenly on May 12, without giving any advance warning.[36] Ernst Schäfer's second wife, Ursula, was sitting in her family's private apartment, chatting with her mother, when the SS leader suddenly walked in through the door. She was shocked to see one of the most powerful men in the Reich cross the room. Attentive as he often was to social pleasantries, he kissed her mother's hand, but he offered no such gallantry to Ursula Schäfer. Himmler had already conducted his own silent racial test on her. He thought her too Slavic in appearance, with her high, broad cheekbones.[37]

When Schäfer arrived to greet Himmler, the two men headed off on a tour of the castle. Himmler seemed pleased by what he saw, remarking with genuine interest on the institute's racial work.[38] Later in the evening, Schäfer accompanied Himmler on a stroll along the castle walls at Mittersill. Himmler gazed down contentedly at the quiet Austrian town below. Although it was still early in the evening, most of the lights in the houses below were out—a good sign, in Himmler's view, that the townspeople were busy conceiving more sons for the Reich.[39] Soon after this, Himmler left.

Schäfer later told his wife that he thought the SS chief had paid them a visit in order to size up Mittersill as a possible hiding place for the end of the war. Tucked away in the Austrian Alps, the castle must have seemed an appealing lair.

IN THE EARLY morning of June 6, 1944, Allied troops slid into the cold waters off the Normandy coast, ducking heavy enemy fire. The German military had prepared for months for just such an invasion in Normandy, planting deadly underwater mines and an assortment of treacherous metal obstacles in the shallow waters. Along the shore, the Germans had installed a ribbon of pillboxes and machine-gun nests, intent on annihilating Allied troops as they waded through the shallows. The Normandy beaches were terrible death traps, and some of the German infantry divisions stationed nearby knew exactly how to use them. But despite their heavy and very precise fire, American, British, and Canadian forces managed to clamber to shore, and by eleven in the morning, a few German defenders were spotted abandoning their posts and surrendering to American troops.[40] That evening, the heaviest fighting was over on the shores of Normandy, and the Allied armies had established their beachheads.

Over the next three months, Allied troops liberated much of France, rapidly advancing toward the green valleys of Alsace. Unable to stop them, the SS evacuated the concentration camp at Natzweiler, marching the feeble survivors eastward toward Dachau. In Strassburg, August Hirt paced anxiously across the floor of his office. He was a worried man. His assistant Otto Bong had yet to finish making casts of the prisoners killed for the Jewish Skeleton Collection. Worse still, he had failed to deflesh most of their corpses, and the Allied forces were now heading toward Strassburg. For Hirt, this posed a great dilemma. He was keen to complete his work on the skeleton collection. But he knew that if the Americans or the French found readily identifiable corpses, he risked arrest for war crimes. What was he to do?

Unable to see his way clear, he took up the matter with Sievers, who referred the problem to Himmler's personal administrative officer on September 4, 1944. Should the collection be preserved? Or should Hirt render the bodies virtually unidentifiable by removing all the soft tissue? Or would Himmler prefer more drastic action—the complete destruction of the bodies?[41] Himmler and his staff seem to have dithered over the problem, but in mid-October, a certain SS-Hauptsturmführer Berg issued orders to destroy all the bodies "if Strassburg should be endangered because of the military situation."[42] To soften the blow, Sievers assured Hirt that it was only a temporary setback. He and his team would be able to repeat the study, Sievers promised, if they were permitted "to work and research peacefully."[43]

Thus mollified, Hirt instructed his assistants to dissect the remaining cadavers, place them in coffins, and consign them to the incinerator, just as the corpses used in anatomy lessons were. But before he released the bodies to his assistants, he committed a final indignity. He pried loose the mouths of the dead prisoners and pocketed their gold teeth.[44] Then he prepared to flee eastward to the German city of Tübingen, just across the Rhine. A few days later, on October 21, Sievers notified Himmler's staff that Hirt had complied with the orders, completely destroying the Jewish Skeleton Collection.[45]

French troops liberated Strassburg a month later, and it was not long after this that French authorities learned that a scientific institute at the Reich University of Strassburg had been in constant contact with the Natzweiler concentration camp. Investigators quickly descended upon Hirt's anatomy institute, combing the offices and laboratories for evidence of war crimes. In one of the labs, they discovered sixteen cadavers of young and relatively healthy-looking men and women floating naked in containers filled with an alcohol solution. They also found remains from another seventy bodies, including fifty-four glass microscope slides containing human testicular tissue.[46]

Suspecting the worse, the investigators fished the bodies from the tanks, one by one, examining them carefully for clues to their identities. Someone had cut off a patch of skin from the left arms of fifteen of the bodies. But along the arm of one male, a tattooed concentration-camp number could be clearly seen.

IN EARLY JANUARY 1945, French and British journalists began filing the first newspaper stories on Hirt's atrocities at Natzweiler. Sievers was stunned to read their reports. For weeks, he had tried keeping up the old appearances at Waischenfeld, sitting in his office, reading official letters and dictating replies to his secretary as if everything were under control. But the charade had become increasingly difficult to maintain.[47] Telephone calls from the outside world had tapered off noticeably, and the courier brought fewer letters to the office in Waischenfeld. Food was growing scarce, and soon there would be no fuel left for the vehicles. The Ahnenerbe's connection to the world was growing more tenuous by the day. Sievers hated it. He had pinned everything on Himmler and Hitler and their seemingly boundless power, and he was beginning to see just how foolish he had been.

The Reich's foreign affairs ministry proposed fighting the ugly stories coming out of Natzweiler, hoping even then to paste together the tattered facade of Nazi respectability.[48] It requested a statement from Hirt that would somehow explain away the evidence of the atrocity, and Sievers dutifully relayed the request to the anatomist, who was safely lodged, along with many of his former university colleagues, in Tübingen. There was talk of resurrecting the Reich University of Strassburg just as soon as military conditions permitted, and the administration had appointed Hirt as its new dean of medicine, the former dean having been captured. As a result, Hirt was keeping busy, searching for accommodations for the exiled faculty, but he still hoped to resume the experiments at the earliest opportunity.

Sievers believed that the newspaper reports of Hirt's work were mere propaganda, based solely on rumors and suspicion. Hirt, after all, had informed him that all material evidence of the skeleton collection was destroyed.[49] So he encouraged Hirt to pen a strong denial. The anatomist followed the advice, for he feared that an international scandal would damage his scientific reputation in Germany and abroad.[50] In his statement, he described an article in the *Daily Mail* disparagingly as a "typical atrocity story."[51] The corpses discovered in the anatomical institute, he declared, were simply bodies used to teach medical students the practice of dissection. They had been obtained from the same legal sources that French anatomists had previously used to obtain cadavers. He also observed that he had conducted only animal experiments at Natzweiler, and he completely denied any involvement in *Rassenkunde* studies. "I do not know anything about racial research and have never received such an order," he stated. "The only thing which has to do with race in my institute is the large anthropological collection of skulls which was built prior to the First World War."[52]

Sievers thought Hirt's blatant lies were "excellent."[53] But the experience seems to have instilled in the official a new sense of caution. He ordered the staff at Mittersill to destroy all correspondence, photographs, and other materials related to "the matter Auschwitz/Prof. Dr. Hirt Strassburg."[54] He then began burning boxes of incriminating documents in the courtyard of the Steinhaus itself.[55]

21. THOR'S HAMMER

IN ALL THE YEARS Hitler had known Himmler, he had always counted upon the younger man's deep, unwavering sense of loyalty. Since the two men had met in Bavaria in the mid-1920s, Hitler had come to recognize his subordinate's remarkable skills as an administrator and his zealous dedication to Nazi party doctrine. But what finally bound the two men irrevocably together was Himmler's personal allegiance. He had guarded Hitler's life assiduously from assassins and built a personal bodyguard service, the SS, which was second to none. He had daily carried out Hitler's dirty work—the mass executions, the liquidations, the slaughter—without objection. His fealty seemed beyond question—so much so in fact that Hitler had entrusted one of the most dangerous weapons of the Nazi state, the police, to Himmler alone.

Moreover, as the war dragged on, and as the early victories turned sour, Hitler had heaped new powers on *der treue Heinrich,* "loyal Heinrich." He named Himmler Germany's new interior minister in the summer of 1943. The following year, he placed him in charge of the Reserve Army, then called the People's Militia, which proceeded to conscript all men between the ages of sixteen and sixty not already in military service. His faith in Himmler's ability to get a tough job done seemed unshakable. In November 1944, he had handed him a major military post, appointing him commander in chief of the Army Group Upper Rhine, responsible for holding a key German bridgehead south of Strassburg and west of the Rhine.

By early 1945, the clerkish-looking SS leader had emerged as the second most powerful man in the Reich, far outstripping Göring, Goebbels, and Bormann. With each new appointment, Himmler's hopes for personal advancement grew and he dreamed of one day inheriting Hitler's empire. As such, he stubbornly searched for a way—any way—of snatching victory from the jaws of defeat.

HIMMLER HAD NEVER forgotten the old Norse legends that his father had once read to him, the tales of Thor and his magical weapon, a deadly hammer that struck like lightning. He had persuaded himself long ago that the *Edda* was literal truth and that Thor's hammer was in fact a sophisticated piece of electrical engineering developed by Aryans to vanquish their enemies. So in November 1944, Himmler directed his staff to examine a bizarre plan to build a modern version of Thor's hammer—a mammoth electrical weapon capable of shutting down all the electrical systems of Allied troops, from their radio communications and radar to the ignitions of their tanks.[1]

A great irony lay in this last-minute quest to develop a wonder weapon to save Germany from its enemies. Some two and a half years earlier, in February 1942, the Reich Research Council had organized a private daylong presentation in Berlin on uranium. One of the speakers, Nobel Laureate physicist Werner Heisenberg, had described for the audience the process of nuclear fission and the devastating potential of an atomic bomb. Pure uranium-235, Heisenberg observed, is "an explosive of quite unimaginable force," something that had occurred to many nuclear physicists. To harness this power, he noted, American scientists were apparently "pursuing this line of research with particular urgency."[2]

Building a German atomic bomb would cost billions of reichsmarks, however, and require the expertise of many thousands of German scientists and technicians. To gather support for the idea, the Reich Research Council had sent Himmler and many other high-ranking Nazis personal invitations to attend the February 1942 meeting. But a secretary at the council had mistakenly tucked a program for another, far more technical meeting into the envelope. As a result, Himmler and his colleagues turned down the invitation, fearing a tedious day of it, and Heisenberg's ideas for a potent new explosive failed to win broad support in the Nazi leadership.[3] When Albert Speer raised the idea of such a bomb with Hitler a few months later, he shrugged it off, seeing little to be gained.[4] The idea, noted Speer, "quite obviously strained his intellectual capacity."[5] And the project suffered a further heavy blow over the next year and a half, as Allied saboteurs and bombers incapacitated a heavy-water plant in Norway needed to produce weapons-grade raw material for the bomb. So it was that Germany failed to develop a major atomic weapon.[6]

As a result, Himmler turned at the end of the war to a far-fetched scheme for a bizarre electrical weapon. An obscure German company, Elemag, had put the plan forward in mid-November 1944. According to the firm's engineers, existing technology could be used to transform the earth's atmosphere into a giant remote-control device capable of flipping the switch on the Allies' electrical equipment. "It is established," explained the Elemag engineers, "that by removing the insulating effect of the atmosphere, one makes it impossible for any electrical device of a familiar construction and implementation to function. The present state of technology offers the possibility of influencing the insulating material of the atmosphere for the task at hand. It is well known that ultra-shortwave electrical vibrations of certain frequencies also develop the ability to ionize the atmosphere they permeate, thus causing a reverse electrical reaction. In other words, they transform the insulating material of the atmosphere into a voltage conductor."[7]

Himmler eagerly referred this proposal to his SS technical office. Aware no doubt of the Reichsführer's peculiar interests, the SS scientists examined the proposal carefully. As Himmler waited, he could not refrain from boasting to associates such as Adolf Eichmann and others about a new weapon that would render the Allies literally powerless. His personal physician, Felix Kersten, recalled one such occasion:

> When I returned to Himmler's headquarters in December I found him singularly optimistic. Once again he was prophesizing a German victory!
>
> Germany was in ruins, the bombardment was ever more intense, Germany was almost encircled—and Himmler was talking of victory! I could hardly believe my ears.

One day, in a particularly expansive mood, he hinted at the reason for this presumably unfounded optimism. . . . "Very soon we shall put our last secret weapon into use. And that will change the war situation entirely."[8]

The SS technical office mulled over the blueprints for the giant voltage conductor for weeks, searching, it seems, for some way to deliver the bad news to Himmler. But on January 8, it was forced to issue its verdict. The mammoth remote control was simply a fantasy—one well beyond Germany's capabilities. Indeed, the SS staff warned, "all the means to be made available [to the research] would have to be designated as lost in advance."[9]

Himmler, however, refused to believe it. Despite all his heavy responsibilities as a military commander, all the briefings and field reports, all the tactical and strategic planning with his officers, he immediately dispatched the proposal to Dr. Werner Oseberg, head of the planning office of the Reich Research Council, requesting another opinion. Oseberg, accustomed to dealing with serious physicists, must have been astounded by the document crossing his desk. But he clearly understood the importance that Himmler placed on the plan. He referred the matter to two prominent German scientists—one of whom was an expert on electromagnetic waves.

All three men submitted their reports in early February. The idea of such an electrical weapon, noted Oseberg dryly, "is unrealizable, given the present stage of technology. Elemag's statements themselves lack any deep understanding of the technological and physical processes involved."[10]

BY THE END of March 1945, Allied bombers had reduced much of Berlin and Munich to smoking heaps of debris, leaving inhabitants to fend as best they could. Total destruction seemed only a few weeks away. In the west, Allied troops had bridged the Rhine and were now driving hard toward Germany's industrial heartland, the Ruhr Valley. Toward the east, the Soviet army was preparing to cross the Oder River, less than one hundred kilometers from Berlin. The city's inhabitants shuddered to think what would happen then. For weeks, they had seen and heard terrible reports of the vengeance that Soviet troops were taking as they moved eastward, burning and looting German homes, beating and gang-raping German women as a payback for the atrocities Hitler's forces had committed on Soviet soil. Some of these reports were frighteningly accurate.[11]

Safe for the time being in his new field headquarters—a convalescent hospital just outside Berlin whose roof was conveniently painted with red crosses—Himmler had begun examining his own options. As a military commander, he had failed to live up to Hitler's expectations, and much to his humiliation, he had been relieved of his army group command—not, however, before receiving the full brunt of Hitler's fury. This failure and the hard words that had followed had greatly strained the old understanding between the two men. Hitler had found Himmler seriously lacking as a subordinate, and now that defeat was looming, Himmler had begun to seriously question Hitler's judgment, his military acumen, and his fitness to serve as Germany's leader. In defeat, they had fallen out like a pair of wolves.

Hitler, after all, was prepared to go down in flames like an old Norse god, taking all Germany with him if need be. He angrily refused to capitulate to the enemy, no matter what terms might be offered. Himmler, however, wanted to live and he had already coldly turned his mind to a future without Hitler.[12] Indeed, he hoped to play a pivotal role in postwar Europe. He believed he could persuade the Allies to overlook his terrible crimes by handing over to them some of his SS divisions, thereby bringing the war to a speedier end. Himmler had grown accustomed to power, and he desperately hoped to hold on to it, perhaps as the new anti-Communist leader of Germany.[13] If that meant betraying Hitler, the man he had served so obediently for nearly two decades, then he was fully prepared to do so. Already, he had dispatched a secret proposal to the western Allies.[14] They had briskly turned it down, but he intended to try again.

As he stewed over the future in his field headquarters, however, Himmler followed the reports from the battle lines carefully. In the final days of March 1945, he realized with grim certainty that the Americans would soon capture Wewelsburg, the SS castle where Wiligut had once pontificated on ancient runes. He was loath to let the castle, his Nordic academy for senior SS officers, pass into enemy hands. So on March 30, he ordered Wewelsburg's staff to evacuate, leaving behind only a detachment of SS guards.[15] A day later, at three in the afternoon, a member of Himmler's personal entourage, SS-Hauptsturmführer Heinz Macher, appeared in the village with a small demolition squad.

Macher ordered his men to place dynamite in the castle's west and south towers, as well as in two other adjoining staff buildings.[16] But the squad lacked what was needed to blow up the entire building, since the SS was running short of explosives. So Macher instructed his men to set fire to the curtains and other flammables. As flames began to dart from the windows, the SS officer set off the charges. The two towers buckled and sagged, then slumped in a dense cloud of dust and debris. From the safety of their homes, residents of the adjacent village of Wewelsburg peered out at the spectacle. Macher and his men then finished off the work, firing antitank grenades at the stronghold, and after leveling as much as they could, they and the remaining SS guards departed for Paderborn. Just two hours had elapsed since they arrived.

The villagers waited until Macher and his men were safely out of sight. Then they began pillaging the burning castle. In the last hours of daylight and over the next two days, they scrambled across halls and up burning staircases to plunder guest rooms, libraries, workshops, storage rooms, kitchens, and, of course, Himmler's own personal study. They helped themselves to expensive carpets and carved chairs, inlaid tables and china plates decorated with Wiligut's faux runic symbols. Some broke into the castle's sprawling wine cellar, which was reputed to stock nearly forty thousand bottles, many almost certainly stolen from the finest cellars of Europe. As Wewelsburg burned, the revelers grabbed bottles of champagne, Bordeaux, port, and schnapps, breaking so many on the way back up the narrow staircase that they had literally to slop through puddles of wine, their pant cuffs stained a deep red.[17]

Others stumbled upon what remained of the castle's museum—a glorified ancestor room to educate the SS leadership in the ways of their Aryan forebears.[18] Since the beginning of the war, Himmler had added dramatically to its collections, acquiring crates of plundered artifacts— from Viking swords and golden helmets to Bronze Age belts, mirrors, Scythian bronze arrowheads, and classical Greek terra-cotta figurines.[19] In addition, foreign governments had learned that the surest way to curry favor with Himmler was to present him with a rare antiquity. When the new Fascist regime of Spain wanted to court his support, for example, the Spanish foreign minister bestowed upon him a pair of artifacts excavated from an old West Goth grave in Segovia.[20]

Wewelburg's staff had taken pains to hide some, if not all, of these antiquities, knowing that the Allies were closing in. They ordered false walls to be built in a former Augustinian monastery known as Gut Böddeken, located a few kilometers from Wewelsburg, and stowed an unknown quantity of prehistoric treasures behind them.[21] But in the chaos that followed the abrupt departure of Himmler's troops from the region, someone seems to have divulged the location of this secret cache. Looters soon descended.

The first American troops reached the charred ruins of Wewelsburg on April 2 and occupied the village the following day. They picked up a few odd souvenirs from inside the razed castle—a box of silver rings here, an SS typewriter there—and saw the splintered remains of a large reptilian fossil in the courtyard. But there was little sign of the treasure that the castle had once housed. And there were few valuables remaining at the old monastery of Böddeken. When Allied art experts arrived to inventory what was left from Himmler's plunder in hopes of returning it to its rightful owners, they found only a small pile of swords, shields, crossbows, and ancient helmets, all that remained of Himmler's prize collection of prehistoric weapons.[22]

EARLY IN THE morning of May 1, Himmler received word that Hitler and his new bride, Eva Braun, had committed suicide in the bunker below the Chancellery building in Berlin. Although Himmler had once hoped to be named Hitler's successor, the German leader had coldly passed him over. Hitler had learned from a series of foreign media reports a few days earlier that Himmler had betrayed him. The SS chief had entered into private negotiations with the western Allies, offering to bring the war to a quick end—something Himmler believed he could enforce.[23] At first Hitler could scarcely believe the news. But as he mulled over the way that he had been mollified and gulled by Himmler, he had exploded with fury, calling it "the most shameful betrayal in human history."[24] Soon after, he had named Admiral Karl Dönitz as his successor.

Dönitz, the commander in chief of the German navy, was as surprised as any by this choice, but unlike his predecessor, he did not intend to stand back watching Germany's annihilation. He transferred his headquarters to a naval cadet school near Flensburg, not far from the Danish border, and dispatched his representatives to France to see General Dwight Eisenhower.

He wanted to sue for peace. On May 7, 1945, his representatives signed Germany's official unconditional surrender, and the following night at eleven o'clock all hostilities officially ceased.

In the last months of the war, the Allies had agreed among themselves to establish an international military tribunal to try Germany's major war criminals. Although Hitler had escaped arrest, the other senior Nazi leaders were still at large, and the Allies were determined to bring them to justice. At the top of their list was the man who had meticulously planned, down to the last finicky detail, the industrialized slaughter of Jews and others categorized as enemies of the Reich. Well aware of this, Himmler began making plans to escape. He had already pocketed the identification card of a field police officer, Heinrich Hitzinger. He now shaved off his mustache, removed his famous glasses, slung a black patch over his right eye, and assumed Hitzinger's identity.

For the first few days, he kept to himself at Flensburg, where he had brought his young mistress Hedwig Potthast and their two children.[25] On May 10, however, he decided to strike out south with a small entourage. Several of his senior SS officers had already departed for the Alps, where they planned to launch a new Nazi guerilla movement, code name Werwolf. Himmler, it seems, had decided to cast his lot in with Werwolf. He donned civilian clothes and set off by automobile with his personal administrative officer, Dr. Rudolf Brandt, and a few other senior SS officers.

The small group crossed a bridge over the Elbe River without attracting the suspicion of guards, then lit out on foot. Over the next week or so, the men lived on the lam, a novel experience for Himmler. They joined the churning floodtide of people on the roads—refugees and former soldiers, newly liberated slave laborers and recently freed concentration-camp prisoners—and at night they made their beds in farmers' haylofts or slept on benches in railway stations. Meanwhile, Allied intelligence officers searched for Himmler. In the House of Commons in London, a member of the opposition asked Winston Churchill pointedly where the SS leader was. Churchill expertly deflected the question. "I expect he will turn up somewhere in this world or the next, and will be dealt with by the appropriate local authorities. The latter of them would be the more convenient to His Majesty's Government."[26]

On May 21, British troops stopped three German field police at a bridge outside of Bremervörde, a small town west of Hamburg.[27] They failed to recognize Himmler, but they were under orders to arrest all German police officers and so they dispatched the trio to a nearby internment camp. Interrogators there found Himmler's papers suspicious. The documents had been issued in Flensburg, where Dönitz and senior Nazi advisors had fled in the final days of the regime and where, according to recent reports from the Danish resistance, the SS leader had gone into hiding.[28] So the British sent him on to a second camp at Lüneburg.

By this time, however, Himmler had grown tired of all the skulking about. He had come up with a plan to offer Churchill the services of the Werwolf guerillas to aid in the coming fight against Communism in the East.[29] So after arriving at the Lüneburg camp on May 23, 1945, he told the guards who he was and asked to see the commanding officer, Captain Tom Selvester, who was apparently at lunch. As Himmler waited, he doffed the eyepatch and dug a pair of spectacles from his pocket and put them on. The transformation was startling. The British camp commander knew at once that he was speaking to Heinrich Himmler.

Selvester called in an intelligence officer to strip-search the prisoner. The examination produced two curiosities—small brass cylinders, each one-half inch long and about the diameter of a cigarette.[30] One of the cylinders was empty; the other contained a small blue-glass vial, most likely of poison. Two days earlier, a senior SS officer had killed himself by swallowing a hidden vial of cyanide. The intelligence officer then compared the prisoner's signature to a known sample from Himmler, and asked Himmler a number of questions. Convinced that he was looking at one of the most notorious figures of the Nazi regime, the officer proceeded to show Himmler photographs of emaciated concentration-camp prisoners and the huge mounds of corpses that Allied troops had recently discovered at Buchenwald. Himmler merely glanced at the pictures and shrugged. "Am I responsible for the excesses of my subordinates?" he asked.[31]

That evening, Colonel Michael Murphy, chief of intelligence on the staff of British Field Marshal Bernard Montgomery, took custody of the prisoner. He shoved Himmler roughly into a car, and drove him to an interrogation center just outside Lüneberg. There he curtly ordered Himmler to strip and directed an army medical doctor to conduct a second examination in hopes of retrieving a missing poison vial. Himmler undressed meekly enough, but the display of photos from Buchenwald and Murphy's abusive manner seem to have unsettled him, revealing all too clearly the futility of offering Werwolf's services to Churchill. He had at last run out of options. There was no future but ignominy at the hands of his captors, no escape but death.

The army doctor peered into Himmler's mouth. He thought he saw something dark lodged in the lower molars. He beckoned Himmler toward better light and positioned his head for a better look. But Himmler suddenly wriggled free, snapping his mouth shut and grinding the cyanide capsule between his teeth. He quickly lost consciousness. Horrified by the turn of events, Murphy and an associate hoisted Himmler upside down by his legs, and shook him frantically. Then the army doctor tried pumping his stomach. But the prisoner did not regain consciousness. Heinrich Himmler, the chief architect of the Holocaust, had stopped breathing. He was pronounced dead at 11:04 P.M.[32]

FOR TWO DAYS, Himmler's body lay on the floor in Lüneburg. Russian and American officials traipsed by and peered down at it to confirm the identification, then a medical person appeared and made plaster casts of Himmler's head and removed his brain and took it away.[33] Finally, a British officer wrapped Himmler's body up in camouflage netting knotted with army telephone wire, and hauled it out to a truck with the help of a few soldiers.[34] Together they drove to the nearby wilderness of Lüneberg Heath, where they dug a hole and secretly buried Himmler in an unmarked grave.

They left nothing behind for Nazi loyalists to turn into a shrine.

22. NUREMBERG

In APRIL 1945, A prominent American journalist wrote a story for *Collier's* magazine describing a series of conversations with ordinary Germans at the end of Nazi rule. Martha Gellhorn was a veteran war correspondent who had spent six years penning dispatches from the front lines. An elegant chain-smoker who customarily dressed for battle in her Saks Fifth Avenue best, she had long made a point of avoiding the honeyed lies of generals in favor of the plain, unvarnished words of soldiers. She hungered for the truth, and on D-Day, she braved enemy fire to walk upon the beaches of Normandy, while her famous husband—fellow war correspondent Ernest Hemingway—contented himself with what he could see from the bridge of a landing craft.

In the spring of 1945, Gellhorn had traveled with American forces as they pushed their way into Germany's Ruhr Valley. En route, she had chatted daily with German citizens she encountered. She was stunned by what she heard. Not one person she met would confess to being a Nazi. No one, moreover, would admit to knowing any Nazis, past or present. And no one she talked to had a bad word to say against the Jews. On the contrary, nearly everyone had a story about saving a Jewish neighbor or acquaintance from the camps. Gellhorn was deeply dismayed: she heard the same lies over and over again in a terrible refrain. "I hid a Jew for six weeks. I hid a Jew for eight weeks. (I hid a Jew, he hid a Jew, all God's chillun hid Jews.) We have nothing against the Jews; we always got along well with them."[1]

This seeming amnesia distressed the plainspoken writer. She had traveled to Germany in the early years of the Third Reich with a group of French pacifists and seen for herself the feverish worship of Hitler and the country's appalling depths of anti-Semitism. And in the spring of 1945, she found it disturbing that no one she talked to was prepared to own up to the past or admit the terrible error of Nazism, much less shoulder any of the blame for the war that had destroyed much of Europe. "To see a whole nation passing the buck," she observed, "is not an enlightening spectacle."[2]

Such brazen attempts to conceal the past posed a serious problem for Allied forces intent on bringing Nazi war criminals to justice and rebuilding democratic institutions in Germany. Nazism had deliberately cultivated a culture of lies, equivocation, and fantasy, and at Potsdam in July 1945, Allied leaders agreed that the occupation forces had to publicly expose the errors and crimes of the former regime in order to convince the German people "that they cannot escape responsibility for what they have brought on themselves."[3] To succeed in this, the Allies had to denazify Germany. They had to eradicate the Nazi party and root out Nazism from the German courts, press, and schools. First, however, they had to bring all those guilty of war crimes to justice, beginning with the most senior offenders.

To try the accused, the four major Allied powers established an international military tribunal at Nuremberg, a city replete with symbolism. Before the war, Hitler had presided over annual Nazi party rallies and mass parades each year at Zeppelin Field in Nuremberg, and it was at one of these rallies that party officials had enacted the Nazis' notorious racial legislation, the Nuremberg Laws. In preparation for the trials, the tribunal's document division set to work in the city's Palace of Justice, sorting through the towering piles of government records in search of evidence. The sheer quantity of documentation was overwhelming, but on November 14, 1945, guards led twenty-two of the Nazi regime's senior leaders—from Hermann Göring and Joachim von Ribbentrop, the former foreign minister, to Julius Streicher, the former publisher of the anti-Semitic newspaper *Der Stürmer*, and Ernst Kaltenbrunner, the former head of the Reich Security Office—into the courtroom.

Over the next eleven months, as reporters from around the world flocked to cover the trial, American investigators began preparing for twelve more major prosecutions of Nazi war criminals at Nuremberg. Top on the docket was the Medical Case. Investigators had uncovered shocking evidence of the complicity of senior German doctors in Nazi atrocities committed in mental hospitals and concentration camps. Some of the most chilling evidence came from a thick sheaf of documents describing the Ahnenerbe's skeleton collection and medical experiments.

AMERICAN AUTHORITIES HAD discovered the Ahnenerbe files in a dark cave near the small Bavarian village of Pottenstein. In the final weeks of the war, Wolfram Sievers had resolved to save documents concerning some of the Ahnenerbe's atrocities, hoping, it seems, that the researchers could one day pick up where they had left off. On the advice of a cave expert, Sievers had decided to conceal them and many other Ahnenerbe records in a cave known as Kleines Teufelsloch, or Little Devil's Hole, which was situated near the neighboring village of Pottenstein.[4] He obtained a crew of concentration-camp prisoners to transport the boxes, and he saw to it that the records were safely stowed behind the blast rubble in the cave.[5] Then he and his assistant, Wolf-Dietrich Wolff, set off from Waischenfeld to join a nearby SS detachment.

On April 14, 1945, the first American tanks appeared in Waischenfeld. A young villager on a bicycle came out to greet them, waving a white flag. Most of the locals had fled, so the troops moved into the official-looking Steinhaus, turning it into a temporary administrative center. Soon after, they received a tip-off about Nazi files hidden in the Little Devil's Hole. A small detachment of American troops went out to retrieve them. The boxes contained many thousands of Ahnenerbe documents—letters, personnel records, memoranda, orders, maps, handwritten notes, and official reports, some clearly stamped Geheim, or "secret." They recorded in minute detail the war crimes of the Ahnenerbe—the greedy plundering of museums and private art collections from Poland to the Caucasus; the cold brutality of human medical experiments at Natzweiler and Dachau; and the conspiracy to murder Jewish concentration-camp prisoners for their skeletons.

By this time Sievers had returned to Waischenfeld in hopes of seeing his family. He hid in a village barn for a few days, where he was spotted by some of the local inhabitants. They had never much liked Sievers's high-handed manners, nor the way he had paraded through town in his SS uniform, so they reported him to the new authorities. On May 1, an American patrol captured him and took him into custody. Impressed by his high rank as an SS-Standartenführer, they photographed him, asked him some preliminary questions, then shipped him off to nearby Bamberg for a brief hearing.[6] Soon after that, Allied investigators dispatched him to an internment camp in England.

Over the next eighteen months, American intelligence officers pored over the captured documents, studying their contents and selecting correspondence of evidentiary value. A team of translators then rendered into English clinical reports of freezing experiments, mustard-gas tests, and skeleton collections, and passed the finished documents on to the team of prosecutors, who were stunned by what they read. "Their own reports illustrated with pictures are far better than any of the studies we have compiled on the persecution of Jews, crimes against humanity, etc.," noted one prosecutor. "The Germans certainly believed in putting everything in writing."[7]

With such detailed evidence before them, prosecutors charted the senior chain of command responsible for the Ahnenerbe's atrocities at Natzweiler and Dachau. They were particularly interested in four of the Ahnenerbe's staff: Walther Wüst, the superintendent; Sievers, the director of the Institute for Military Scientific Research and the official who had directly overseen the experiments; and the institute's two senior researchers, Dr. August Hirt and Dr. Sigmund Rascher.

The two physicians, however, lay well beyond the reach of the Allies' justice. In February 1945, just shortly before the war ended, German police had arrested Rascher for his role in a bizarre child-abduction scheme. According to the original investigators, Rascher's wife Nini had kidnapped a series of infants after discovering that she was unable to conceive children of her own.[8] Rascher himself had gone along with the scheme, proudly representing three of the stolen babies as his own newborn sons to Himmler. But in 1944, Munich investigators uncovered the truth. When Himmler learned of the deception, he sent Nini Rascher to Ravensbrück concentration camp and eventually dispatched her husband to Dachau, the camp where Rascher had once reveled in conducting his experiments. Just a few short days before American forces liberated Dachau, an SS officer shot and killed the physician in his cell.[9]

Hirt was similarly unavailable. In February 1945, he had suddenly left his office in Tübingen and journeyed secretly to the Black Forest. There he cached food and hid out in a hut in the woods. From time to time, he slipped down to a local farm, hungry for news of the war, until the farm's inhabitants eventually invited him to stay. And it was there that Hirt heard the news of Germany's surrender. Fearing arrest by the Allies, he borrowed a pistol from the farmer, then returned to the forest. He shot himself on June 2, 1945. The farmer who had sheltered him recovered his body and reported the death to authorities.[10]

That left the Nuremberg prosecutors with just two senior Nazi officials to try for the Ahnenerbe's atrocities—Wüst, the scholar who had overseen all of the Ahnenerbe's scientific research, and Sievers, the organization's managing director. But in all the thick files of Ahnenerbe correspondence, they could find little clear evidence of Wüst's complicity in the experiments. They regretfully struck his name from their list. That left them with just one senior official: Sievers.

ON DECEMBER 9, 1946, twenty-three defendants dressed in civilian clothes and military uniforms carefully stripped of all badges of rank filed into a courtroom in Nuremberg. Taking their places along two long wooden benches, they listened in silence to the charges laid against them. Twenty of the defendants were medical doctors accused of war crimes ranging from planning the mass murder of the mentally handicapped and others deemed "unworthy of life" to forcibly performing medical experiments on concentration-camp prisoners. The remaining three were Nazi officials.[11] Sievers numbered among them. Prosecutors had indicted him on four counts, including unlawfully, willfully, and knowingly committing war crimes and crimes against humanity. The indictment specifically detailed his role in aiding and abetting the skeleton collection and the human medical experiments at Dachau and Natzweiler.

The chief of counsel, Brigadier General Telford Taylor, outlined the critical importance of the trial in his opening statement:

> The defendants in the dock are charged with murder, but this is no mere murder trial. We cannot rest content when we have shown that crimes were committed and that certain people committed them. To kill, to maim, and to torture is criminal under all modern systems of law. These defendants did not kill in hot blood, nor for

personal enrichment. Some of them may be sadists who killed and tortured for sport, but they are not all perverts. They are not ignorant men. Most of them are trained physicians and some of them are distinguished scientists. Yet these defendants, all of whom were fully able to comprehend the nature of their acts, and most of whom were exceptionally qualified to form a moral and professional judgment in this respect, are responsible for wholesale murder and unspeakably cruel tortures.

It is our deep obligation to all peoples of the world to show why and how these things happened. It is incumbent upon us to set forth with conspicuous clarity the ideas and motives which moved these defendants to treat their fellow men as less than beasts. . . .

This case and others which will be tried in this building, offer a signal opportunity to lay before the German people the true cause of their present misery. The walls and towers and churches of Nuremberg were, indeed, reduced to rubble by Allied bombs, but in a deeper sense Nuremberg had been destroyed a decade earlier, when it became the seat of the annual Nazi Party rallies, a focal point for the moral disintegration of Germany, and the private domain of Julius Streicher. The insane and malignant doctrines that Nuremberg spewed forth account alike for the crimes of these defendants and for the terrible fate of Germany under the Third Reich.[12]

Over the next nine months, the tribunal examined nearly fifteen hundred documents and listened to the nightmarish testimony of sixty-two witnesses, many of whom recounted in hideous detail the experiments performed on prisoners in German concentration camps. Throughout the trial, Sievers sat impassively on the defendants' bench. On the witness stand, he insisted that he was innocent of all crimes. "I always tried to prevent the Ahnenerbe from becoming involved in medical research," he asserted under oath.[13]

In the evenings in a small six-foot-by-twelve-foot cell at Nuremberg, Sievers taught himself English and French and read widely.[14] He reassured his wife that he would be released when all the facts of his case were made public.[15] He based this belief on a bizarre set of circumstances. Since 1929, he had enjoyed a close personal relationship with Dr. Friedrich Hielscher, the leader of a secret resistance group that had fought Nazism in Germany.[16] Indeed, Sievers declared, Hielscher had recruited him as a member of the group, and during the ten years that he had served as the managing director of the Ahnenerbe, he had led a double life, covertly supplying information to the resistance fighters.

Hielscher was an immensely complex, charismatic man, a bisexual who seems to have once written love poetry to Sievers.[17] He agreed to testify on Sievers's behalf and took the stand on April 15, 1947. Under oath, he explained that he had sent Sievers to work for Herman Wirth in April 1932, believing that Wirth was a rising star in the Nazi firmament.[18] He also testified that he had later encouraged Sievers to join the Ahnenerbe as a Trojan horse, so that he could funnel secret information from SS and police contacts to the resistance group and eventually help assassinate Himmler.

Hielscher then recounted some of Sievers's successes as a resistance worker. He had, Hielscher testified, passed on vital information about the movements of the Waffen-SS during the war. In addition, he had assisted several prominent people to elude arrest and imprisonment by the Gestapo. These included the Nobel Prize–winning physicist Niels Bohr. Moreover, he had proven to be such a valuable asset to the resistance group that Hielscher had insisted that Sievers keep his post at the Ahnenerbe even after the medical experiments began. Sievers had been desperate to quit, Hielscher claimed. And in the summer of 1944, the two men had plotted an assassination attempt on Himmler, which had unfortunately come to naught.

Hielscher, however, could produce little evidence for these remarkable assertions, and it later emerged that he had greatly exaggerated the truth.[19] He was very fond of Sievers and owed the Nazi official a large debt of gratitude. Sievers had obtained his release from a Gestapo prison during the war and Hielscher clearly wanted to return the favor. While Sievers had supplied some of Hielscher's group with travel documents, and had even allowed them to meet from time to time in the Ahnenerbe offices—a cooperativeness that likely arose from his old sense of loyalty to Hielscher—most of the witness's other claims were fictional.[20] Sievers played no part in Bohr's escape.[21] And he and Hielscher had never conspired to assassinate Himmler.

The truth was far more sinister. While Sievers had assisted a minor resistance group from time to time, ladling out small favors in order to impress an old friend, he had planned and organized some of the most heinous crimes of the war, and served as a very lethal instrument of the Nazi state. He was, as the prosecution concluded in its final statement, "an unresisting member of a so-called resistance group."[22]

On August 21, 1947, the tribunal judges handed down their verdict: they pronounced Sievers guilty on all four counts. Soon after, they sentenced him to death. Ten months later, Sievers climbed the stairs of the gallows in the courtyard of Landsberg Prison, only a few paces away from where Hitler had written the first volume of *Mein Kampf.*

23. SECRETS

The TRIALS AT NUREMBERG riveted the world with their gruesome testimony of slave-labor factories, mobile gas chambers, forced deportations, and massacres in the East. They resulted in the convictions of several hundred senior Nazi officials and succeeded in exposing some of the most repellant crimes of the Nazi regime. But as dramatic as the trials were, they failed to convince many Germans of the great error of Nazism. Some 40 percent still looked favorably in 1951 upon the "Hitler period of German history."[1] The Allies' great goal of rooting out Nazi culture remained little more than a lofty ideal. Many of the guilty went free.

In the months following the German surrender, Allied forces had attempted arresting all leading party and government officials. In the American zone, the new military government passed a law requiring each public-office holder or each person aspiring to public office to fill out an extensive questionnaire concerning his or her past Nazi party involvement.[2] On the strength of the answers, American troops broadened their net, arresting tens of thousands of party officials and members of the SS, the Gestapo, and other suspicious organizations and imprisoning them in internment camps.[3] In Bavaria alone, American authorities fired or arrested one hundred thousand employees, leaving schools without teachers, telephone exchanges without operators, and post offices without postal workers.[4] Life quickly ground to a halt, while the internment camps bulged at the seams.

Thinking better of this approach, the new military governments established local denazification tribunals across Germany to sift through the vast numbers of suspects and identify the most dangerous Nazis. The local tribunals in turn hired nearly twenty-two thousand people to assess the information from the questionnaires and classify those who had filled them out into one of five categories, depending on the degree of their involvement in the former regime—from "exonerated" and "fellow traveller," to "lesser activist," "activist," and "major offender." For the three most serious categories, the tribunals were free to impose penalties ranging from modest fines all the way up to ten years' imprisonment in a work camp and the seizure of all personal property.[5]

The new system, however, was riddled with holes that allowed the great majority of senior Nazi officials to wriggle free. Some falsified their questionnaires, knowing that many government records had gone up in flames during the bombing attacks.[6] Or they offered discreet bribes—a few pounds of butter or a burlap sack filled with flour—to tribunal members known to accept such things.[7] Often they traded favors or cash for the support of witnesses willing to testify that they had once helped a member of a resistance group or a Jewish family evade arrest—a practice so common that cynical Germans soon coined a new word, *Persilscheine,* or "whitewashing certificates," for such statements.[8] Faced with such ploys, tribunal members exonerated both war criminals and those who had fostered and intensified the poisonous atmosphere of racism in Nazi Germany. The tribunals themselves became known as *Mitläuferfabriken,* factories for massproducing "fellow travellers."[9]

Moreover, with the onset of the Cold War, the western Allies lost much of their fervor for finding and prosecuting Nazi war criminals, focusing their attentions instead on the new Communist threat behind the Iron Curtain. As a result, many important Nazis escaped virtually unscathed from denazification. They resumed their former jobs and picked up the pieces of their lives again, as if nothing had ever happened.

At the end of the war, this is precisely what most of the Ahnenerbe's senior researchers attempted to do.

SOON AFTER AMERICAN troops captured the city of Marburg, an eccentric, sixty-year-old savant began besieging American authorities with petitions for their assistance. Dr. Herman Wirth, who had helped Himmler create the Ahnenerbe in the mid-1930s and who publicly claimed to have discovered the cradle of a superior Aryan civilization in the high Arctic, was frantic to recover a large library of books and hundreds of plaster casts he had made of the Swedish rock art.[10] Wirth had been forced to leave them behind in the Ahnenerbe headquarters in Berlin after he was ousted as the research organization's first president, and he desperately wanted them back. He planned on resuming his research on ancient Aryan writings.

American officials in Marburg had no idea who Wirth was. They were unaware that he had once lectured widely in northern Europe, spreading the myth of Aryan supremacy to anyone willing to listen. They politely agreed to search for the material. Eventually, however, a neighbor denounced him and American intelligence officers clapped him in an internment camp while they decided what to do with him.[11] Wirth became a barracks leader there. He changed his last name to Wirth-Roeper Bosch and reinvented his Nazi past. He told authorities that he had fiercely opposed Nazism and the SS and had even been dismissed from his professorship at the University of Berlin because of it.[12] He neglected, however, to mention that he had repeatedly pleaded during the war for permission to give propaganda lectures in the Netherlands and Sweden—only to be turned down. "Cultural propaganda," wrote Sievers, "is a delicate matter which Wirth is not capable of performing in a skilful way."[13]

Wirth's disavowal of his Nazi past was masterful, however. American authorities classified him as a "political victim of the Third Reich and displaced person" and released him from internment in 1947.[14] Free again, he bundled up his wife and children and set off first to the Netherlands, the country of his birth, then to Sweden, the land of the rock art that so enchanted him. There he changed his name twice more—to Felix Bosch and later to Heinrich Bosch—and found work briefly at a private photographic institute in Lund.[15] But despite his change of address, name, and employment, he was still the same old Herman Wirth. One visitor to his home in Lund reported that a large oil painting of Wirth dressed in an SS uniform hung in the family's private library.[16]

Eager to be reunited with his large circle of admirers, he moved back to Germany with his wife in the 1950s.[17] But he continued to study and make casts of the rock art of Bohuslän. His bumbling fieldwork infuriated Swedish officials. In 1964, they accused Wirth and his assistants of permanently damaging two of the country's most important petroglyph sites.[18] They contemplated levying a fine of 5,000 Swedish crowns for damages, but eventually settled on officially banning the seventy-nine-year-old Wirth from cleaning, drawing, casting, "or in any other manner altering the conditions of the rock art of Bohuslän or any other place in all of Sweden."[19]

Wirth, however, was not so easily deterred. He continued to stage exhibitions and lectures, attracting a large group of followers who avidly lapped up his theories about an ancient Ice Age civilization in the far north.[20] With the backing of powerful supporters, he drew up plans in the late 1970s for a new museum to showcase his collection of rock-art casts and so charmed officials in Rhineland-Palatinate that they agreed to put up 1.1 million marks for the project, which was to be installed in a castle in the small town of Thallichtenberg.[21] In 1980, at the age of ninety-five, Wirth was poised on the brink of a remarkable resurrection, but just at this moment of triumph, a *Spiegel* reporter came snooping around. The resulting article exposed Wirth's Nazi past and ridiculed his befuddled ideas, dashing any chance of success.[22]

Wirth died a year later, with hardly a penny to his name.[23] His admirers, however, stubbornly refused to let his ideas die. German publishers continued to produce pseudo-scholarly books on ancient symbol research, and some German documentary filmmakers touted the notion of a primeval high civilization in northern Europe, even going so far as to borrow footage from old Nazi films to illustrate these ideas. And a former Ahnenerbe colleague acquired part of Wirth's collection for a rock-art museum tucked away in the small Austrian town of Spital am Pyhrn.[24] Today, fourteen of Wirth's large casts hang in a bright, well-lit room there. The exhibition makes no mention of Wirth or his dark past. Even so, the casts remain a shrine to Wirth's ideas and, as one Austrian scholar notes, "a clandestine Nazi memorial."[25]

IN THE SUMMER of 1946, a distinguished-looking Finnish nobleman paced the floor of a cell in Åkershus Prison in Oslo. Yrjö von Grönhagen, the scholar who recorded the magical spells of Finnish sorcerers for Himmler, and sought to bring Finland and Nazi Germany closer together during the war, insisted he was guilty of no crime. At the start of the war, he had joined the Finnish army, eager to defeat the invading Soviet forces, and when his homeland sued for peace in 1940, he returned to Berlin. There, as he later explained to his son, he worked for a time as an "extraordinary representative" of the Finnish foreign ministry, furthering "German-Finnish cultural exchange."[26] As part of this mission, he wrote a series of books on Finland for German readers; produced a German radio program on Viipuri, the old capital of Karelia; and finished a German propaganda film entitled *Freedom for Finland*.[27] All stressed the strong cultural links that bound the two countries together.

When the war ended, the Finnish foreign ministry transferred Grönhagen to Oslo, where he began repatriating prisoners of war.[28] In the midst of this work, however, the Finnish Security Police began investigating the diplomat for his political activities. They arranged for him to be arrested and detained in Åkershus Prison, where the British Security Service held those suspected of collaborating with the Nazis. The British, however, uncovered nothing damaging against Grönhagen: "He appears to have no connection with the Finnish or German SS," wrote one British officer blithely.[29] The Finnish Security Police were much more suspicious. When Grönhagen returned to Helsinki, the investigation continued. Finally in early 1947, the courts set him free.[30]

Grönhagen had hoped to return to his studies of Karelia and its rich folklore after the war. He was still keenly interested in the ancient songs of *The Kalevala*, although he seems to have lost the valuable sound recordings he had made in Karelia.[31] To his dismay, leading folklorists in Finland shunned him.[32] They were quick to snub a man who had accepted the tenets of Nazi *Rassenkunde* and alleged that the Finns were Aryans. And many were horrified by his close relationship to Himmler and the Ahnenerbe.[33] One prominent Finnish scholar, Dr. Kustaa Vilkuna, publicly accused Grönhagen of being a German spy who merely masqueraded as a folklorist.[34] One by one, the doors to Finnish academic life closed.

For a time, Grönhagen worked as a Russian interpreter in a dockyard; then he purchased a tourist hotel in Lapland.[35] In 1964, he became the general secretary of the Order of St. Constantine the Great, a Christian ecumenical organization dedicated to keeping alive the intellectual heritage of the ancient Greek and Byzantine civilizations. The former folklorist spent his summers in Lapland and his winters in Greece for more than thirty years, then finally returned to Helsinki in 2000. He lived quietly there until his death in 2003 at the age of ninety-two.

DR. FRANZ ALTHEIM, the charming classical historian whose mistress Erika Trautmann was a close personal friend of Göring, was on the faculty of the University of Halle when the Soviet army arrived in the city. After the troops had established control, intelligence officers led the university's faculty and staff down into a basement for interrogation.[36] During questioning, someone seems to have mentioned that Altheim was closely associated with the Ahnenerbe and that some of his books even bore a preface written by Himmler himself. For battle-hardened Soviet officers who would as soon string up an SS officer as look at him, this would have been a damning bit of news.

Altheim, however, stuck firmly to the story that he was just a scholar. He said nothing of his work as an intelligence agent both before and during the war.[37] In 1938, he and Trautmann had gathered intelligence for Himmler in Romania and Iraq, two countries possessing oil reserves that would prove of critical importance during the war. And they had continued to collect and pass on secret information during the war.[38] Before the invasion of the Soviet Union in 1941, for example, they had submitted a detailed report outlining three routes for smuggling weapons to pro-Nazi supporters in Iraq.[39] In addition, they also seem to have analyzed political, economic, and social conditions in Iran, supplying intelligence useful for a possible German invasion.[40]

Altheim made no mention of this work to the Soviets. He did, however, confess to straying into the political realm on two important occasions, mixing a little truth into his lies.[41] Once, he declared, he had used his Ahnenerbe connections to try to free the president of Oslo University from a German concentration camp; another time he had tried to win the release of a woman prisoner of mixed Jewish ancestry, the daughter of one of his closest friends.[42] Both of these assertions were true, and they seem to have swung the balance in his favor. The Soviets arrested two of the university staff and deported them to Siberia. But they gave Altheim a clean bill.[43]

The historian returned gratefully to his teaching. But he was not ready to settle back into the life he had previously known with Trautmann. He seems to have found her too old to be desirable, and it may have occurred to him that a mistress with high Nazi party connections was more of a liability than an asset to a man with a past to hide.[44] As it happened, he had met someone else. At Halle he had discovered a beautiful young student,

Ruth Stiehl, with flowing red hair and a brilliant gift for languages and classical history. He fell in love with her, and when he escaped to the West in 1949 at the age of fifty-one to take up a professorship at the Free University in Berlin, Stiehl accompanied him.[45]

Even so, Altheim did not abandon Trautmann. He found a house on a quiet street in Grunewald forest in Berlin, a lovely spot tucked away in the woods. There, in one part of the house, he lived with his new young mistress, collaborating on books and academic publications, just as he had once done with Trautmann. In the other part, Trautmann lived quietly, doing photography work for her former lover and cooking dinners and keeping house for him and his new mistress.[46] It was a curious ménage à trois, one that puzzled Altheim's friends and students. But the historian undoubtedly had his reasons. He loved to shock bourgeois sensibilities.[47] And Trautmann was likely the only person alive who knew about his past as a Nazi intelligence agent. _- VERY PART-TIME_

Altheim went on to a brilliant career at the Free University, and an equally brilliant retirement in Münster. He authored or coauthored more than 250 publications, on subjects as diverse as Asian feudalism, the Arab world before Muhammad, and the history of the Huns.[48] He dazzled students with his wit and the vast sweep of his knowledge, and he rose to the top of his profession, becoming, as one obituary writer later recorded, "one of the best known scholars of German antiquity studies."[49] Even in his final years, he was a legendary figure. He drove to the University of Münster each day in a white Mercedes 220 S coupe with Stiehl, a beautiful woman dressed in a leopard-skin coat and cradling a Pekinese in her arms. "They looked like two mythical beings," recalled a former student with a laugh.[50]

Trautmann died in October 1968 and Altheim bought her an old Germanic style gravestone, in keeping with her last wishes.[51] He died eight years later of cancer. Neither his friends nor his former students had any inkling of his past as an intelligence agent. As fond as Altheim was of attracting attention, he had never breathed a word of his clandestine activities. He filled the walls of his study with photographs of famous people he had known during his career. But there were no photographs of Himmler or Göring or colleagues from the Ahnenerbe. "That was a cleansed picture," noted one of his former students. "He wanted to wipe it away, I think."[52]

IN HIS LABORATORY at the University of Groningen, Dr. Assien Bohmers continued to pore over the strange stone spearheads of Mauern during the late 1940s. The scholar who had sought the origins of the "Cro-Magnon race" in Germany and who had promoted Nazism along the northeastern Dutch coast, in hopes of one day becoming the Gauleiter of a Nazi Friesland, had landed neatly on his feet. He had convinced his Allied captors to release him from prison after just nine months of internment by brazenly claiming to have been part of the same resistance group as Sievers.[53] He had then managed to land a position as a research worker at the renowned Biological-Archaeological Institute at the University of Groningen in the Netherlands, where he had taught as a lecturer during the war.

Just how Bohmers managed to finagle his way into a teaching position in a country that reviled the Nazis is very unclear. "Bohmers," explains one Dutch scholar who met the archaeologist after the war, "was a very enigmatic man, and I think even his nearest colleagues didn't know him well."[54] But a former colleague of Bohmers who has examined his case closely believes the prehistorian may have blackmailed the head of the institute, Dr. Albert van Giffen.[55] During the Nazi occupation of the Netherlands, Bohmers persuaded van Giffen, one of the country's foremost archaeologists, to remove his signature from a petition opposing the dismissal of Jewish university professors, and it seems likely that he threatened to expose van Giffen's action and portray him as an anti-Semite.[56]

At Groningen, Bohmers went about his archaeological work with suitable professionalism, but his personal behavior grew increasingly furtive and erratic. During the saber-rattling of the Cold War, he became obsessed with the idea that the Soviet Union would invade the Netherlands. He was particularly worried about being captured. He sold his house and bought a large seaworthy yacht so that he could sail to Scandinavia should the Soviet Army suddenly arrive.[57] He stocked the yacht with a bristling arsenal of firearms, all purchased illegally from an assistant who was stealing artifacts from the institute. After a police investigation, the University of Groningen suspended Bohmers, and in January 1965, the archaeologist asked to be dismissed.

He then turned his energies to marketing the crafts of traditional Dutch wood-carvers. In 1972 he gave a talk to a group of Frisian university students about his years in the Ahnenerbe. By that time, he had reinvented

Even before Warsaw surrendered in 1939, Himmler put Peter Paulsen (right) in charge of stealing Poland's treasures for the SS. *Bundesarchiv, RuS Paulsen, Peter*

Nazi researchers were fascinated by the rune-covered spear of Kowel (shown in two views). Paulsen stole it in Warsaw and sent it to Himmler. *From* Die Krise der Alten Welt

Warsaw's prince-bishop protested passionately against Paulsen's theft of the famous Veit Stoss altar (a detail shown right). *ullstein-Poss*

Bundles and trunks filled with stolen currency and looted art lie neatly on the floor of a German salt mine, where the Nazi government stored them. During the war, Ahnenerbe scholars played a key role in several looting missions. *USHMM, courtesy of National Archives*

Herbert Jankuhn (third from right) poses with his wife and colleagues at a German dig in the early 1930s. As the head of the Ahnenerbe's prehistory and excavations department in 1940, he became "the most powerful archaeologist in the Third Reich." *Harald Jankuhn*

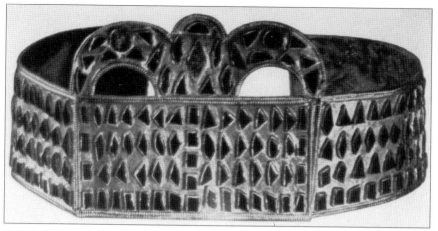

To justify the German invasion of Russia, many Nazis claimed the Gothic tribes of the Crimea as their ancestors. In 1942, Himmler sent Jankuhn to the Crimea and Caucasus to find Gothic jewels to rival the crown of Kerch (above). *From* Geschichte 6.Klasse: Oberschulen/Gymnasium

Himmler loved boasting that Gothic tribes once founded a great empire along the Black Sea shores. But Ahnenerbe archaeologists could find little proof of this. Instead of a major metropolis, they found only cave cities such as Chufut-Kale (above). *John Noble/Lonely Planet Images*

A woman wails as she finds her son among corpses left by the German forces in the Crimea. *Getty Images*

Himmler inspects a cotton crop in the Crimea. When the war ended, he planned to expel and murder most of the region's inhabitants and replace them with ethnic Germans and SS farming colonies. In his maps, he renamed the region Gotengau or "Goth region." *USHMM, courtesy of James Blevins*

Menacing racial exhibits on the Jews sprang up across the Reich in the late 1930s. But German scholars were unable to scientifically define the Jewish race, a fictional construct. One survey showed that 11 percent of Jewish children were blue-eyed and blond. *Naturhistorisches Museum Wien, Abteilung für Anthropologie*

Ernst Schäfer displays a trophy at Mittersill. In 1942, he took command of a mission to "racially explore" tribes in the Caucasus. *NARA, Still Pictures Unit, RG-319*

Schäfer picked Bruno Beger to racially assess Caucasus tribes, such as the Mountain Jews (left). If deemed Jewish, they would be slated for execution. *From* Jewish Encyclopedia *(1901–1906)*

An anatomist by training, August Hirt headed an Ahnenerbe project in 1943 to murder 86 Jewish prisoners for their skeletons. *Privatarchiv Hans-Joachim Lang*

Natzweiler concentration camp possessed an evil reputation. It was there that Hirt and Beger performed research on 86 Jewish men, women and children, who were then gassed. A driver delivered the corpses to Hirt's lab for defleshing. *USHMM, courtesy of Abraham M. Muhlbaum*

Wolfram Sievers and other defendants in the Medical Trial listen to testimony at Nuremberg. Sievers (lower right-hand corner) insisted that he joined the Ahnenerbe to aid a German resistance group. Unconvinced, the judges sentenced him to death.
USHMM, courtesy of Hedwig Wachenheimer Epstein

In October 1970, Bruno Beger (right) was brought to trial in Frankfurt for aiding and abetting in the murder of 86 Jews at Natzweiler concentration camp. He was convicted of this crime, receiving a sentence of three years on probation.
ullstein–dpa (85)

the entire history of the research organization. "According to Dr. Bohmers," one of the audience members later recalled, "there was a group, a core in the Ahnenerbe, of highly qualified scholars who formed a kind of secret society opposed to National Socialism, or maybe [more accurately], to Hitler and his ideas."[58] Throughout the presentation, Bohmers spoke of a certain "Heini" with admiration. At first audience members had no idea who he was talking about, but they eventually realized he was referring to Himmler. "I was shocked to meet a man who had been so near to Heinrich Himmler," explains one of the audience members, Dr. Oebele Vries, today.[59]

Disenchanted with life in the Netherlands, Bohmers moved to Sweden, and died in Gothenburg in 1988. Since then, his professional reputation has sunk lower still in the eyes of Dutch archaeologists. A prominent Groningen scientist, Dr. Tjalling Waterbolk, who once headed the Biological-Archaeological Institute, has since completed an extensive investigation of an archaeological forgery ring that tried to pawn off fake stone hand axes as Middle Paleolithic treasures in the mid-1960s. He concluded that Bohmers was most likely the ringleader.

DR. ERNST SCHÄFER departed for the cloud forests of Venezuela in 1950, eager to return to his old life as a hunter and naturalist. He had served Himmler faithfully, searching for Aryans in the mountains of Tibet and touring occupied Europe during the war as one of the exemplars of Nazi science. He had put concentration-camp prisoners to work at Mittersill and accepted the command of the Caucasus mission, which was designed in part to racially diagnose the Mountain Jews prior to their liquidation. One knowledgeable witness, Walther Wüst, even declared after the war that Schäfer had sat on the board of trustees for the Institute for Military Scientific Research, which had presided over the Ahnenerbe's medical experiments.[60]

But in the final days of the war, Schäfer had erased the official record of his deeds as thoroughly as he could. He carefully burned incriminating documents at Mittersill and destroyed other key pieces of evidence, including Gabel's plaster casts of the Auschwitz prisoners, which could have been used to identify the victims.[61] During his denazification hearing, his attorney presented affidavits from more than forty witnesses who stated that Schäfer had cooperated with resistance groups and assisted Jews and Polish scientists persecuted by the Nazis. On the strength of this evidence, the local denazification tribunal cleared him in June 1949, awarding him an "exonerated" classification.[62]

By then, Schäfer was anxious to put Germany far behind him. Through a friend, he lined up a job as the director of Rancho Grande, a biological research station in Venezuela, and flew there with his wife and three children.[63] The station amounted to little more than a bizarre-looking concrete bunker set at the top of a windy hairpin road in the middle of dense forest, but Schäfer was enchanted. The forest and the lowlands that surrounded it teemed with life—howler monkeys, fer-de-lance, coral snakes, tapirs, pumas, giant anteaters, and more than five hundred species of birds. Ornithologists had yet to study or describe many of these species, so Schäfer spent the next few years observing the birds in their natural habitat and hunting specimens for the station's research collection.

From time to time, foreign ornithologists dropped in at the station, keen to see the forest's rich fauna, and it was on one of these visits that Schäfer met the former Belgian king, Leopold. Leopold had been forced to abdicate the throne after his swift surrender to the Wehrmacht during the war; many Belgians thought him a traitor. To fill his empty hours, the former king had taken up ornithology and natural history. At Rancho Grande, he and Schäfer became fast friends, and he insisted that the German zoologist come to work as his scientific advisor in Belgium. Schäfer accepted the offer. He wanted to educate his children in Europe.

Leopold lodged the family in luxury and gave Schäfer a big workspace in Castle Villers-sur-Lesse. He then commissioned him to make a film to commemorate the fiftieth anniversary of Belgium's annexation of the Congo. Leopold had raised sixty million Belgian francs—the equivalent of $7.59 million today—to make it.[64] Schäfer jumped at the chance. He roamed the Congo for months with a large film crew, shooting *The Lords of the Ancient Forest*. From time to time, Leopold and his wife Lillianne flew in to see how things were going, and together they planned a grand premiere. Just before the film opened, however, veterans of the Belgium resistance announced that the director was a former Himmler man. The bombshell sparked a furor in Brussels. Leopold asked the Schäfers to leave the country quietly. He feared the scandal would bring down the royal house. Schäfer, however, countered with a lawsuit, eventually accepting a reported settlement of two million Belgian francs, the equivalent today of some $252,800.[65]

In Hannover, Schäfer landed a position as a curator at a prominent museum. He was immensely happy to be back on German soil, but during his lengthy absence, strange stories had begun circulating in high society about his wartime activities at Mittersill. Mittersill's owners had restored the property to its previous grandeur and opened it again as a sport and shooting club where the old European nobility could mingle happily with the likes of Clark Gable and Gina Lollobrigida.[66] But according to one popular story, the castle's owners had found on their return "thousands of skulls from Tibet, India and China eerily stored on new constructed shelves in all of the rooms."[67] The tale aroused much comment and particularly intrigued one Mittersill guest, Ian Fleming.[68] After hearing the story, the creator of James Bond transformed Mittersill into the fictional scientific research station of Piz Gloria, the Alpine lair of Bond's archenemy Ernst Blofeld.

Schäfer, however, paid no attention to these tales. He continually reshaped his past to suit the times, editing out any parts that cast him in an unflattering light. He retired to the quiet spa town of Bad Bevensen in Lower Saxony, and remained an avid birder and naturalist until the end of his days. He died on July 21, 1992, at the age of eighty-two.

EDMUND KISS NEVER did lead a great scientific expedition to the Bolivian Andes. The architect who searched for proof of an ancient Aryan high civilization at Tiwanaku and who penned fantasy and adventure novels to popularize Hitler's racial theories, remained in active military service throughout the war, first as an officer and later as the commander of an antitank gun division in Norway, East Prussia, and Poland. Near the end, he assumed command of the SS men at Wolfschanze, guarding the dank bunker complex where Hitler had once mesmerized his dinner guests with table talk about the Crimea.[69]

Arrested and interned, Kiss entertained his fellow prisoners with stories of his travels in Bolivia and Libya. By the end of the war, however, the fifty-nine-year-old officer was suffering severely from diabetes, so American authorities released him from Darmstadt camp in June 1947.[70] Under the terms of his discharge, he was not allowed to take any form of work, except menial manual labor, and the denazification tribunal proposed classifying him as a "major offender," noting that Kiss was a member of

Himmler's personal staff and had received an honorary SS dagger. Alarmed by this, Kiss hired an expensive lawyer and countered the charge, pointing out that he had never taken out membership in the Nazi party.[71]

At his denazification hearing in 1948, Kiss tried to present himself as a reformed man. He explained, for example, that he had developed serious doubts about the World Ice Theory. Colleagues in England and the United States had mocked the theory, regarding him as "the complete idiot from Germany."[72] Moreover, the "new nuclear theory" had given him pause for thought: he was revising his ideas accordingly.[73] But Kiss refused to completely renounce other Nazi notions. On the stand, he declared his belief in Rassenkunde, stating "there is something to racial theory, no question about it."[74]

The tribunal obligingly reclassified Kiss as a "fellow traveller" and fined him 501 marks.[75] Kiss paid in full. Then, with the hearing behind him, he retired as a writer. In December 1960, he died. His fantasy novels of Tiwanaku and Atlantis were almost immediately forgotten, gathering dust on the storage shelves of libraries. But his research on the World Ice Theory and his crackpot theories on the sun gate of Tiwanaku stubbornly lived on. In the 1950s and '60s popular authors such as H. S. Bellamy and Denis Saurat published Kiss's ideas in a series of books. These in turn gave birth to a new generation of fabulists, whose numbers include such modern writers as Graham Hancock. One of Hancock's most popular books, Fingerprints of the Gods, devoted three chapters to wild speculations on the origins of Tiwanaku, selling more than three million copies after its publication in 1995.

ART EXPERTS ATTACHED to the staffs of the Allied military commanders hunted determinedly after the war for the treasures plundered from Poland and other European nations. They questioned German museum directors, combed government records, and searched thousands of possible hiding places, from remote salt mines and barns to dusty medieval cellars. They inventoried rooms brimming with loot—Old Masters, ancient gold coins, classical Chinese bronzes, ancient Scythian jewels—and dispatched their contents to Allied collection points so that they could be returned to their rightful owners. They became sleuths on a very grand scale; the war years had seen the largest plundering of artworks in history.[76]

Dr. Peter Paulsen, the archaeologist who stole the Veit Stoss altar from Kraków and orchestrated the looting of Poland's archaeological treasures, did his best to bury his past. When the war ended, he spent three years in internment, then took a series of teaching jobs. In 1961, he landed a prestigious position as a medieval expert at the provincial museum in Württemberg.[77] He became an expert on the Allamanni, an ancient German tribe, and reestablished his scholarly reputation. "He was well-regarded," recalled prominent German archaeologist Achim Leube. Indeed Paulsen remained active in archaeology until he developed eye problems in his old age. He died in 1985, at the age of eighty-three. His obituary, like so many others of the time, skipped lightly over his wartime activities. It made no mention of his work for the SS and the Ahnenerbe.[78]

Paulsen's colleague in the plunder, the Austrian museum director Eduard Tratz, enjoyed an even smoother ride back to respectability. During the war, Tratz had personally ransacked the Natural History Museum in Warsaw, stealing specimens to mount Nazi racial exhibits in the museum he founded, Haus der Natur. A department head in the Ahnenerbe and a major supporter of the Nazi regime in Salzburg, Tratz thought it important to educate the Austrian public in the theories of *Rassenkunde*. After the war, he spent two years in internment, receiving the "lesser activist" classification.[79] It amounted to no more than a slap on the wrist, for by 1949 he had resumed his old job as director of the Haus der Natur.[80]

Allied investigators forced the museum to return many of its looted artifacts and zoological specimens, but they could not induce Tratz or his successors to discard all the museum's Nazi exhibits. As late as the 1990s, the Haus der Natur exhibited racial casts of the supposed Nordic and Jewish races.[81] And even today, tours of schoolchildren gaze unknowingly at Tibetan mannequins created from the head casts made by Bruno Beger, one of the collaborators in the Jewish Skeleton Collection.

Tratz, a highly respected member of Salzburg society, died in 1977. A bronze bust of the zoologist now adorns the museum's main foyer.

DR. HERBERT JANKUHN, the Ahnenerbe archaeologist who fraternized with leaders of Einsatzgruppe 11 in the northern Caucasus, and who sought evidence of a Gothic empire in the Crimea to bolster Nazi claims to the future Gotengau, spent the last years of the war as an intelligence

officer with Viking Division on the eastern front. Like most German soldiers, he dreaded capture by the Soviet army, and was much relieved when
his commanding officer decided in the dying days of the war to lead the
Viking Division on a hurried retreat westward. In this way, Jankuhn and his
fellow Viking Division officers managed to surrender to American troops
entering Bavaria.[82]

The archaeology professor was interned in Allied camps for three
years. When he was released, his eleven-year-old son did not know who he
was. He was a bitter man. His family possessed little money for the *Persilscheine* needed to win a lighter sentence. As a result, the local denazification tribunal barred him for several years from holding a university
teaching position, and Jankuhn was forced to hustle for work.[83] He sought
out and received annual scientific research grants to continue his excavations at the old Viking site at Haithabu, near the Danish border, one of the
most important sites in Germany.[84] In the winter, he analyzed the data,
published his findings, and gave guest lectures at Hamburg and Kiel.[85]

By such means, he supported his family and managed to advance his
scientific reputation. In 1956, the University of Göttingen offered him a
teaching position, which Jankuhn gratefully accepted.[86] His career at the
university was meteoric. Ten years after his appointment, the university
named Jankuhn the dean of the philosophy faculty.[87] While a few of his
students and colleagues knew of his SS background, most at Göttingen
were unaware of his work in the Ahnenerbe and of the influence he once
wielded as the most powerful archaeologist in Nazi Europe.[88]

Scholars elsewhere, however, were a good deal more knowledgeable
and far less forgiving. In 1968, for example, Jankuhn planned a trip to Norway and offered to give a public lecture at the University of Bergen. The
university turned the suggestion down cold, for faculty members still remembered his imperious manner as an SS archaeologist in Oslo during the
war, and the way he had once denounced a Norwegian archaeologist, Anton Brøgger. "Jankuhn was not welcome," recalls another Norwegian archaeologist, Anders Hagen, who was then on faculty at Bergen. "He was a
really intelligent man, but he couldn't understand these things."[89]

Jankuhn later told his sons that he could not explain his behavior during the Nazi years, even to himself, but he never apologized. Indeed, he
staunchly defended his old comrades. During an interview with historian
Michael Kater in 1963, he declared that the SS was largely innocent of

genocide. Only SS concentration-camp guards, he stated, had truly persecuted Jews. It was an immensely cynical attempt at whitewashing—given
his association with the leaders of Einsatzgruppe 11 in the Caucasus.[90]
Moreover, he never abandoned his ideas about German territorial claims in
the East. In later life, he told his son Dieter that "the world doesn't stop at
the Iron Curtain. And anyway, these parts have been the settlements of our
ancestors."[91]

Jankuhn died in 1990, honored, respected, and eulogized as one of the
deans of German archaeology.

WALTHER WÜST, THE Aryan studies expert who led the Ahnenerbe for
eight years, from 1937 to 1945, and whom the *Völkischer Beobachter* once described as "among the most loyal and dependable men to the Führer," was
interned for forty months at the end of the war.[92] During this time, American intelligence officials attempted to build a case against him for trial at
Nuremberg, but they found it extremely difficult going. Wüst had been
very cautious in his handling of the Institute for Military Scientific Research, delegating operational responsibility for the medical experiments
to Sievers. Adding to the difficulty of the case was the fact that investigators lacked Wüst's own correspondence files from the Ahnenerbe bureau
in Munich. The office had been hit in an aerial attack near the end of the
war and the files seem to have been destroyed or lost in the rubble.[93]

Under interrogation, Wüst claimed to know nothing at all about the
human experiments or the skeleton collection, and insisted stubbornly that
the Institute for Military Scientific Research was completely separate from
the Ahnenerbe, although this was patently false.[94] Moreover, he repeatedly
claimed that the Ahnenerbe was merely a conventional research organization, devoted to the scholarly pursuit of truth and knowledge. He stuck
firmly to this story, much to the frustration of investigators, who made
note of his "pathological attempts to whitewash himself of his clearly established responsibilities."[95] In the end, prosecutors at Nuremberg chose
not to put him on trial.

The local denazification tribunal classified Wüst in 1950 as a "fellow
traveller," and the University of Munich took him back as a professor-in-
reserve, paying him a monthly salary of 494 marks.[96] But Wüst did not return to the classroom or the administration. He found work instead at the

Bavarian State Library in Munich, where he could often be seen in the 1950s, a small man working on slips of paper for what he hoped would be his magnum opus, a dictionary of Old Indoaryan. Many of his former colleagues were baffled by his apparent inability to reconstruct his academic career, when so many other Nazi scholars were slipping back effortlessly into their former posts. "It is as if the earth had swallowed you and the 'research society,'" marveled one colleague, Gustav Freytag, who hoped that the "disinfection" would soon be over and that Wüst would find a position suitable to his "great skills."[97] But Wüst's isolation and the withdrawal from academic life may have been self-imposed. Some former colleagues thought that the scholar had gone a little "funny" after the war, working on strange new ideas, such as the role of bears in Paleolithic societies.[98]

In 1958, the German federal states founded a new agency, the Zentrale Stelle, or Central Office, in Ludwigsburg, for investigating Nazi war crimes committed in concentration camps, Jewish ghettos, and other places that had no relation to military activities. The new agency had its hands full of crimes to examine, mounting nearly four hundred major investigations in the first year of its operation.[99] But from the start, its staff took a keen interest in the Ahnenerbe and its former head, Walther Wüst. Document experts sifted through all available records of Wüst's activities during the Nazi regime. Among the Ahnenerbe papers, they found the office diaries of Sievers. These delineated Wüst's responsibility for the Institute for Military Scientific Research, and revealed that the scholar had even attended official meetings with Himmler concerning the institute's medical experiments.[100]

But investigators from the Central Office were unable to find a smoking gun to link Wüst clearly and directly to the crimes.[101] In 1972, the prosecuting attorney was forced to suspend the investigation.[102] Current archival research, however, reveals that Wüst was well aware of the nature of at least some of the medical experiments, while they were under way. In 1944, for example, as Hirt continued to subject concentration-camp prisoners to lethal mustard-gas experiments, Wüst recommended that the anatomist be promoted to the rank of SS-Sturmbannführer, referring specifically to his work on "secret experiments on poison gases, etc."[103] In this letter, Wüst observed that "Hirt has taken on this task with selflessness and diligence, in such a manner that it has badly compromised the condition of his health, not the least of which is due to the use of poison during the experiments."

To the end of his life, however, Wüst proclaimed his innocence. He died on March 21, 1993, at the age of ninety-two, never having finished his Indoaryan dictionary.

AMERICAN FORCES CAPTURED Bruno Beger in Italy at the end of April 1945. The anthropologist who had collaborated with August Hirt and selected and measured Jewish prisoners for the Ahnenerbe's skeleton collection, had spent the last part of the war conducting racial research on a Muslim division of the Waffen-SS.[104] He mulled over his future in a series of Italian and German prisoner-of-war camps, then was interned at Darmstadt for fourteen months.[105] Intelligence officers interrogated him several times. In February 1948, a local denazification tribunal classified the anthropologist as "exonerated," unaware it seems of his exact role in the skeleton conspiracy.[106] Beger returned home to his wife and five children a free man.

He had few job prospects, however. His academic specialty— *Rassenkunde*—had disappeared from university syllabuses across Germany, and research grants for the moribund field had completely dried up. Beger was forced to look for another line of work. He and an old friend, Dr. Ludwig-Ferdinand Clauss, another racial researcher, found positions at a publishing house in Oberursel specializing in educational books.[107] After that, Beger became a sales representative for a large paper company and went into the paper business.[108]

All the time, however, he seems to have yearned to return to scientific studies. He traveled on a private research expedition in 1954 to Algeria and Morocco with Dr. Volkmar Vareschi, an old colleague from the Ahnenerbe institute at Mittersill, and on a second trip to the Middle East in 1958 and 1959 with Clauss.[109] A year later, he tried publishing a serialized version of his Tibetan diaries in a local Frankfurt newspaper, the *Nordwest Spiegel*. Less than a third of the diary appeared, however. Reader complaints seem to have forced the editors to cancel the rest. "Not only did I have to defend myself against unjustified reproaches," complained Beger later, "I had to concentrate on the earning of money and my worries about my large family."[110]

In 1960, the Central Office in Ludwigsburg began a preliminary investigation of all those involved in the skeleton collection conspiracy. The staff carefully pored through archival records and attempted to track down witnesses who knew about the collection or had observed the events that took place at Auschwitz, Natzweiler, or Strassburg. On March 30, 1960, the police took Beger into custody. Four months later, he was released. Nevertheless, investigators continued to amass thick folders of evidence on Beger and two colleagues—Dr. Hans Fleischhacker, the anthropologist who had assisted Beger in performing the measurements at Auschwitz; and Wolf-Dietrich Wolff, the SS officer who had worked for Sievers. Eight years later, the Central Office had gathered sufficient evidence to press charges. So it handed over its files to prosecuting attorneys in Frankfurt, where Beger then lived.[111]

The trial opened on October 27, 1970. Throughout the proceedings, Beger insisted that Sievers and Hirt had kept him in the dark about the ultimate fate of the selected Jews until after he left Auschwitz. When he had learned the truth, he claimed, he was horrified and had journeyed to Natzweiler with the intention of confronting Hirt and dissuading him from killing the prisoners. Witnesses and surviving Ahnenerbe documents, however, poked holes in this story. Evidence from Sievers's diary showed that the racial specialist had discussed a Jewish skull collection as early as 1941, two years before the murders. Moreover, Sievers's former personal secretary and mistress, Gisela Schmitz, testified that it was Beger, rather than Hirt, who had written most of the notorious letter proposing that a Jewish skull collection be made from the heads of "Jewish-Bolshevik commissars." More damaging was the official expense claim that Beger submitted for the trip to Natzweiler. On it, he had noted that the purpose of the trip to the French concentration camp was to perform blood tests and take X-rays of the prisoners.[112] To refute this damaging evidence, the defense presented testimony from several old associates and friends of Beger, who recalled that the anthropologist had been disturbed by conditions he had observed in the camp.

It was a long and complex trial, but it did not rouse the public furor that might have been expected. By the early 1970s, many Germans had wearied of newspaper articles about war-crimes trials and Nazi atrocities. Moreover, a surprising number of Germans still sympathized with the Nazi cause. Indeed, one 1971 German survey showed that 50 percent of the population still held that "National Socialism was fundamentally a good idea which was merely badly carried out."[113]

In early March 1971, the Frankfurt court dismissed the charges against

Fleischhacker. The prosecution was unable to prove that he had known ahead of time about the plan to murder the prisoners.[114] A month later, it suspended the charges against Wolff, noting that the former administrative assistant was guilty of little more than naïveté.[115] But the prosecuting attorney claimed a victory in the case against Beger. On April 6, the court in Frankfurt convicted the anthropologist of being an accomplice in the murder of eighty-six Jews in the gas chamber at Natzweiler.[116]

The judge then proceeded to deliberate on a suitable sentence.[117] He observed that the anthropologist had fallen as a youth under the influence of Nazi party doctrine, which had clouded his critical judgment. He noted that Beger had performed the research at Natzweiler against his will, convinced that he could not prevent the slaughter. He also recalled that Beger had written a letter to Sievers after his trip to Auschwitz, favorably mentioning four prisoner-assistants by name. This, the judge stated, demonstrated that the anthropologist felt compassion for those imprisoned in the camp, although it is highly unlikely that any of the four listed prisoner-assistants was Jewish.[118] Finally the judge noted that Beger had been forced to wait ten years for the trial, which had placed him under significant psychological stress, and had already endured internment after the war and four months in custody in 1960. On the strength of these factors, the court was inclined to clemency. It sentenced him to a three-year prison term and ordered that he pay the costs of the trial.[119]

Beger's friends were much disturbed by this sentence. They thought it unduly harsh and wrote letters of protest to the local newspapers, pointing out that the former SS officer was guilty merely of following orders. Beger's lawyer filed an appeal.

In 1974, a German appellate court reduced his sentence to three years on probation.[120]

24. SHADOWS OF HISTORY

In FEBRUARY 2002, I learned that Bruno Beger was still alive. He was ninety years of age and living at the time with a daughter in the small German town of Königstein, so I wrote him a short letter, requesting an interview. Nearly sixty years had passed since Beger had stood in front of the concentration-camp barracks at Auschwitz, selecting men, women, and children for his research on Jews, and in that time, science had utterly repudiated the foundations of Nazi racial doctrine. Indeed, leading researchers from around the world had convened several times to set the scientific record straight on the concept of racial difference.

At the invitation of the United Nations Educational, Scientific and Cultural Organization, for example, twelve of the world's leading physical anthropologists and geneticists had gathered in Paris in 1951 to review the scientific literature and to issue an official statement on race, the second since the war had ended. While the twelve scientists struggled to define precisely what race was—noting that many populations could not be easily fitted into any existing racial classification—they had no trouble at all in delineating what it was not. "Moslems and Jews," they observed pointedly, "are no more races than are Roman Catholics and Protestants."[1]

They then proceeded to examine the basic tenets of *Rassenkunde*, finding them severely flawed. "There is no evidence for the existence of so-called 'pure' races," they concluded, adding that "available scientific knowledge provides no basis for believing that the groups of mankind differ in their innate capacity for intellectual and emotional development."[2] The notion of superior or inferior races was mere fiction. And as for the perils of miscegenation—something that had greatly troubled Nazi scholars—the researchers simply shook their heads. "There is no evidence that race mixture produces disadvantageous results from a biological point of view."[3]

By carefully reviewing the scientific literature, the scientific community unmasked Germany's racial experts as mere bigots. It stripped away their scholarly pretenses and vanities, revealing the true nature of their work—the advancement of racism and the dissemination of anti-Semitism. And as I reflected on this, I wondered what the senior staff of the Ahnenerbe, men who had constructed their research on the treacherous ground of *Rassenkunde*, might truthfully say about their research today. Could they admit to themselves and to others the errors of their scholarship? Could they explain why they had abandoned real science and scholarship for such a skein of lies? And would a distance of sixty years bring some clarity and light into the dark shadows of their lives?

I could think of no better person to answer these questions than Bruno Beger. Interviewing the former Nazi racial expert would bring the entire story of the Ahnenerbe—from its mad and seemingly harmless global quest for ancient Aryans to its final years of barbarous cruelties—out of the shadowy world of the past and place it firmly, immediately in the present. I was immensely curious to hear what Beger might say today about his racial research in Tibet, his stint in the Ahnenerbe, and his involvement in a major war crime, but I did not hold out much hope that he would agree to talk to me. It seemed highly unlikely that he would be willing to discuss such disturbing matters. So I was surprised when I eventually received word that he would see me.

On a cool, damp April afternoon, a colleague and I took a train from Frankfurt to the small town of Königstein. I felt very uneasy about the appointment that lay ahead, and by the time we arrived my spirits were as leaden as the skies outside the train windows.

KÖNIGSTEIN IS A pretty little Hessian town, a quiet suburban retreat for the health-conscious and the moneyed. In summer, Königstein's residents hike the nearby wooded trails and take the clean, bracing air. They wander the romantic ruins of Königstein Castle and plan day trips to the rustic vineyards of the Rhine to sample the latest vintage of their favorite Riesling or Müller-Thurgau. Königstein, in other words, is a very German kind of place. Just slightly off the main tourist path, it is a town where English is not often heard on the streets and where foreigners are unlikely to intrude, unless they happen to be relatives visiting from America or Britain. And the people of Königstein seem to like it that way. Certainly Bruno Beger did.

Just a month short of his ninety-first birthday, Beger kept a low profile. He refrained from listing his name, as his neighbors did, beneath the row of door buzzers at the main entrance of his building—a fact he omitted to mention, along with his suite number, when he agreed to see me. To enter past the building's security system and find Beger, my colleague and I were forced to wait patiently for a resident to happen by in the middle of the afternoon, someone who was willing to check for Beger's name on the mailboxes inside. Only then were we able to proceed up the stairs to the fourth floor and down a narrow corridor to Beger's apartment. And it was just at that last moment, when I stood at the doorstep, hand poised to knock, that Beger subtly began to reveal his presence.

Alongside his doorway, someone had tacked up four photographic prints—glimpses, it seems, of Beger's favorite mental landscapes. Two of these showed typically German scenes, rustic images of handsome farm folk going about their chores in Lower Saxony—images that could have come straight from the pages of an SS calendar sixty years ago. The remaining two, however, showed something far more exotic, something strange and thrilling and rather mysterious, particularly for this stolid suburban apartment building. They were photographs taken half a world away of Tibetans staring into the camera's eye—curious, puzzled, and bemused.

Beger's daughter, a snowy-haired woman in her sixties, answered the door and led the way into a cozy private study, a place that resembled a miniature Tibetan museum. Black-and-white photographs of shy Tibetan women and children stared down from the wall, just a few of the thousands of photographs that Beger had taken during his travels across Tibet in 1938 and 1939. Stacks of papers about Tibet cluttered the top of an old wooden desk. Rich Oriental rugs that might have been Tibetan hung from the walls. And some of the souvenirs from Beger's famous trek to Lhasa were placed artfully around the room. One that caught my eye was a small Tibetan container. Its top was shaped like a whitened human skull.

Beger eased himself down a little stiffly into a straight-backed chair that his daughter had drawn up to a table. She sat down protectively beside him. I had expected him to be frail and withered and quite deaf, for this was the impression he had given on the phone. But in this I was mistaken: he was not at all ground down by time. Indeed, he looked almost preternaturally young—hale, hearty, alert, and sharp, with a full head of hair and an almost roseate hue to his still full cheeks. He was without a doubt the fittest ninety-year-old that I had ever met or seen. Moreover, there was something kindly and avuncular, something friendly and even a bit ingratiating about his manner, as if he had just stepped from a Norman Rockwell painting.

Beger seemed eager to see us, as if he had been looking forward to the meeting. Like many old people, he enjoyed talking about his youth. He relished chatting about his ancestors and his university years, and when we eventually brought out an assortment of photographs of the Tibet expedition that we had located in a German archive, he flipped through them with delight, describing some of the adventures that had befallen him in Tibet. To our astonishment, he casually boasted that he still possessed the calipers and sliding compasses that he had used to perform racial measurements in Tibet. Then, with no prompting at all, he asked his reluctant daughter to fetch one of the notebooks he had used to record racial measurements in Tibet. Such notebooks, he explained, were often used by anthropologists of the day. Indeed, he had received advice about using them from Theodor Mollison, the racial expert who had trained Josef Mengele.

Beger, it transpired, was not hesitant to talk about old racial theories. He pulled out a still-treasured copy of one of Hans Günther's books on *Rassenkunde* from a book cabinet. He flipped open the pages and proudly pointed to a map that he had designed for the book as a young student. Nor did he particularly mind talking about his ideas on the origins of the Jewish "race." Indeed, he still believed—as he and other German *Rassenkunde* specialists had in the 1930s—that the Jews were a mongrel race, "with strong Mongoloid elements, very strong Mongoloid characteristics."

Only when the conversation finally turned and reached at last its inevitable conclusion, the Jewish Skeleton Collection and his particular role in that crime, did Beger's memory begin to falter. But he did not stop talking. He insisted, in a patter that sounded exceptionally well rehearsed, that he was an innocent dupe, a stooge really who had been much too trusting and naïve in his dealings with Sievers and Hirt. When he finally learned the real plans for those he had measured, he explained, "I was of course very angry. What a bad joke to be suddenly pulled into this." But in the awkward silence that followed, he expressed no regret, no sign of sympathy or compassion for the eighty-six Jewish men, women, and children whom he had helped consign to a gas chamber at Natzweiler. He seemed to regard them as minor supporting characters in a greater tragedy.

In the three and a half hours I spent with Beger, his emotions only got the better of him once and that was when he began to describe his trial. It was, he said pointedly—as if this would explain everything—"started by a Jewish lawyer." And it was obvious that he was still baffled by his conviction, unable to fathom how anyone could think him a criminal. Indeed he seemed to see himself as the real victim of the tragedy, a man much wronged by the judicial system and the politics of the day. "They felt the need to convict someone," he muttered darkly. His lawyer had warned him about "how the law is biased."

This hideous self-pity was terrible to witness. We stood up and prepared to leave. But he refused to let us go until he had told one final story about the trial. When he was convicted, he explained, the court had ordered him to pay the costs of the six-month-long proceeding, a heavy financial blow to Beger. But despite months of waiting, he had never received the bill. The reason, he claimed, lay with the trial judge, someone whom he had come to regard as a secret sympathizer. He paused significantly, and permitted himself a smile. The judge, he said, was the son of a German official who attended the Wannsee Conference in 1942 to discuss the coordination of the Final Solution.

BACK IN FRANKFURT, I turned the key to my small hotel room and sank down upon the bed, emotionally drained by my meeting with Beger. I felt exhausted and worn down to the bone, but I could not manage to sleep. Nor could I put Beger's words out of my mind. As I lay awake that night, I thought about those who had once worked for the Ahnenerbe and the motives they had for doing so. Some of the scientific staff, including Beger, were almost certainly true believers, susceptible men who had been swept up in the fervor of Nazism in their youths and who were later incapable of shaking off its hold—even after witnessing for themselves the concentration camps, the terrified prisoners, the spiraling smoke of the gas ovens. These researchers had not flinched from hard acts, believing that they served a higher purpose and that their cruelties were part of the natural order of things. They were precisely the kind of men that Himmler had counted upon to carry out his bidding.

But the others were much more of an enigma. What had driven men like Ernst Schäfer, the Tibet adventurer, Walther Wüst, the Orientalist, or August Hirt, the anatomist—all highly intelligent men capable of seeing beyond the Nazi rhetoric—to cross such a terrible moral line? Why had they lent their names and their reputations to the Ahnenerbe, cloaking it in scholarly credibility? What had persuaded them to forsake the traditional scholarly pursuit of the truth, distorting their research to fit Nazi party doctrine? And what had led them to serve a political system that they must have known to be rotten at the heart—corrupt, cruel, and murderous?

Such questions troubled me all that night in Frankfurt, keeping me restless and on edge until daylight, when I rose at last to prepare for another day of interviews. And even now, as I am about to close the last chapter of this book, they plague me still. After scrutinizing the personal files of these men and poring over the details of their life stories, after contemplating their academic work and talking to their families and friends, I still do not understand why they did what they did, why they willingly contributed to such evil.

No one, after all, forced them to join the SS and its research arm, the Ahnenerbe. They had other choices. They could have contented themselves, for example, with modest university posts, avoiding the limelight and leaving the grander promotions and the important research grants to others. Many intellectuals survived the Nazi years in this way. Or they could have left public life entirely, as some German writers and artists did, embarking on something known today as "inner emigration."[4] Such dissidents made their homes in remote parts of Germany, declining to place their talents at the service of the Reich and retiring to the private sphere. Or, most admirably of all, they could have worked actively against the Nazi state, although such work would have placed their own lives and that of their families in grave peril.

But few, if any, of the Ahnenerbe's senior researchers chose to conduct their lives with such integrity, and today I wonder why. What was Schäfer thinking when he accepted command of the Caucasus mission, whose goal in part was to racially diagnose Jewish tribes, all the better for their extermination? What was Wüst's response when he first learned of Himmler's plan to forcibly conduct medical experiments on human beings? What thoughts passed through Hirt's mind as he first contemplated murdering Jewish prisoners for their skeletons? Were they conscious of crossing some great moral chasm, of leaving behind the familiar world of ethics, decency, and human compassion? Or had they become so numb to the great evil that surrounded them that they failed to notice where the next step would lead?

History, no matter how thoroughly researched and carefully pondered, has its strict limits, beyond which we cannot go. We cannot know in the end what these men were thinking at the moment they willingly relinquished their humanity and crossed over the divide to barbarism. And it is impossible to say exactly why they did what they did, though some combination of fatal ambition, moral weakness, and unthinking prejudice seems the most likely explanation. While many other historians have written about the dire consequences of such personal failings in the Third Reich, I believe that the terrible power of science, and the manner in which science was manipulated to justify some of the worst atrocities of the Holocaust, is a little-known story. We like to think today that science is immutable, the gold standard of human knowledge, but as the history of the Ahnenerbe has shown us, it can be bent and warped to catastrophic ends. We cannot afford to forget this lesson.

x my view :
"HATE CONQUERS ALL"

"FEAR CAUSES HATE"

NOTES

1. FOREIGN AFFAIRS

[1]Hans A. Halbey, "Klingspor, Karl," in *Neue Deutsche Biographie*, ed. Historische Kommission bei der Bayerischen Akademie der Wissenschaften (Berlin: Duncker & Humblot, 1979); Karl Klingspor, *Über Schönheit von Schrift und Druck* (Frankfurt am Main: Georg Kurt Schauer, 1949).

[2]Evidence abounds for Himmler the bookworm. He kept a detailed, annotated list of all the volumes he read between September 4, 1919, and January 6, 1934. See Nachlaß Himmler, BAK, NL Himmler, N 126/9. For an interesting description of at least part of Himmler's private library, see Felix Kersten, *The Memoirs of Doctor Felix Kersten*, ed. Herma Briffault (Garden City, N.Y.: Doubleday, 1947), pp. 11–12. And for a description of a few of the books he gave to others, see Richard Breitman, *The Architect of Genocide* (New York: Alfred A. Knopf, 1991), p. 39. See also Sievers to Koehler & Amelang, 09.12.1937, BA, NS 21/166.

[3]Joachim Köhler, *Wagner's Hitler*, trans. Ronald Taylor (Cambridge, U.K.: Polity Press, 2000), p. 13. For a complete itemized list of the manuscripts, see Albert Speer, *Spandau* (New York: Macmillan, 1976), p. 164. According to Speer, these rare manuscripts were lost at the end of the war.

[4]"Herr Hitler's Birthday," *Times Digital Archive 1785–1985*, April 20, 1939. http://web1.infotrac.galegroup.com/.

[5]For further details on this present, see Steven Lehrer, *Hitler Sites* (Jefferson, N. C.: McFarland, 2002), pp. 160–166.

[6]Dietrich Eichholtz and Kurt Pätzold, eds., *Der Weg in den Krieg* (Cologne: Pahl-Rugenstein, 1989), p. 327.

[7]Michael Kater, *Das 'Ahnenerbe' der SS 1935–1945* (Stuttgart: Deutsche Verlags-Anstalt, 1974), p. 110.

[8]The Klingspor-Museum in Offenbach possesses a copy of this portfolio in its collections. Personal communication, Stephanie Ehret, Klingspor-Museum.

[9]Berlin–Brandenburgischer Akademie der Wissenschaften, *Deutsches Wörterbuch von Jacob Grimm und Wilhelm Grimm*, rev. ed. (Stuttgart: S. Hirzel Verlag, 1998), s.v. "Ahnenerbe." During the Nazi era, German writers concerned with racial purity often used *Ahnenerbe* in a biological sense to mean "inherited characteristics from the ancestors." But Nazi writers also used the word to mean "cultural traditions and customs passed on from the ancestors." For more details, see Cornelia Schmitz-Berning, *Vokabular des Nationalsozialismus* (Berlin: Walther de Gruyter, 1998), pp. 17–18.

[10]*Das Ahnenerbe* (Offenbach am Main: Gebrüder Klingspor, n.d.), p. 4. My citations to this portfolio refer to the copy held in the Klingspor-Museum. The title of the portfolio itself is missing from this copy. However, I found it in a copy held in private hands today.

[11]"Bericht über die bei der Forschungs-und Lehrgemeinschaft 'Das Ahnenerbe' in Berlin-Dahlem vorgenommene Prüfung des Jahresabschlusses zum 31.März 1940," NARA, RG242, T580/199/569.

[12]Adolf Hitler, *Mein Kampf,* trans. Ralph Manheim (Boston: Houghton Mifflin, 1971), p. 290.

[13]In their famous book, *We Europeans,* Julian Huxley and A. C. Haddon observed that "the Jews of different areas are not genetically equivalent. . . . The word Jew is valid more as a socio-religious or 'pseudo-national' description than as an ethnic term in any genetic sense." See Julian Huxley and A.C. Haddon, *We Europeans* (London: Jonathan Cape, 1935), pp. 96–97.

[14]Adolf Hitler, *Mein Kampf,* p. 305.

[15]The 1939 edition of *Encyclopaedia Britannica,* for example, defined Aryans as follows: "This word is used by some of the 'Satem' speakers of Indo-European languages with the meaning 'noble' and is the name of one of the tribes of these people. As Sir George Grierson points out, 'Indians and Iranians who are descended from an Indo-European stock have a perfect right to call themselves Aryans but we English have not'." And this view was strongly upheld by scholars such as J. R. R. Tolkien. In 1938, a German publishing house approached Tolkien with the idea of publishing his book *The Hobbit* in Germany. First, however, the publishers wanted to know whether Tolkien was *arisch* or Aryan. The British author was incensed by this inquiry. "I regret that I am not clear as to what you intend by *arisch,*" he noted. "I am not of Aryan extraction: that is Indo-Iranian; as far as I am aware none of my ancestors spoke Hindustani, Persian, Gypsy, or any related dialects. But if I am to understand that you are enquiring whether I am of Jewish origin, I can only regret that I appear to have *no* ancestors of that gifted people." For further details, see

J. P. Zmirak, "Tolkien, Hitler and Nordic Heroism," *Frontpage Magazine,* December 20, 2001.

[16]*The Universal Jewish Encyclopedia,* ed. Isaac Landman (New York: The Universal Jewish Encyclopedia, 1939–1943), s.v. "Jewish Nobel Prize Winners."

[17]"Rede vor den SS-Gruppenführern zu einer Gruppenführerbesprechung im Führerheim der SS-Standarte 'Deutschland' am 8.11.1938," in *Heinrich Himmler Geheimreden 1933–1945,* ed. Bradley F. Smith and Agnes F. Peterson (Frankfurt am Main: Propyläen, 1974).

[18]See, for example, Himmler to Wüst, 28.05.1940, BA, NS 21/227.

[19]Himmler to Wüst, 25.10.1937, NARA, RG 242, T580/186/366.

[20]The book was Wijnand van der Sanden's classic scientific study, *Through Nature to Eternity.* (Amsterdam: Batavian Lion International, 1996).

[21]Wijnand van der Sanden, *Through Nature to Eternity,* p. 167. Also "Nazi Leader Heinrich Himmler on the 'Question of Homosexuality,'" in *Homosexuals: Victims of the Nazi Era 1933–1945* (Washington: United States Holocaust Memorial Museum, n.d.), http://www.ushmm.org

[22]As cited in "Nazi Leader Heinrich Himmler on the 'Question of Homosexuality,'" in *Homosexuals: Victims of the Nazi Era 1933–1945.*

[23]Harry Osterhuis, "Medicine, Male Bonding, and Homosexuality in Nazi Germany," *Journal of Contemporary History* 32, no. 2 (April 1997): 194.

[24]Jankuhn borrowed this idea from Tacitus, a Roman historian who wrote a famous book on the ancient German tribes. In chapter 12 of *Germany and Its Tribes,* Tacitus notes that "penalities are distinguished according to the offence. Traitors and deserters are hanged on trees; the coward, the unwarlike, the man stained with abominable vices, is plunged into the mire of the morass, with a hurdle put over him. This distinction in punishment means that crime, they think, ought in being punished, to be exposed, while infamy ought to be buried out of sight." There is no evidence, however, that Tacitus ever visited Germania, the homeland of the Teutonic tribes. Nevertheless, Herbert Jankuhn, who became the Ahnenerbe department head for archaeology, argued in one of his publications that the male bog bodies were either punished cowards or men who had taken part in "perverse sexual offences." Wijnand van der Sanden, *Through Nature to Eternity,* p. 167.

[25]Ibid, p. 167.

[26]*Homosexuals: Victims of the Nazi Era 1933–1945.*

[27]Rüdiger Lautmann, "The Pink Triangle: The Persecution of Homosexual Males in Concentration Camps in Nazi Germany," *Journal of Homosexuality* 6, no. 1/2 (Fall/Winter 1980/81): 157.

[28]Wolfgang Kaschuba, "Am Ort der Geschichte," in *Prähistorie und Nationalsozialismus,* eds. Achim Leube and Morten Hegewisch. (Heidelberg: Synchron, 2002), pp 13–16.

[29]Kater himself was subjected to threats, particularly from Dr. Walther Wüst, the former scientific director of the Ahnenerbe. In 1966, Wüst and his lawyer attempted to force the University of Heidelberg to prohibit the publication of Kater's dissertation. Wüst failed.

[30]Achim Leube, "Einleitung," in *Prähistorie und Nationalsozialismus*, pp. ix–xiv.

[31]Some of the most tantalizing clues appeared on a large, rather plainly drawn map in the Klingspor portfolio that Himmler planned to give Hitler in the spring of 1939. Sprawling across two large pages, the map seems to be a visual summary of the chief projects of the Ahnenerbe. Small blue triangles mark the locations of the brain trust's major archaeological digs, while orange dots chart the locations of its research centers. Beyond these, however, the cartographer drew six colored lines that ricocheted and careened across Europe and Asia. According to the map's legend, they depict the routes of six *Forschungsreisen*, or "research trips," from Sicily and Serbia to Finland and Iraq. In addition, six black arrows soar across the pages. They point toward six other destinations: Iceland, Bolivia, the Canary Islands, Libya, Iran, and Tibet.

[32]Michael Kater, *Das "Ahnenerbe"*, p. 113

[33]The map featured in the Klingspor portfolio depicts the state of the Ahnenerbe expeditions around January 1, 1939. (We can deduce this date from the fact that the portfolio was to be presented to Hitler on April 20, 1939: to meet this deadline, the map would have had to be drawn several months in advance.) Five of the six "research trips" depicted on the map were completed before January 1, 1939. (In addition, it seems very likely that the sixth research trip—to Greece—also took place before January 1, 1939, but its dates are currently unknown.) The remaining six trips were either in progress or still in the planning stage as of January 1939. Therefore the cartographer did not have complete trip route information: he or she merely indicated the general route of these expeditions with a straight black line. From April to August 1939 researchers completed two more foreign trips, to Libya and Tibet. The remaining four trips languished, however, due to the outbreak of war.

[34]"*Das Ahnenerbe*," p. 8.

2. THE READER

[1]Gregor Strasser characterized Himmler in this way during a conversation with his brother Otto, who later published the account. As quoted in Peter Padfield, *Himmler* (London: Cassell, 2001), p. 80.

[2]Bradley Smith, *Heinrich Himmler* (Stanford: Hoover Institute Press, 1971), p. 80.

[3]Heinrich Himmler's reading list. BAK, NL Himmler, N 1126/9.

[4]Wijnand van der Sanden, *Through Nature to Eternity* (Amsterdam: Batavian Lion

International, 1996), p. 171; N.G.L Hammond, and H.H. Scullard, *The Oxford Classical Dictionary* (Oxford: Oxford University Press, 1978), p. 1,034.

[5]Heinrich Himmler's reading list. BAK, NL Himmler, N 1126/9, no. 218.

[6]*The Compact Edition of the Oxford English Dictionary* (Oxford: Oxford University Press, 1971), s.v. "Philology."

[7]Gerald Reitlinger, *The SS: Alibi of a Nation, 1922–1945* (Melbourne: Heinemann, 1956), p. 18.

[8]In 2002, the Staatliches Museum für Völkerkunde in Munich mounted a superb exhibit entitled "Prinzessin Therese von Bayern: Eine Bildungsreise zu den Indianern im Jahre 1893." The exhibition included artifacts that the princess had collected, as well as a photograph of her private collection at the Wittelsbach palace.

[9]Ernst Hanfstängl was a student of Gebhard Himmler in 1898. In a formerly classified report he wrote for the American government on Heinrich Himmler, he included a description of the elder Himmler and his behavior in the classroom. See NARA, RG 226, Entry 171, Box 8.

[10]George Hallgarten, as quoted in Peter Padfield, *Himmler*, p. 20.

[11]Alfred Andersch, *The Father of a Murderer* (New York: New Directions, 1994). In a recent detailed study of the relevant school records, historian Walter Habersetzer confirms the factual basis of this story. [See Walter Habersetzer, *Ein Münchner Gymnasium in der NS-Zeit.* (Munich: Verlag Geschichtswerkstatt Neuhausen, 1997).] Other students also came forward later with disturbing tales of Gebhard Himmler's methods, see "Mr. & Mrs.," *Time*, June 16, 1947. Andersch, it is interesting to note, later went on to become one of the leading German writers of the postwar period.

[12]Walther Habersetzer, *Ein Münchner Gymnasium*, p. 12.

[13]Peter Padfield, *Himmler*, p. 69.

[14]The story of Siegfried and the fall of the Nibelung dynasty was an immensely popular one in medieval northern Europe. Bards of the age told several versions of the story, including those preserved today in the Volsungsaga and the Thidreksaga. Wagner, who was entranced by the old epics, drew on disparate elements of the *Nibelungenlied, Volsungssaga,* and *Thidreksaga* to create his famous opera cycle, *Der Ring des Nibelungen.*

[15]So enamored was Himmler with the *Nibelungenlied* that he picked up the book again as an adult in 1923, gushing in his booklist about its "incomparable eternal beauty in language, depth, and all things German." See Heinrich Himmler's reading list. BAK, NL Himmler, N 1126/9, no. 180.

[16]As a young man, Himmler gave a copy of the *Edda* as a present to a friend, a practice he reserved for books he much admired. See Marianne to "Heini" (Heinrich Himmler) 08.09.1926, BAK, N 1126/17.

[17]See Alfred Andersch, *The Father of a Murderer*, pp. 28–29.

[18]Peter Padfield, *Himmler*, p. 22.

[19]Examples of these rooms can be seen in noble residences in Germany. The Wittelsbach winter palace in Munich, for example, has a very similar *Ahnengalerie*.

[20]Willi Frischauer, *Himmler* (London: Odhams, 1953), p. 17.

[21]Giles MacDonogh, *The Last Kaiser* (New York: Times Books, 1977), pp. 295–296.

[22]Willi Frischauer, *Himmler*, p. 15.

[23]Peter Padfield, *Himmler*, p. 22.

[24]Rüdiger Lautmann, "The Pink Triangle." *Journal of Homosexuality* 6, no. 1/2 (Fall/Winter 1980/81): 147.

[25]Willi Frischauer, *Himmler*, p. 16.

[26]Guido Knopp, *Hitler's Henchmen*, trans. Angus McGeoch (Phoenix Mill, England: Sutton, 2000), p. 119.

[27]Ernst Hanfstängl, "Heinrich Himmler," NARA, RG 226, Entry 171, Box 8.

[28]Willi Frischauer, *Himmler*, p. 20.

[29]"Mr. & Mrs.," *Time*, June 16, 1947.

[30]Peter Loewenberg, "The Unsuccessful Adolescence of Heinrich Himmler," *The American Historical Review* 76, no. 3 (June 1971): 624.

[31]Peter Padfield, *Himmler*, p. 36.

[32]Indeed, some fraternities in Germany still keep this tradition today.

[33]Heinrich Himmler's personal diary. HIA. Heinrich Himmler Collection. Box 15. See for example, the entry dated 13.11.1921.

[34]Ernst Hanfstängl, *Hitler* (London: Eyre & Spottiswoode, 1957), p. 33.

[35]David Clay Large, *Where Ghosts Walked* (New York: W. W. Norton, 1997), p. 147.

[36]Heinrich Himmler's reading list. BAK, NL Himmler, N 1126/9 no. 47. Under the heading "Artur Dinter. Die Sünde wider das Blut," for example, Himmler noted that the book "introduces one to the Jewish question with shocking clarity and causes one to approach this situation with extreme distrust."

[37]Heinrich Himmler's reading list. BAK, NL Himmler, N 1126/9, no. 107.

[38]Peter Padfield, *Himmler*, p. 59.

[39]David Clay Large, *Where Ghosts Walked*, p. 203.

[40]Konrad Heiden, "Introduction," in Adolf Hitler, *Mein Kampf*, trans. Ralph Manheim (Boston: Houghton Mifflin, 1971), p. xviii.

[41]Heinrich Himmler's reading list. BAK, NL Himmler, N 1126/9, no. 276.

[42]Heinrich Himmler's reading list. BAK, NL Himmler, N 1126/9. In this document, Himmler lists 276 books that he read from September 4, 1919, when he started the booklist, to February 19, 1927, when he finished the second volume of *Mein Kampf*. Researcher Charlotte Stenberg classified the subject matter of each book by three methods. She searched initially for the book in the Library of Congress classification system. She also compared the title to the historical novels listed in the database of the University of Innsbruck's Projekt Historischer Roman. And she examined Himmler's own comments on the book for indications of its subject. If none of these methods yielded results, then she looked for similar books by the

same author in order to determine classification. By these methods, she was able to categorize 211 of the books on the list. Sixty-seven of these books, or 32 percent, were either fictional or nonfictional explorations of the past.

[43]Adolf Hitler, *Mein Kampf,* p. 290.

3. ARYANS

[1]C. Loring Brace, "Race Concept," in *History of Physical Anthropology,* ed. Frank Spencer (New York: Garland, 1997), vol. 2, p. 861.

[2]*The Compact Edition of the Oxford English Dictionary,* vol. 2 (Oxford: Oxford University Press, 1971), s.v. "Race."

[3]Julian Huxley and A.C. Haddon, *We Europeans* (London: Jonathan Cape, 1935), p. 18.

[4]Robert Watt, "Parsons, James," in *Bibliotheca Britannica,* vol. 1 (Edinburgh: Constable, 1824). This entry contains a detailed list of his eclectic papers and books.

[5]J.P. Mallory, *In Search of the Indo-Europeans* (New York: Thames & Hudson, 1989), p. 10.

[6]James Parsons, *The Remains of Japhet, being historical inquiries into the affinity and origins of the European languages,* as quoted in J.P. Mallory, *In Search of the Indo-Europeans,* p. 10.

[7]J.P. Mallory, *In Search of the Indo-Europeans,* Table 3, pp. 12–13.

[8]Ibid., p. 12.

[9]Julian Huxley and A.C. Haddon, *We Europeans,* p. 147.

[10]*The Compact Edition of the Oxford English Dictionary,* s.v. "Aryan."

[11]Linguists have now placed nearly forty major modern languages into the Indo-European family group, from English, French, German, Danish, Swedish, and Irish Gaelic to Serbo-Croatian, Yiddish, and Romany. As two experts on this Indo-European family have noted, "nearly half the world's population speaks an Indo-European language as a first language." Thomas Gamkrelidze and V.V. Ivanov, "The Early History of Indo-European Languages," *Scientific American* 262, no. 3 (March 1990): 110–116.

[12]Schlegel to Ludwig Tieck, 15.12.1803, as quoted in Léon Poliakov, *The Aryan Myth,* trans. Edmund Howard. (London: Sussex University Press in association with Heinemann, 1974), p. 191.

[13]Léon Poliakov, *The Aryan Myth,* p. 192.

[14]Ibid.

[15]Ibid.

[16]Most scholars today agree that a society known as the Indo-Europeans did indeed speak the Indo-European protolanguage. But after more than a century of investigation, scholars are unable to agree on the original homeland of this society and do not ascribe any racial characteristics to this ancient group.

[17]Léon Poliakov, *The Aryan Myth,* p. 191.

[18]Ibid.

[19]*The Jewish Encyclopedia*. (New York and London: Funk and Wagnalls, 1902), s.v. "Benfey, Theodor."

[20]One of the rules that early German linguists relied on heavily was first developed by fairy-tale collector Jacob Grimm in 1822. It was known as Grimm's law of *Lautverschiebung* ("sound shifting"). This rule establishes a number of regular correspondences between early Germanic stops and fricatives (two different types of consonant sounds) and the stop consonants of other Indo-European languages.

[21]John V. Day, "Aryanism," in *History of Physical Anthropology*, ed. Frank Spencer (New York: Garland, 1997), pp. 109–111.

[22]Scholars in recent years have continued to use words from the reconstructed Proto-Indo-European vocabulary to support their own theories about the ancient homeland. They have arrived at very different conclusions. Modern scholars such as Thomas Gamkrelidze and V. V. Ivanov have noted, for example, that some words in the reconstructed vocabulary are for animals not found in northern Europe—most notably leopards, lions, monkeys, and elephants. See Thomas Gamkrelidze and V.V. Ivanov. "The Early History of Indo-European Languages," p. 110. The scholarly search for the homeland of the Indo-Europeans continues today. Currently, researchers have narrowed the search to four possible locations: north-central and eastern Europe, Anatolia, the Balkans and the basin of the Danube River, and the Pontic-Caspian steppes of the Ukraine and Russia. See J.P. Mallory, "Human Populations and the Indo-European Problem," *The Mankind Quarterly* 33, no. 2 (Winter 1992): 131–154.

[23]Tacitus, "Germany and Its Tribes," *The Complete Works of Tacitus*, ed. Moses Hadas (New York: The Modern Library, 1942), p. 710.

[24]For an interesting synopsis of some of the serious errors in Tacitus' book, "Germany and Its Tribes," see Julian Huxley and A.C. Haddon, *We Europeans*, pp. 33–37.

[25]J.P. Mallory, "Human Populations and the Indo-European Problem," p. 135–137.

[26]Today, biologists define a race as a population that has a distinct lineage and that differs in a significant genetic way from other populations—to such a degree, in fact, that it may be considered a subspecies.

[27]Scientists are still unable to agree on a classification system for human races. Many researchers today prefer to talk about race in terms of geographically defined groupings—such as the African geographic race, European geographic race, or Polynesian geographic race. But there is no clear agreement on this system.

[28]*Encyclopaedia Britannica* (London: the Encyclopaedia Britannica Company, 1939), s.v. "Anthropology. The Study of Race."

[29]Elvira Weisenburger, "Der 'Rassenpapst' Hans Friedrich Karl Günther, Professor für Rassenkunde," in *Die Führer der Provinz*, ed. Michael Kissener and Joachim Scholtyseck et al. (Konstanz: Universitätsverlag Konstanz, 1999), p. 170.

[30]Hans F.K. Günther. *The Racial Elements of European History*, trans. G.W. Wheeler. (Port Washington, N.Y.: Kennikat Press, 1970), pp. 10–23.

[31]Hans F. K. Günther, *The Racial Elements*, pp. 51–55.

[32]Günther's first racial book, *Rassenkunde des deutschen Volkes*, or "Racial Studies of the German People," sold 400,000 copies in Germany by 1945.

[33]One of the first scholars to do this was Gustav Kossinna, one of Germany's most famous archaeologists. See Julian Huxley and A.C. Haddon, *We Europeans*, pp. 66–67. Many other younger German archaeologists and scholars then hastened to follow his example in the 1920s and 1930s.

4. DEATH'S-HEAD

[1]Heinz Höhne, *The Order of the Death's Head*, trans. Richard Barry (New York: Coward-McCann, 1970), p. 52; Gerald Reitlinger, *The SS: Alibi of a Nation, 1922–1945* (Melbourne: Heinemann, 1956), p. 25.

[2]Peter Padfield, *Himmler* (London: Cassell, 2001), p. 89.

[3]Heinz Höhne, *The Order of the Death's Head*, pp. 51–52. The photographic studio belonged to Heinrich Hoffman, who introduced his assistant Eva Braun to Hitler. Steven Lehrer, *Hitler Sites* (Jefferson, North Carolina: McFarland & Company, 2002), p. 57.

[4]Gudrun Schwarz, *Eine Frau an seiner Seite* (Hamburg: Hamburger Edition, 1997), p. 82.

[5]Peter Padfield, *Himmler*, p. 83.

[6]Ibid.

[7]Verbal information from a Himmler relative. As quoted in Heinz Höhne, *The Order of the Death's Head*, p. 49.

[8]Undated memorandum written by Heinrich Himmler. As quoted in Heinz Höhne, *The Order of the Death's Head*, p. 44.

[9]Michael Kater, "Die Artamanen," *Historische Zeitschrift* 213 (1971): 576–638.

[10]One strong supporter of the Artamanen was Hans F. K. Günther, the philologist who had classified the German people into five distinct races for the publisher Julius Lehmann.

[11]Michael Kater, "Die Artamanen," p. 613.

[12]Clifford R. Lovin, "Blut und Boden: The Ideological Basis of the Nazi Agricultural Program," *Journal of the History of Ideas* 28, no. 2 (April–June 1967): 280–281.

[13]R.Walther Darré, *Das Bauerntum als Lebensquell der Nordischen Rasse* (Munich: J.F. Lehmann, 1933), pp. 367–369.

[14]Clifford R. Lovin, "Blut und Boden," p. 283.

[15]Ibid., p. 285.

[16]Ibid.

[17]As quoted in Heinz Höhne, *The Order of the Death's Head*, p. 49.

[18]Hans F. K. Günther, *Kleine Rassenkunde Europas* (Munich: J. F. Lehmanns Verlag, 1925), p. 82.

[19]Elvira Weisenburger, "Der Rassepapst: Hans Friedrich Karl Günther, Professor für Rassenkunde," in *Die Führer der Provinz*, ed. Michael Kissener and Joachim Scholtyseck (Konstanz: Universitätsverlag Konstanz, 1997), p. 174.

[20]This is from a speech Himmler gave on January 18, 1943. The quote is taken from Heinz Höhne, *The Order of the Death's Head*, p. 52.

[21]Isabel Heinemann, *"Rasse, Siedlung, deutsches Blut"* (Göttingen: Wallstein, 2003), p. 60.

[22]Isabel Heinemann, *"Rasse, Siedlung, deutsches Blut,"* p. 61.

[23]The remark was made after the war by Gottlob Berger, who had served as the head of SS recruitment in 1938. See Gerald Reitlinger, *The SS: Alibi of a Nation*, p. 16.

[24]Peter Padfield, *Himmler*, p. 377. The original source for this story was a memoir written by Dr. Werner Best in a Danish prison and dated September 18, 1949. This memoir was given to historian Schlomo Aronson. For further details, see Shlomo Aronson, *Reinhard Heydrich und die Frühgeschichte von Gestapo und SD* (Stuttgart: Deutsche Verlag Anstalt, 1971), p. 266, footnote 81.

[25]Geschäftsanzeige der Firma Hugo Boss, 1933, *Alb-Neckar-Zeitung. 26./27*, August 1933. As posted online in Elisabeth Timm, "Hugo Ferdinand Boss (1885–1948) und die Firma Hugo Boss," Eine Dokumentation, 18.04.1999. Available at http://www.metzingen-zwangsarbeit.de/.

[26]Peter Padfield, *Himmler*, p. 86.

[27]Christian Bernadac, *Le Mystère Otto Rahn* (Paris: Éditions France Empire, 1978), p. 248.

[28]Isabel Heinemann, *"Rasse, Siedlung, deutsches Blut,"* p. 51.

[29]"Führerbesprechung der SS-Gruppe Ost," as quoted in Peter Padfield, *Himmler*, p. 101.

[30]Barbara Distel and Ruth Jakusch, eds., *Concentration Camp Dachau 1933–1945* (Munich: Comité International de Dachau, Brussels Lipp GmbH, 1978), p. 46.

[31]Peter Padfield, *Himmler*, p. 87.

[32]Isabel Heinemann, *"Rasse, Siedlung, deutsches Blut,"* p. 66.

[33]Adolf Hitler, *Hitler's Table Talk, 1941–1944*, ed. H.R. Trevor-Roper (London: Phoenix Press, 2000), no. 138, p. 289. Hitler made these remarks on February 4, 1941, when Himmler was his dinner guest.

[34]Ibid.

[35]Ibid., no. 253, p. 566.

[36]The intensity of Himmler's feeling on this subject can be seen clearly in the comments he writes for his booklist. After finishing Gustav Freytag's book *Aus der Römerzeit*, he noted, "If only we were still like the ancient Germanic tribes with their traditions and their healthy establishments!" And after putting down Werner

Jansen's historical novel on the Amelungen, he enthused, "The great heroic saga of the Amelungen, in which every Germanic person is a great person who behaves himself as fate and honor demand. . . . I feel as if I belong to these Germanic tribespeople, however, these days, I am quite alone in this feeling." Heinrich Himmler's reading list. BAK, NL Himmler, N 1126/9. no. 217, 259.

[37]Isabel Heinemann, *"Rasse, Siedlung, deutsches Blut,"* p. 67.

[38]Darré, "Aktenvermerk," 16.[?].1934, BA, NS 2/277.

[39]Harm, "Aktenvermerk," 02.06.1936, BA, NS 2/135. SS educators even tested officer candidates on their knowledge of prehistory. As one surviving examination reveals, officers were expected to memorize a series of dates, including that of the Early Stone Age, Bronze Age, and Iron Age, as well as the routes of later Germanic migrations and details of Arminius's victory over the Romans in Teutoberg Forest. See "Prüfungsfragen für SS-Führer," NARA, BDC, A3345 B, R127/334.

[40]Isabel Heinemann, *"Rasse, Siedlung, deutsches Blut,"* p. 95, footnote 129.

[41]*Deutsche Geschichte. Lichtbildvortrag: Erster Teil: Germanische Frühzeit "Das Licht aus dem Norden,"* ed. Reichsführer-SS, IfZ, DC 25.10. Intriguingly this copy of the booklet bears the stamp of "Leibstandarte SS. 'Adolf Hitler' Abtlg. Schulung." This was the educational department of the SS Bodyguard Regiment Adolf Hitler. See also *Deutsche Geschichte. Lichtbildvortrag. Zweiter Teil: Die Grossgermanische Zeit "Eiserne Zeit—Germanen marschieren!"* BA, NS 31/163.

[42]See for example "Genau beachten!" in the *SS-Leitheft,* 29.05.1936, NARA, BDC A3345 B, R122/1462.

[43]It would appear such columns were highly necessary, as some SS men did not understand what the SS was looking for in terms of brides. In one memorable column, a fictional SS-Mann is looking for permission to marry "a girl whose stature is dwarfish. She is 143 cm tall (4.69 feet). In addition, she suffers from heart and eye problems, so that for health and racial-biological reasons alone, she does not come into question as a potential wife for a SS-member." See "Beispiele aus der Praxis des Sippenamtes," *SS-Leitheft,* 15.08.1937, NARA, BDC, A3345 B, R122/676.

[44]*SS-Leitheft,* November 1944, NARA, BDC, A3345 B, R124/1039.

[45]Karl Hüser, *Wewelsburg 1933 bis 1945* (Paderborn: Verlag Bonifatius Drückerei, 1982), pp. 27–29.

[46]"Reichsführerschule der SS auf der Wewelsburg," *Bürener Zeitung,* January 24, 1934, reprinted in Karl Hüser, *Wewelsburg 1933 bis 1945,* p. 162.

[47]"Lebenslauf," NARA, BDC, A3343, SS0 Weisthor, Karl-Maria (10.12.1866).

[48]"Abschrift, 06.01.1925, Bericht über die krankhafte Gemütsart des Herrn Oberst Wiligut," as published in Hans-Jürgen Lange, *Weisthor* (Engarda: Arun, 1998), p. 114.

[49]"Salzburger Landesheilanstalt für Geistes- und Gemütskranke. Die Krankheitsgeschichte, 29.11.1924," as published in Hans-Jürgen Lange, *Weisthor,* p. 102.

[50]"Dr. Schweighofer, Handschriftlicher Vermerk, 04.12.1914" (sic. most likely 1924), as published in Hans-Jürgen Lange, *Weisthor,* p. 107.

[51]Ibid.

[52]"Beschluss über die Anhaltung in einer geschlossenen Anstalt, 11.12.1924," as published in Hans-Jürgen Lange, *Weisthor*, p. 111.

[53]"Fortsetzung der Krankengeschichte. 15.11.1925," as published in Hans-Jürgen Lange, *Weisthor*, pp. 129–130.

[54]Nicholas Goodrick-Clarke, *The Occult Roots of Nazism* (London: I. B. Taurus, 1992), p. 183. For more information on Wiligut's trances, see Rüdiger Sünner, *Schwarze Sonne* (Freiburg: Herde, 1999), p. 50.

[55]Nicholas Goodrick-Clarke, *The Occult Roots of Nazism*, pp. 186–187.

[56]Himmler's former chief of staff Karl Wolff, who was an eyewitness to part of this meeting, recounted the story in an article that appeared in *Neue Illustrierte*, April/May 1961. As excerpted in Christian Bernadac, *Le Mystère Otto Rahn*, p. 348.

[57]Christian Bernadac, *Le Mystère Otto Rahn*, pp. 348–349.

[58]Indeed, it seems highly likely that Himmler was alluding to Arminius and his battle against the Romans by mentioning the "old road of the German heroes." So sacred had this history become to German nationalists that in 1875 they had built a popular monument to Arminius at a supposed site of the battle in Teutoberg Forest. For further details, see Malcolm Todd, *The Early Germans* (Oxford, UK; Cambridge, MA, USA: Blackwell Publishers, 1992), p. 266.

[59]Harald Meyer, *Wewelsburg. SS-Burg, Konzentrationslager, Mahnmal, Prozess* (Paderborn: Selbstverlag, 1982), p. 5.

[60]Stuart Flowers and Stephen Cook, *Heinrich Himmler's Camelot* (Andrews, N.C.: Kressmann-Backmeyer, 1999), p. 22.

[61]Karl Hüser, *Wewelsburg 1933 bis 1945*, p. 214.

[62]Hans-Jürgen Lange, *Weisthor*, pp. 276–278.

[63]Karl Hüser, *Wewelsburg 1933 bis 1945*, p. 72–73.

[64]Ibid., p. 33, footnote 44.

[65]Otto Rahn, *Luzifers Hofgesind* (Leipzig: Schwarzhäupter, 1937), pp. 44–51.

[66]"Wewelsburg Kr.Paderborn.-Museumeinrichtung 1935–1944," Nach Erinnerungen von Wilhelm Jordan, 29.12.1979, KWA 70/1/2/14.

[67]Ibid.

[68]Isabel Heinemann, *"Rasse, Siedlung, deutsches Blut."* pp. 88–89.

[69]Michael Kater, *Das 'Ahnenerbe' der SS 1935–1945* (Stuttgart: Deutsche Verlags-Anstalt, 1974), p. 27.

5. MAKING STONES SPEAK

[1]Georg Fritz and Walter Puttkammer, *Berlin. Die Alte und die Neue Stadt* (Berlin: Klinkhardt & Biermann, 1936), p. 24.

[2]*Berlin and Potsdam, Grieben's Guidebooks*, vol.108 (Berlin: A. Goldschmidt, 1931), p. 78.

[3]Wirth to Reichserziehungsministerium, 16.02.1935, BA (ehem.BDC) REM: Wirth, Herman (06.05.1885).

[4]"Bericht über die Prüfung der Buchführung der Gemeinschaft 'Das Ahnenerbe,'" 31.03.1937, NARA, RG242, T580/199/569.

[5]Wirth to Museumdirektion Oslo, 10.08.1930, KHM, Inkomne Skriv 1930; E. to Wirth, 23.01.1935, KHM, Kopibok 1935; Sievers to Direktor Universitetes Oldsaksamling, 22.04.1936, KHM, Inkomne Skriv 1936; Engelhardt to Ahnenerbe, 04.05.1936, KHM, Kopibok 1936; Metropolitan Museum of Art to Wirth, 04.05.1936, NARA, RG242, T580/198/560.

[6]Yrjö von Grönhagen, "Ungefährer Plan für die Arbeit der Abteilung Pflegestätte für Indogermanisch-Finnische Kulturbeziehungen," 25.02.37, NARA, RG242, T580/206/716.

[7]Michael Kater, Das 'Ahnenerbe' der SS 1935–1945 (Stuttgart: Deutsche Verlags-Anstalt, 1974), p. 38.

8These museum shows were, to say the very least, eclectic and highly eccentric. For a description of one such show, see "Der Heilbringer: Urreligionsgeschichtliche Ausstellung." n.d. Deutsche Allgemeine Zeitung. UAR, Personalakte alt. Wirth, Herman.

[9]It was Dr. Lisa Schroeter-Bieler who told me about Wirth's ability to quote Sanskrit texts. Others, such as Paul Rohkst, who knew Wirth after the war, attested to his powers as a speaker. According to Rohkst, he once lectured to an audience in Augsburg for four hours, taking only one break. As Rohkst recalls the occasion, no one left the hall.

[10]"Voices of Notable Scholars to the Investigations of Herman Wirth." North America: The New or the Ancient World? (n.p., n.d.), NARA, RG242, T580/143/167.

[11]Personal communication, Prof. Dr. Helmut Arntz.

[12]During a lengthy interview with historian Michael Kater in 1963, Wirth recalled that he met Himmler at the von Leers' and that this evening took place in late fall of 1934. (See "Gedächtnisprotokoll. Unterredung Prof. Dr. Herman Wirth und Michael H. Kater," 22.06.1963. IfZ. ZS/A-25, vol.2.) But two letters that Wirth wrote to Darré in 1934 [see BA, (ehem.BDC) WI: Wirth, Herman (06.05.1885)] clearly suggest that he had met Himmler several months earlier than this. In the first letter from Wirth to Darré, written on June 15, 1934, Wirth invited both Darré and Himmler to attend one of his forthcoming lectures. In the second letter, dated July 7, 1934, Wirth referred to tasks "as discussed with RFSS." This strongly suggests that he had met Himmler sometime before July 7, 1934.

[13]Robert Wistrich, Who's Who in Nazi Germany (London: Weidenfeld and Nicolson, 1982), s.v. "Johann von Leers."

[14]Anna Bramwell, Blood and Soil: Richard Walther Darré and Hitler's Green Party (Bourne End, Buckinghamshire: Kensal Press, 1985), pp. 49–50.

[15]"Gedächtnisprotokoll. Unterredung Prof. Dr. Herman Wirth und Michael H. Kater." 22.06.1963. IfZ. ZS/A-25, vol. 2.

[16]Himmler clearly thought of himself as an expert on racial matters. He liked to tell people that he personally conducted the racial assessments of SS applicants. When he conducted tours of concentration camps, he sometimes pointed out the purported "criminal physiognomies" of certain individuals.

[17]For an interesting discussion of this, see Richard Steigmann-Gall, *The Holy Reich* (Cambridge: Cambridge University Press, 2003), pp. 86–114.

[18]Ibid., pp. 106–108.

[19]*Völkischer Beobachter,* July 1, 1935. As quoted in Richard Steigmann-Gall, *The Holy Reich,* p. 129.

[20]Felix Kersten, *The Memoirs of Doctor Felix Kersten,* ed. Herma Briffault (Garden City, N.Y.: Doubleday, 1947), p. 61.

[21]Nils Åberg, "Herman Wirth: En germansk kulturprofet," *Fornvännen* 28 (1933): 247–249.

[22]Herman Wirth, "Lebenslauf," NARA, RG242, A3343, SS0 Wirth, Herman: (06.05.1885).

[23]Herman Wirth, *Der Aufgang der Menschheit* (Jena: Eugen Diederichs, 1928), p. 14.

[24]Ibid.

[25]"Erläuterungen zu den Abbildungen: Abb. 2," in *Verzeichnis der Schriften, Manuskripte und Vorträge von Herman Felix Wirth Roeper Bosch* ed. Eberhard Baumann (Toppenstedt: Uwe Berg, 1995), p. 349.

[26]"Die Meinung andere Leute über Herman Wirth," Gesine von Leers to C.W. Mack, 10.08.1932, BA (ehem. BDC) Ahnenerbe: Wirth, Herman Felix (06.05.1885).

[27]Ingo Wiwjorra, "Herman Wirth—Ein gescheiterter Ideologe zwischen 'Ahnenerbe' und Atlantis," in *Historische Rassismusforschung,* ed. Barbara Danckwortt and Thorsten Querg (Hamburg: Argument Verlag. 1995), p. 93.

[28]As archaeological historian Luitgard Löw points out, Wirth was likely influenced by the ideas of the renowned German historian Oswald Spengler.

[29]For a photograph of these gable decorations, see *Verzeichnis der Schriften,* ed. Eberhard Baumann, p. 357.

[30]Herman Wirth, *Der Aufgang der Menschheit,* pp. 15–17.

[31]Herman Wirth, *Der Aufgang der Menschheit,* pp. 622–626.

[32]*Encyclopaedia Britannica,* 14th ed. (London: Encylopaedia Britannica Co. Ltd., 1938), s.v. "Inscriptions."

[33]The oldest rune alphabet consisted of twenty-four letters and is known as Futhark. But runic script evolved considerably over time and place in Europe. Anglo-Saxon rune-masters, for example, added new letters, bringing the total to over thirty, while Scandinavian writers reduced the number to just sixteen.

[34]*Encyclopaedia Britannica,* 14th ed, s.v. "Runes."

[35]Herman Wirth, *Der Aufgang der Menschheit,* p. 22.

[36]Julian Huxley and A.C. Haddon. *We Europeans* (London: Jonathan Cape, 1935), p. 94.

[37]Herman Wirth, *Der Aufgang der Menschheit*, pp. 55–63.

[38]Ingo Wiwjorra, "Herman Wirth," pp. 97–99.

[39]Herman Wirth, *Der Aufgang der Menschheit*, p. 126–127.

[40]Wirth later brought some of these young acolytes with him into the Ahnenerbe. These included Wolfram Sievers, Otto Plassmann, and Otto Huth.

[41]Plassmann to Galke, 27.01.1937, BA (ehem.BDC) Ahnenerbe: Plassmann, Otto: 12.06.1895.

[42]James Webb, *The Occult Establishment* (La Salle, IL: Open Court, 1976), p. 322. Wirth even posted a rather unusual sign in his home: "Please walk softly and don't smoke: a deep breather lives here."

[43]Herman Wirth, "Hochschulsiedlung—Ein Weg zum Aufbau," *Berliner Lokal-Anzeiger*, February 19, 1933, BAK, 73/11853.

[44]"Gedächtnisprotokoll. Unterredung Prof. Dr. Herman Wirth und Michael H. Kater," 22.06.1963, IfZ. ZS/A-25, vol. 2.

[45]James Webb, *The Occult Establishment*, p. 322.

[46]Herman Wirth, "Das Geheimnis Arktis = Atlantis," *Die Woche*, Nr.38, 29.04.1931. p. 1144–1156.

[47]Fritz Wiegers, "Herman Wirth und die Geologie," in *Herman Wirth und die deutsche Wissenschaft* (Munich: J.F. Lehmanns, 1932), p. 8.

[48]Herman Wirth, *Der Aufgang der Menschheit*, pp. 105–109.

[49]Ibid.

[50]Ibid.

[51]Fritz Wiegers, "Herman Wirth und die Geologie," p. 12.

[52]"Die Meinung anderer Leute über Herman Wirth," Gesine von Leers to C.W. Mack, 14.05.1933, BA (ehem. BDC) Ahnenerbe: Wirth, Herman Felix (06.05.1885).

[53]Arn Strohmeyer, *Der gebaute Mythos. Das Haus Atlantis in der Bremer Böttcher-strasse. Ein deutsches Missverständnis* (Bremen: Donat, 1993).

[54]Hans Müller-Brauel described the collection during a walking tour of Haus Atlantis in 1932. See "Haus Atlantis—Die Sammlung Väterkunde," *Die Böttcherstrasse in Bremen* (2001 by Deutsches Rundfunkarchiv/Radio Bremen).

[55]Gero von Merhart, *Zu dem Buche Herman Wirth, "Der Aufgang der Menschheit,"* BA, (ehem. BDC) REM: Wirth, Herman (06.05.1885).

[56]Professor Köfler to Archivrat, 09.10.1935, BA (ehem. BDC) REM: Wirth, Herman (06.05.1885).

[57]Nils Åberg, "Herman Wirth: En germansk kulturprofet," *Fornvännen* 28, 1933: 247–249.

6. FINDING RELIGION

[1]Adrianus Corten, "A Proposed Mechanism for the Bohuslän Herring Period," *ICES Journal of Marine Science* 56 (1999): 207–220.

[2]Personal communication, Camilla Olsson. Also John Coles in association with Lasse Bengtsson, *Images of the Past: A Guide to the Rock Carvings and Other Ancient Monuments of Northern Bohuslän* (Uddevalla: Bohusläns museum och Bohusläns hembygdsförbund, 1990), p. 14.

[3]Ibid., p. 39. Swedish archaeologists have managed to date the rock carvings of Bohuslän by a variety of painstaking methods. They have compared certain distinctive objects portrayed in the engravings, such as spectacle broaches, with real examples found in graves that can be accurately dated. They have also studied superimposed images on the rocks, to determine which styles are the oldest, and examined changes in shorelines for additional dating clues. For a more detailed discussion of the dating of the Bohuslän art, see Anne-Sophie Hygen and Lasse Bengtsson, *Rock Carvings in the Borderlands* (Gothenburg, Sweden: Warner Förlag, 2000), pp. 172–184.

[4]The inhabitants of Bohuslän had long been intimately familiar with the carvings, but they first took a serious interest in them during the nineteenth and early twentieth century when scholars from the outside began to study these remarkable works of ancient art. Personal communication, Camilla Olsson.

[5]Anne-Sophie Hygen and Lasse Bengtsson, *Rock Carvings in the Borderlands*, p. 186.

[6]Indeed, in 1994, UNESCO classified the rock-art sites in the parish of Tanum in Bohuslän as a World Heritage Site, noting that "The rock carvings of Tanum region constitute an outstanding example of Bronze Age art of the highest quality." For further details, see Anne-Sophie Hygen and Lasse Bengtsson, *Rock Carvings in the Borderlands*, pp. 210–211.

[7]Sievers to Himmler, 07.05.1936, BA (ehem. BDC) Ahnenerbe: Wirth, Herman Felix (06.05.1885).

[8]Herman Wirth, "Bericht über die Hällristningar-Expedition des 'Deutschen Ahnenerbe' vom 27.08 bis 03.09.1935," BA (ehem. BDC) REM: Wirth, Herman (06.05.1885). Characteristically Wirth makes a number of errors in the report. The title itself is clearly incorrect: the expedition ended in the beginning of October. See Wirth to Reichsantiquar Stockholm, 14.07.1936, "Bericht über die erste Hällristningar-Expedition des 'Deutschen Ahnenerbes,' Ende August-Anfang Oktober 1935," ATA, Stockholm.

[9]Herman Wirth, "Bericht über die Hällristningar-Expedition des 'Deutschen Ahnenerbe' vom 27.08 bis 03.09.1935." BA (ehem. BDC) REM: Wirth, Herman (06.05.1885).

[10]"Gedächtnisprotokoll Unterredung Prof. Dr. Walther Wüst und Michael H. Kater," 04.04.1963. IfZ, ZS/A-25, vol.3.

[11]Albert Speer, *Inside the Third Reich*, as quoted in Bettina Arnold, "The Past as Propaganda: Totalitarian Archaeology in Nazi Germany," *Antiquity* 64 (1990): 469.

[12]"Gedächtnisprotokoll Unterredung Prof. Dr. Walther Wüst und Michael H. Kater," 04.04.1963, IfZ, ZS/A-25, vol.3.

[13]Ibid.

[14]Himmler did indeed use the casts in at least one speech he gave. See Sievers to Kotte, 07.04.1937, BA (ehem. BDC) Ahnenerbe: Kottenrodt, Wilhelm (11.11.1904).

[15]Jack Tresidder, *Dictionary of Symbols: An Illustrated Guide to Traditional Images, Icons, and Emblems* (San Francisco: Chronicle Books, 1998), s.v. "Swastika"; Hans Biedermann, *Dictionary of Symbolism*, trans. James Hulbert (New York: Facts on File, 1992), s.v. "Swastika."

[16]Wirth to Himmler, 07.04.1936, BA, NS 21/693.

[17]Wirth to Schindler, 10.06.1929, BA, NS 8/125. See also "Die Meinung anderer Leute über Herman Wirth," Roselius to Wirth, 10.12.1933, BA (ehem. BDC) Ahnenerbe: Wirth, Herman Felix (06.05.1885).

[18]"Die Meinung anderer Leute über Herman Wirth," Roselius to C.W. Mack, 27.07.1932, BA (ehem. BDC) Ahnenerbe: Wirth, Herman Felix: (06.05.1885).

[19]Himmler to Galke, 28.10.1936, BA (ehem. BDC) Ahnenerbe: Wirth, Herman Felix (06.05.1885).

[20]From comments made during a speech Himmler gave to SS-Gruppenführer, February 18, 1937. See Peter Padfield, *Himmler* (London: Cassell, 2001), p. 193.

[21]One of the negotiators was Wolfram Sievers, who was the managing director of the Ahnenerbe. See "Bericht," 13.12.1935, BA, NS 21/693.

[22]Exactly how much Wirth owed in late 1935 is unclear, as correspondence regarding his debts is likely incomplete. In all probability, 84,000 reichsmarks was the bare minimum. See "Bericht," 13.12.1935, BA, NS 21/693. Also see Galke to Amelang & Koehler, 31.03.1936, BA, NS 21/693.

[23]In a speech Himmler gave to the SS-Gruppenführer on February 18, 1937, he described the conditions of this financial assistance. See Peter Padfield, *Himmler: Reichsführer-SS*, p. 193.

[24]Sievers, "Bericht über die Forschungsfahrt 1936," 12.06.1936, NARA, RG242, T580/203/686.

[25]Sievers, "Personalfragebogen," 24.10.1935, IfZ, SSO Sievers, Wolfram (10.07.1905).

[26]"RuSHA Ärztlicher Untersuchungsbogen: Wolfram Sievers," BA (ehem.BDC) RS: Sievers, Wolfram (10.07.1905).

[27]Sievers to Bousset, 30.07.1936, NARA, RG242, T580/203/686.

[28]Sievers to Himmler, 07.05.1936, NARA, RG242, T580/203/686. See also Galke to Reichsführer-SS, 09.05.1936, NARA, RG242, T580/203/686.

[29]This scientific funding agency was founded by the five German academies of science, the union of German universities, the Kaiser Wilhelm Society, and other German scientific organizations in 1920. On December 19, 1929, the organization underwent a name change to *Deutsche Gemeinschaft zur Erhaltung und Förderung der Forschung (Notgemeinschaft)*. In October 1937, it was rechristened *Deutsche Forschungemeinschaft*, or the German Research Foundation.

[30]Sievers to Himmler, 14.07.1936, NARA, RG242, T580/203/686.

[31]Sievers to Galke, 17.07.1936, NARA, RG242, T580/203/686.

[32]Bousset, "Stammrollenauszug," BA (ehem.BDC), SSO: Bousset, Helmut (08.12. 1902).

[33]Ahnenerbe to Landesfinanzamt, 22.07.1936, BA, NS 21/556; See also Luitgard Löw, "Herman Wirth und die Suche nach der germanischen Geistesurgeschichte in Skandinavien," unpublished paper.

[34]Wirth to Reichsantiquar Stockholm, 14.07.1936, ATA Stockholm.

[35]Ibid. See also Wirth, "Bericht über die erste Hällristningar-Expedition des 'Deutschen Ahnenerbes,' Ende August-Anfang Oktober 1935," ATA, Stockholm.

[36](Galke) to Sievers, "Betr. Forschungsreise nach Skandinavien," 30.07.1936, NARA, RG242, T580/203/686.

[37]Wirth. "Bericht," 25.08.1935, BA (ehem. BDC) Ahnenerbe: Wirth, Herman Felix: (06.05.1885).

[38]In the late summer of 2002, I visited Backa and several other sites in Bohuslän that Wirth had studied. I traveled with German archaeologist Luitgard Löw and Swedish archaeologist Camilla Olsson, who were kind enough to bring me up to date on current scientific knowledge of the sites. Many of the descriptive details in this section are taken from that trip.

[39]Anne-Sophie Hygen and Lasse Bengtsson, Rock Carvings in the Borderlands, pp. 150–151.

[40]Indeed, Wirth once told a Swedish antiquity official that he intended to expand on Almgren's work. Wirth to Curman, 12.01.1939, ATA Stockholm.

[41]Wirth dates the Swedish engravings to both the Neolithic and Bronze Age. See Wirth (to Reichsantiquar Stockholm), "Bericht über die zweite Hällristningar-Expedition des 'Ahnenerbes.' Berlin 1936." This report arrived in Stockholm in December 1938. ATA Stockholm. For the dates that Wirth ascribes to the Neolithic and Bronze Ages, see Herman Wirth, Der Aufgang der Menschheit (Jena: Eugen Diederichs, 1928), p. 57.

[42]Herman Wirth, Die Heilige Urschrift der Menschheit (Leipzig: Koehler & Amelang, 1936), pp. 84–93.

[43]Herman Wirth, Der Aufgang der Menschheit, p. 90.

[44]Herman Wirth, "Die Heiligen Zeichen der Weihenacht unserer Ahnen," Nationalsozialistische Landpost, 21.12.1934. See also Herman Wirth, Die Heilige Urschrift der Menschheit, p. 518.

[45]Wirth to Himmler, "Bericht über die Arbeiten der 2.Expedition des Deutschen Ahnenerbes," 04.09.1936, NARA, RG242, T580/143/167.

[46]"Tysk professor avgjuter Tanums många hällristningar," Bohusläningen, 23.09.1935, Tanum Hällristningsmuseum. I am indebted to Luitgard Löw for sending me a copy of this clipping.

[47]Wirth (to Reichsantiquar Stockholm), "Bericht über die zweite Hällristningar-Expedition des 'Ahnenerbes.' Berlin 1936." This report arrived in Stockholm in December 1938. ATA Stockholm.

[48]Ibid.

[49]Personal communication, Ms. Camilla Olsson.

[50]"Tysk nazistungdom till Halland," *Nytid,* 19.08.1934. Tanum Hällristningsmuseum. It is unclear who this German professor was, but it could have been Wirth on an earlier trip.

[51]Wirth (to Reichsantiquar Stockholm), "Bericht über die zweite Hällristningar-Expedition des 'Ahnenerbes.' Berlin 1936." ATA Stockholm.

[52]Wirth to Himmler, "Bericht über die Arbeiten der 2.Expedition des Deutschen Ahnenerbes," 04.09.1936, NARA, RG242/143/167.

[53]Personal communication, Dr. Jan Brøgger.

[54]"Arkeologien og samfundenes åndelige balanse," *Morgenbladet,* 29.07.1936, KHM Brøgger, A.W.

[55]A.W. Brøgger, editor's comment, *Viking; tidskrift for norrøn arkeologi* (November 1937): n.p.

[56]Wirth to Brøgger, 07.09.1936, KHM, Inkomne Skriv. 1936.

[57]Wirth to Brøgger, 21.09.1936, KHM, Inkomne Skriv. 1936.

[58]Ibid. In his letter, Wirth refers to the island as Rødøy, but it is clear from the information in the letter that he is actually writing about Rødøya in Alstadhaug County in Norway. It is on Rødøya that the engraving of the skier is found.

[59]So famous is the Rødøya skier in Norway that organizers of the 1994 winter Olympics in Lillehammer chose it as the basis for the distinctive logos used for the various sporting events. Personal communication, Ms. Mette Hide.

7. ENCHANTMENT

[1]Fritz Bose, "Law and Freedom in the Interpretation of European Folk Epics," *Journal of the International Folk Music Council* 10 (1958): 31.

[2]John Martin Crawford, "Preface," in *The Kalevala: The Epic Poem of Finland* (New York: The Columbian Publishing Co., 1891), p. xxxvii.

[3]At least one of these Finnish critics set aside his objections, however, after learning that the famous German linguist and story collector Jacob Grimm praised *The Kalevala* highly, ranking it with the Greek myths. Juha Y. Pentikäinen, *Kalevala Mythology,* trans. Ritva Poom (Bloomington: Indiana University Press, 1989), pp. 22–26.

[4]John Martin Crawford, "Preface," *The Kalevala,* pp. xliii.

[5]Nicholas Goodrick-Clarke, *The Occult Roots of Nazism* (New York: New York University Press, 1992), pp. 154–163.

[6]"J.R.R. Tolkien Dead at 81; Wrote 'The Lord of the Rings,'" *The New York Times*, September 23, 1973.

[7]In the early 1920s, Gorsleben collaborated with the notorious anti-Semite Julius Streicher, who later edited the Nazi newspaper *Der Stürmer*. For further details see Nicholas Goodrick-Clarke, *The Occult Roots of Nazism*, pp. 156–159.

[8]Nicholas Goodrick-Clarke, *The Occult Roots of Nazism*, p. 156.

[9]As quoted in Nicholas Goodrick-Clarke, *The Occult Roots of Nazism*, p. 159.

[10]Adolf Hitler, *Hitler's Table Talk, 1941–1944*, ed. H.R.Trevor-Roper (London: Phoenix Press, 2000), p. 249.

[11]Heinrich Himmler's reading list. BAK, NL Himmler, N 1126/9. no. 180.

[12]Marianne to "Heini," 08.09.1926. From Himmler's private correspondence with friends. BAK, NL Heinrich Himmler, N1126/17.

[13]Indeed, three whole issues of the Edda Society's publication in 1934 were devoted to Wiligut's ideas. See Nicholas Goodrick-Clarke, *The Occult Roots of Nazism*, p. 160.

[14]According to Wiligut's personal assistant, Gabriele Winckler-Dechend, Himmler often invited Wiligut and her for a dinner that consisted of a sandwich, a glass of wine, and a conversation that sometimes stretched on for hours. Under Wiligut's influence, Himmler created a department in the Ahnenerbe called the Teaching and Research Center for Folktales, Fairytales and Sagas, whose work it was to collect ancient Germanic tales. The role of this center is clearly spelled out in the Klingspor portfolio. "Fairytales, sagas and comic fables are all varieties of stories that flow from the soul of the German people. . . . These folktales are to be understood as they are told, without false glorification, in essence to remain true to their words and sounds." *Das Ahnenerbe* (Offenbach am Main: Gebrüder Klingspor, n.d.), p. 13.

[15]Himmler to Wüst, 28.05.1940, BA, NS 21/227. See also Himmler to Wüst, 05.09.1938, in *Reichsführer! . . . Briefe an und von Himmler*, ed. Helmut Heiber (Stuttgart: Deutsche Verlags-Anstalt, 1968), pp. 59–60.

[16]Himmler to Wüst, 28.05.1940, BA, NS 21/227.

[17]"Lebenslauf," Reichskulturkammer Fragebogen, 06.05.1942, BA (ehem. BDC) RKK: Grönhagen, Yrjö von (03.10.1911). See also "Einige Angaben über das Geschlecht Grönhagen," unpublished document in collection of Mr. Juhani von Grönhagen.

[18]Indeed, during the Second World War, Grönhagen wrote a book of German propaganda describing Karelia and its history of battling against Russian Bolsheviks. See Yrjö von Grönhagen, *Karelien: Finnlands Bollwerk gegen den Osten* (Dresden: Franz Müller Verlag, 1942).

[19]"Auskunft über Jury Karlowitsch Grönhagen," 02.10.1937, ARK: A 3860.

[20]Ibid.

[21]Personal communication, Mr. Juhani von Grönhagen.

[22]Yrjö von Grönhagen, private unpublished journal. Collection of Mr. Juhani von Grönhagen. I am indebted to Mr. Juhani von Grönhagen, who kindly brought along the journal to an interview on March 19, 2004.

[23]Yrjö von Grönhagen, "Zum Geleit," in *Finnische Gespräche* (Berlin: Nordland Verlag, 1941).

[24]The newspaper was the *Frankfurter Volksblatt*. Yrjö von Grönhagen, private unpublished journal in the collection of Dr. Juhani von Grönhagen.

[25]Yrjö von Grönhagen, private unpublished journal in the collection of Mr. Juhani von Grönhagen. A copy of this entry also appears in ARK, A 3860, dated 13.10.1935.

[26]Juha Pentikäinen, "Finland as a Cultural Area," in *Cultural Minorities in Finland*, ed. Juha Pentikäinen and Marja Hiltunen (Helsinki: Finnish National Commission for UNESCO, 1995), pp. 11–12.

[27]"Bericht über die Arbeitssitzung der Mitarbeiter des 'Ahnenerbes,'" 25.10.1937, NARA, RG242, T580/128/47.

[28]John Martin Crawford, "Preface," in *The Kalevala*, pp. vi–vii.

[29]"Bericht über die Arbeitssitzung der Mitarbeiter des 'Ahnenerbes,'" 25.10. 1937, NARA, RG242, T580/128/47. Modern scholars have been unable to determine exactly who Tacitus was referring to when he mentioned the Fenni. It is possible, for example, that he was relating stories he had heard of the Sami in Scandinavia and Russia. See Juha Pentikäinen, "Finland as a Cultural Area," pp. 12–13.

[30]Juha Pentikäinen, "Finland as a Cultural Area," pp. 12–13.

[31]Herta and Yrjö von Grönhagen, *Das Antlitz Finnlands* (Berlin: Wiking Verlag, 1942), p. 20.

[32]Ibid., p. 22.

[33]Ibid., p. 30.

[34]Georg von Grönhagen, "Karelische Zauberbeschwörungen," *Germanien* (February 1937): pp. 54–57. "Georg" is the German version of "Yrjö."

[35]Ibid., p. 54.

[36]Rüdiger Sünner, *Schwarze Sonne* (Freiburg: Herder, 1999), p. 234, footnote 152.

[37]Personal communication, Mrs. Gabriele Winckler-Dechend, former personal assistant to Karl-Maria Wiligut.

[38]Personal communication, Mrs. Gabriele Winckler-Dechend. After the war, Grönhagen seems to have revised his opinion, describing Wiligut as an unscientific agitator in his book *Himmlerin salaseura* (Helsinki: Kansankirja, 1948).

[39]Wüst, "Zeugnis," 15.11.1939, NARA, RG242, A3345 DS G119: Grönhagen, Yrjö von (03.10.1911).

[40]Personal communication, Dr. Juha Pentikäinen.

[41]Peter Padfield, *Himmler* (London: Cassell, 2001), pp. 172–174.

[42]Ibid., p. 172.

[43]Himmler to Grönhagen, 19.04.1936, ARK, A3860.

[44]Ibid.

[45]Personal communication, Dr. Juhani von Grönhagen.

[46]Himmler to Grönhagen, 19.04.1936, ARK: A3860.

[47]Pamela M. Potter, *Most German of the Arts* (New Haven & London: Yale University Press, 1998), p. 133.

[48]Hans Heinrich Eggebrecht and Pamela M Potter, "Fritz Bose," in *Grove Music Online Edition*, ed. L. Macy, http://www.grovemusic.com. Indeed, Bose even published a book on this subject in 1934, entitled *Racial Aspects in Music*.

[49]Behrendt, "Bericht: Dr. Fritz Bose," 01.03.1937, NARA, RG242, A3345 DS G113: Bose, Fritz: (28.07.1906.).

[50]Pamela M. Potter, *Most German of the Arts*, pp. 106 and 132.

[51]Heinz Ritter, *An Introduction into Storage Media and Computer Technology* (BASF, 1988), pp. 10–12.

[52]Sievers to Himmler, 17.04.1937, NARA, RG242, A3345 DS G119: Grönhagen, Yrjö von: (03.10.1911).

[53]Personal communication, Dr. Hannu E. Sinisalo, University of Tampere, Finland. Between 1999 and 2001, Sinisalo interviewed Grönhagen, and during one of these interviews, Grönhagen described the route of the field trip of 1936. Viipuri is now known as Vyborg.

[54]Yrjö von Grönhagen, *Finnische Gespräche*, pp. 81–88. We know that Grönhagen went to see Timo Lipitsä during this field trip in 1936 because he presented a photograph he had taken of Lipitsä to Himmler in the autumn of 1936. Interview with Dr. Juha Pentikäinen. See also Fritz Bose, "Typen der Volksmusik in Karelien: Ein Reisebericht," *Archiv für Müsikforschung* 3 (1938): 96–118.

[55]Yrjö von Grönhagen, *Finnische Gespräche*, p. 82.

[56]Ibid., p. 84.

[57]This is what Grönhagen told other informants, such as Juho Hyvärinen. See Juho Hyvärinen to Martti Haavio, 16.10.1936. SKS, Folklore Archives: Correspondences.

[58]Yrjö von Grönhagen, *Finnische Gespräche*. p. 84.

[59]Ibid, p. 86.

[60]It is possible that this is the version of the Kalevala creation story that is preserved today in a collection of Bose's recordings at the Lautarchiv der Humboldt-Universität zu Berlin. See LHUB, LA1512-LA1514, which is described as: "Runengesang Karelisch."

[61]Fritz Bose, "Typen der Volksmusik in Karelien," p. 100.

[62]Ibid., Abbildung 4, p. 108.

[63]Ibid., p. 107.

[64]Ibid.

[65]Scholars continue to debate the origins of the kantele today, pointing out its re-

semblance to instruments as diverse as the Russian *gusli* and the Arabic *quanun*. However, one of the world's greatest historians of Eurasian stringed instruments, Bo Lawergren, suggests that the kantele likely developed from the lyres of Europe. "'To convert a lyre into a zither (like the kantele), all one needs to do is fill the hole in its upper part." Personal communication, Bo Lawergren and Victor Mair.

[66]Yrjö von Grönhagen, *Finnische Gespräche*, p. 54.

[67]Ibid., p. 55.

[68]It is very likely that this magical chant is preserved today in a collection of Bose's recordings at the Lautarchiv der Humboldt-Universität zu Berlin. See LHUB, LA 1519: "Zauber Spruch. Karelisch."

[69]Fritz Bose, "Folk Music Research and the Cultivation of Folk Music," *Journal of the International Folk Music Council* 9 (1957): 20–21.

[70]Yrjö von Grönhagen, *Finnische Gespräche*, p. 41.

[71]LHUB, LA 1504-1505. "Musik-ges."

[72]Fritz Bose, "Typen der Volksmusik in Karelien," p. 102.

[73]Ibid., p. 117.

[74]The Ahnenerbe had entered into an agreement to publish this magazine together with the Society for the Friends of German Prehistory. The first joint issue rolled off the press in March 1936. For further details, see Michael Kater, *Das 'Ahnenerbe' der SS 1935–1945* (Stuttgart: Deutsche Verlags-Anstalt, 1974), p. 105.

[75]Georg von Grönhagen, "Karelische Zauberbeschwörungen," pp. 54–57.

[76]Juho Hyvärinen to Martti Haavio, 16.10.1936, SKS, Folklore Archives: Correspondences.

[77]Personal communication, Dr. Juha Pentikäinen.

[78]Juho Hyvärinen to Martti Haavio, 16.10.1936; Juho Hyvärinen to Martti Haavi, 29.10.1936; Juho Hyvärinen to Martti Haavio, 12.11.1936. SKS, Folklore Archives: Correspondences.

[79]Grönhagen published an account of this extraordinary meeting in a book he wrote after the war, *Himmlerin salaseura* (Helsinki: Kansankirja, 1948). I am indebted to Dr. Juha Pentikäinen, who kindly paraphrased and translated the account into English during an interview.

[80]Personal communication, Dr. Juhani von Grönhagen.

[81]Weisthor to Wolff, 01.07.1937, NARA, RG242, A3345 DS G113: Bose, Fritz (28.07.1906); Sievers to Galke, 31.05.1938, NARA, RG242, A3345 DS G113. Bose, Fritz: (28.07.1906); Sievers to Reichsführer-SS.Persönlicher Stab, 13.03.1941, NARA, RG242, T580/151/229. In one of the primary SS training slide shows on German history, *Das Licht aus dem Norden*, one of the slides depicted two men playing *lurs* during an ancient summer solstice celebration. The accompanying text notes that "A tree with a decorated wreath, a symbol of the sun, is placed on a cliff. Next to it burns the bonfire, which is still customary today, and the lur players greet the sunlight with celebratory sounds." *Deutsche Geschichte. Lichtbildvortrag:*

Erster Tel: Germanische Frühzeit "Das Licht aus dem Norden," ed. Reichsführer-SS, IfZ, DC 25.10.

[82]As an American scholar, Pamela Potter, has pointed out, the *lur* also figured in an important German debate at the time over the origins of polyphony, a style of musical composition that juxtaposes two or more melodies in harmony. German nationalists claimed that ancient Germanic tribes had invented this sophisticated style of composition on their own. As evidence, scholars cited the fact that archaeologists usually found *lurs* in pairs. This, they claimed, proved that Germany's ancestors had always been drawn to consonance and avoided the atonal music deemed typical of Jewish people. For further details, see Pamela M. Potter, *Most German of the Arts,* p. 134.

[83]Himmler to Grönhagen, 28.04.1937, ARK, A3860.

[84]Michael Kater, *Das "Ahnenerbe,"* p. 134.

[85]Grönhagen. "Ungefährer Plan für die Arbeit der Abteilung Pflegestätte für indogermanische-finnische Kulturbeziehungen," 25.02.1937, NARA, RG242, T580/206/716. Two months later, Grönhagen expanded this work plan and gave it a more academic-sounding tone, likely at the insistence of Walther Wüst. See Sievers to Wüst, 21.04.1937. This letter included a copy of Grönhagen's report entitled, "Arbeitsplan der Abteilung 'Pflegestätte für indogermanisch-finnische Kulturbeziehungen,'" 21.04.1937. NARA, RG242, T580/206/716.

[86]Sievers to Grönhagen, 07.10.1937, NARA, RG242, T580/206/716.

8. THE ORIENTALIST

[1]Sievers to Himmler, 24.09.1936, BA, NS 21/302.

[2]Wirth to Hitler, 22.02.1933, BA, R 43II/334; "Die Meinung anderer Leute über Herman Wirth," C.W. Mack to Sievers, 25.02.1933, BA (ehem. BDC) Ahnenerbe: Wirth, Herman Felix (06.05.1885). Mack had discussed Wirth with Hitler in 1929. He noted that Hitler sees "Wirth's prophecies and overestimations of himself as atrocities."

[3]Ian Kershaw, *Hitler 1889–1936.* (London: Allen Lane, The Penguin Press, 1998), pp. 561–562; Ian Kershaw, *Hitler 1936–1945* (London: Allen Lane, The Penguin Press, 2000), pp. 39–40.

[4]Himmler's great fondness for Wirth as a maverick scholar is evident in a letter he wrote shortly after deposing him as the head of the Ahnenerbe. See Himmler to Dr. Wacker, 28.09.1937, BA (ehem.BDC) REM : Wirth, Herman (06.05.1885).

[5]Himmler to Wirth, 08.01.1937, BA, NS 21/703.

[6]Wirth to Himmler, 26.10.1936, BA, NS 21/693.

[7]Himmler to Galke, 28.10.1936, BA (ehem. BDC) Ahnenerbe: Wirth, Herman Felix (06.05.1885).

[8]Sievers to Himmler, 08.06.1936, BA (ehem.BDC) Ahnenerbe: Wüst, Walther

(07.05.1901); SS Stammrollen-Auszug, 08.03.1937, NARA, RG242, A3343, SSO Wüst, Walther (7.05.1901).

[9]Walther Wüst, "Personalbericht," NARA, RG242, A3343, SSO Wüst, Walther: (07.05.1901).

[10]Personal communication, Dr. Henry Hoenigswald.

[11]"SS-Angehörige des 'Ahnenerbe,'" BA (ehem.BDC) 0.879; "Amt Ahnenerbe," BA, NS 19/1630; Wüst, Walther, "Führerstammkarte," NARA, RG242, A3343, SSO Wüst, Walther: (07.05.1901).

[12]Personal communication, Dr. Raimund Pfister.

[13]Indeed, even after joining the party, Wüst complained bitterly about having to rub shoulders with ordinary party members. Wüst to Ahnenerbe, 26.03.1937, BA (ehem. BDC) Ahnenerbe: Wüst, Walther (07.05.1901)

[14]Dietrich Orlow, The History of the Nazi Party. (Pittsburgh: University of Pittsburgh Press, 1973), p. 48.

[15]While the party made some exceptions to this lockout of new members, it did not reopen the membership rolls until May 1937. For further details, see Dietrich Orlow, The History of the Nazi Party, pp. 49–63 and pp. 202–206; Walther Wüst, "Führerstammkarte," NARA, RG242, A3343, SSO Wüst, Walther (07.05.1901).

[16]Wüst, "SS-Stammrollen Auszug," 08.03.1937, BA (ehem. BDC), SSO Wüst, Walther: (07.05.1901).

[17]Sievers to Himmler, 08.06.1936, BA (ehem.BDC) Ahnenerbe: Wüst, Walther (07.05.1901); SS Stammrollen-Auszug, 08.03.1937, NARA, RG242, A3343, SSO Wüst, Walther (7.05.1901).

[18]Hugo and Elsa Bruckmann were early supporters of Hitler and they donated generously to the Nazi party. For further details, see Ian Kershaw, Hitler 1889–1936, pp. 187–188, 282, 299.

[19]Sievers to Plassmann, 07.01.1936, BA, NS 21/351.

[20]Sievers to Himmler, 08.06.1936, BA (ehem. BDC) Ahnenerbe: Wüst, Walther: (07.05.1901).

[21]"Gedächtnisprotokoll Unterredung Prof. Dr. Walther Wüst und Michael H. Kater," 22.04.1963, IfZ, ZS/A-25, vol. 3.

[22]Walther Wüst, "Die indogermanischen Bestandteile des Rigveda und das Problem der 'urindischen' Religion," in Zweites Nordisches Thing, (Bremen: Angelsachsen-Verlag, 1934), pp. 155–164. It should be noted that the state of Pakistan had not yet come into existence at the time of this article. Thus when Wüst uses terms such as "the Northwestern Indian realm," he is largely referring to what is now Pakistan.

[23]Ibid., p. 159.

[24]"Gedächtnisprotokoll Unterredung Prof. Dr. Walther Wüst und Michael H. Kater," 22.04.1963, IfZ, ZS/A-25, vol. 3.

[25]Indeed, of the 346 books that Himmler noted on his booklist, which he kept until January 6, 1934, just ten of the entries—2.8 percent—have Asian subjects. Moreover, just two—or 0.5 percent—pertain to Asian religion. These are Hermann Hesse's *Siddhartha* and Karl Gjellerup's *Der Pilger Kamanita*. Heinrich Himmler's reading list. BAK, NL Himmler, N 1126/9. For an example of extremist writing on the subject of Nordic migrations to the East, see Hans F.K. Günther, *Kleine Rassenkunde Europas* (Munich: Lehmann, 1925), pp. 112–116.

[26]By the time Dr. Felix Kersten met Himmler, which was in 1939, he was reading *Bhagavadgita*, the Vedic poems, the *Rig Veda*, and speeches by Buddha. Indeed his personal administrative officer Rudolf Brandt told Kersten that these books constituted Himmler's main reading for "religious background." See Victor and Victoria Trimondi, *Hitler, Buddha, Krishna* (Vienna: Ueberreuter, 2002), pp. 26–27.

[27]A very early use of this phrase occurs in an article entitled "Walls Within Walls," *The Times*, May 18, 1943.

[28]Michael Kater, *Das 'Ahnenerbe' der SS 1935–1945* (Stuttgart: Deutsche Verlags-Anstalt, 1974), pp. 282–284.

[29]Walther Wüst, "Führerstammkarte," NARA, RG242, A3343, SSO Wüst, Walther: (07.05.1901).

[30]For an example of the clippings he kept, see BStM, Ana 625: Himalaya-Expedition 1929, Dyrenfurth-Expedition 1935, Deutsche Himalaya-Expedition 1931.

[31]"Wer kennt Kafiristan?" *Münchener Zeitung*, January 21, 1935, BA, NS 21/432. See also Walther Wüst, "Die indogermanischen Bestandteile des Rigveda," in *Zweites Nordisches Thing*, p. 161.

[32]Gisela Lixfeld, "Das Ahnenerbe Heinrich Himmlers und die ideologisch-politische Funktion seiner Volkskunde," in *Völkische Wissenschaft. Gestalten und Tendenzen der deutschen und österreichischen Volkskunde in der ersten Hälfte des 20. Jahrhunderts*, W. Jacobeit et al. (Vienna: Böhlau, 1994), pp. 217–255: "Bericht über die Arbeitssitzung der Mitarbeiter des 'Ahnenerbes,'" 25.10. 1937, NARA, RG242, T580/128/47.

[33]"Gedächtnisprotokoll Unterredung Prof. Dr. Walther Wüst und Michael H. Kater," 04.04.1963, IfZ, ZS/A-25, vol.3.

[34]"Bericht über die Arbeitssitzung der Mitarbeiter des 'Ahnenerbes,'" 25.10.1937, NARA, RG242, T580/128/47.

[35]"Lähes 1000 m. taikoja ja kansantapoja," *Laatokka*, 14.09.1937, ARK, A3869.

[36]Wüst to Himmler, 25.10.1937, NARA, RG242, A3345 DS G119: Grönhagen, Yrjö von (03.10.1911).

[37]Sievers to Wüst, 1.11.1937, NARA, RG242, A3345 DS G119: Grönhagen Yrjö von (03.10.1911).

[38]Grönhagen to Himmler via Sievers, 20.12.1937, NARA, RG242, A3345, DS G119: Grönhagen Yrjö von (03.10.1911).

9. INTELLIGENCE OPERATIONS

[1]Ian Kershaw, *Hitler 1936–1945* (London: Allen Lane, The Penguin Press, 2000), pp. 81–85.

[2]William L. Shirer, *The Nightmare Years 1930–1940* (Toronto: Bantam Books, 1985), p. 315.

[3]Indeed, as historian Peter Padfield has noted, Heydrich's "fascination with the English secret service later extended to signing himself 'C' in green ink after the style of Admiral Sir Mansfield Smith-Cumming, founder of MI6." See Peter Padfield, *Himmler: Reichsführer-SS* (London: Cassell, 2001), p. 112.

[4]André Brissaud, *The Nazi Secret Service*, trans. Milton Waldman (London: Bodley Head, 1974), pp. 31–32. See also Richard Rhodes, *Masters of Death* (New York: Alfred A. Knopf, 2002), p. 5.

[5]Peter Padfield, *Himmler*, p. 112.

[6]Ibid.

[7]Ibid, p. 219.

[8]According to one author, these forces arrested an estimated seventy-six thousand people in two days alone in Vienna. See André Brissaud, *The Nazi Secret Service*, p. 195.

[9]William L. Shirer, *The Nightmare Years 1930–1940*, p. 317.

[10]Sievers to Himmler, 13.07.1938, BA (ehem. BDC) Ahnenerbe: Trautmann, Erika (15.04.1897).

[11]Personal communication, Dr. Ruth Altheim-Stiehl. See also Walther K. Nehring, "Erika und Eva Nehring aus Osterwieck," *Familie Nehring*, No. 2, 1973. No stories of this engagement have apparently passed down in the Nehring family, however, and one relative I talked to was skeptical of its authenticity.

[12]Personal communication, Waldemar Nehring.

[13]*The Petroleum Almanac.* (New York: The Conference Board, 1946), p. 3. In 1938, Germany produced 0.19 percent of world crude oil. Romania, by contrast, was the largest producer in Europe excluding the USSR, and it produced 2.44 percent of world crude oil.

[14]From an article in *Pravda*, May 14, 1938. As quoted in Nicholas M. Nagy-Talavera, *The Green Shirts and Others* (Stanford: Hoover Institution Press, 1970), p. 300.

[15]Inge Eichler, "Biographie und künstlerischer Werdegang," in *Wilhelm Altheim: Bilder aus den Volksleben: eine Austellung der Frankfurter Sparkasse von 1822 (Polytechnische Gesellschaft) 1.10-9.11.1979* (Frankfurt: Frankfurter Sparkasse von 1822 [Polytechnische Gesellschaft], 1979), p. 6.

[16]Ibid, p. 14.

[17]Personal communication, Dr. Ruth Altheim-Stiehl.

[18]Otto Rössler, "Kosmopolitische Gelehrsamkeit," *Neue Zürcher Zeitung*, 31.10./ 01.11.1976, Fernausgabe.

[19]Gerhard Radke, "Suche nacht Antwort: zum Tode des Althistorikers Franz Altheim," *Der Tagespiegel*, October 20, 1976.

[20]Personal communication, Dr. Dieter Metzler.

[21]Personal communication, Dr. Ruth Altheim-Stiehl.

[22]Personal communication, Dr. Dieter Metzler.

[23]The Hungarian writer and historian Antal Szerb paints a thinly fictionalized portrait of Altheim, whom he calls Rodolfo Waldheim, in his novel *Der Wanderer und der Mond* (Budapest: Corvina, 1974), first published in 1937. He describes Waldheim showing this trophy from Wilhelm's archaeological working group to the novel's hero. Szerb was a Catholic of Jewish ancestry and perished in a forced labor camp at the end of the Second World War.

[24]Platzhoff to Director of the Universität Erlangen, 25.06.1935, UAF/M, Abteilung 1, Nr. 1; Platzhoff to Neumann, 07.01.1936, UAF/M, Abteilung 1, Nr. 1.

[25]Cordes, "Gutachten," 22.06.1935, UAF/M, Abteilung 1, Nr. 1.

[26]Bäumler to REM, 27.02.1935, IfZ, MA 116/1.

[27]Altheim, "Personalnachweis," entry dated 20.03.1936, BA (ehem.BDC) REM: Altheim, Franz: (06.10.1898).

[28]Joseph Goebbels, *Michael: Ein deutsches Schicksal in Tagebuchblättern*, as excerpted in George L. Mosse, *Nazi Culture: Intellectual, Cultural and Social Life in the Third Reich* (New York: Grosset & Dunlap, 1966), p. 41.

[29]Walther K. Nehring, "Würdigung des Oberbaudirektors Johann Arnold Nehring (Nering) in der Literatur," *Familienblatt des Familienverbandes Nehring-Moek-Wagner* 8, (1966): 26–27.

[30]"Osterwiecker Haus," in *Genealogie des Familienkreises Nehring* 4 (1999): p. 86.

[31]Eva Nehring, "Die Schwestern Eva und Erika Nehring," *Familie Nehring* 3 (September 1975): p. 77.

[32]Ibid.

[33]Ibid.

[34]Walther K. Nehring, "Erika und Eva Nehring aus Osterwieck," *Familie Nehring* 2 (1973): pp. 51–52.

[35]Eva Nehring, "Die Schwestern Eva und Erika Nehring," p. 77.

[36]Johannes Tuchel and Reinhold Schattenfroh, *Zentrale des Terrors* (Berlin: Siedler, 1987), pp. 29–30.

[37]"Zensurenlisten," UAKB, 7/324, SoSe 1923, 88, Nr. 132; 7/325, WS 1923/24, 27, Nr. 86. There is a sad footnote to this story. Otto Ludwig Haas-Heye's daughter, Libertas Schulz Boysen, was arrested in the summer of 1942 by the Gestapo for her involvement in the Red Orchestra group, which smuggled Jews and political dissidents out of Germany. She was imprisoned in the former art school. At one point during her stay there, she reportedly laughed and noted that she was "'sitting' in the art

school where her father had been the Rector." She was executed on December 1 1942. For a literary account of these events see Peter Weiss, *Die Ästhetik des Widerstandes* (Frankfurt am Main: Suhrkamp, 1975).

[38]Walther K. Nehring, "Erika und Eva Nehring aus Osterwieck," p. 51.

[39]Eva Nehring, "Die Schwestern Eva und Erika Nehring," p. 77.

[40]"Die Mitarbeiterstab des Forschungsinstituts für Kulturmorphologie," 11.10.1935, BAK, R73/10112.

[41]"Forschungsreisen des Frobenius-Instituts," in *Das Frobenius-Institut an der Johann Wolfgang Goethe-Universität 1898–1998*. (Frankfurt a. M.: Frobenius Institut, 1998), pp. 36–40.

[42]"Vorgeschichte I (Felsbilder) Juli-August 1934: Südfrankreich und Ostspanien," FIA, LF 514.

[43]Personal communication, Mrs. Katharina Lommel.

[44]"Fotos: Expedition XIV Süd-Frankreich/Ost-Spanien 1934," FIA, Fotoarchiv.

[45]"Vorgeschichte (Felsbilder) 1936: Val Camonica, Italien," FIA, LF 519. See also Franz Altheim, "Forschungsbericht zur römischen Geschichte," *Die Welt als Geschichte* 2, (1936): 68–94.

[46]Personal communication, Mr. Christoph Nehring.

[47]Indeed, educators in the Third Reich even offered this explanation in textbooks used in German schools. One textbook author, Walther Gehl, who was affiliated with the Ahnenerbe, penned a entire chapter entitled, "The Roman Empire as a Nordic creation." See Walther Gehl, *Geschichte* (Breslau: Ferdinand Hirt, 1940), pp. 72–122.

[48]Himmler also believed that Rome's pedigree could be traced back to Germany's ancestors. After visiting Rome in 1937, he wrote to Wüst requesting that he establish a new department in the Ahnenerbe to "find proof that the Romans . . . stem from Aryan Indo-Germanic groups who migrated from the North. . . ." For further details, see Himmler to Wüst, 10.12.1937, NARA, RG242, T580/207/725.

[49]Himmler was well aware of the Italians' attitudes towards this research. He observed to Wüst that the Italians "are solely interested in Caesar's and the Emperor's time. Where they came from is uninteresting to them—and perhaps it is good, in terms of politics, that it is so." Himmler to Wüst, 10.12.1937, NARA, RG242, T580/207/725.

[50]Franz Altheim and Erika Trautmann, "Nordische und italische Felsbildkunst," *Die Welt als Geschichte* 3 (1937): 1–82. The couple argued, in this paper, that the carvings of Val Camonica were several centuries younger than those of Bohuslän in Sweden. This suggested that ancient Scandinavians had developed this art in the north and carried it south to Italy. But rock art is notoriously difficult to date and the techniques for doing so in the 1930s were relatively crude. More sophisticated techniques today—employing several lines of evidence—suggest that the oldest rock art in Val Camonica may date back some nine thousand years. This is six thousand years older than Altheim and Trautmann suggested. The art of Val Camonica is at least as

old—if not older—than the surviving rock art of Sweden. Thus no evidence exists today to support the contention that migrants from northern Europe left the carvings at Val Camonica. Instead archaeologists now theorize that the artists were of local Italian origins and merely shared religious beliefs and ideas common to many peoples across Europe.

[51]Ibid.

[52]Himmler to Höhne, 03.07.1937, BA, NS 21/687.

[53]According to the map prepared for the Klingspor volume, Altheim and Trautmann traveled from Berlin to Munich, Vienna, Zagreb, Split, Cattaro, Dubrovnic, Mitropica, Keztheley, Trieste, Venice, Val Camonica, Bologna, Florence, Rome, Syracuse, and Enna, then returned to Berlin. This route and the nature of the research conducted during the trip are confirmed by several surviving reports and letters, particularly those found in BA, NS 21/687. See Trautmann. "Deutsches Reich Reisepass," BA, NS 21/165; Himmler to Höhne, 03.07.1937, BA, NS 21/687; Höhne to Trautmann, 15.07.1937, BA, NS 21/687; Altheim to Höhne, 24.08.1937, BA, NS 21/687; Altheim RFR 'Personalnachweis,' entry dated 31.08.1937, BA (ehem. BDC) RFR: Altheim, Franz (06.10.1898); Altheim to Höhne, 01.09.1937, BA, NS 21/687; Altheim to Höhne, 09.09.1937, BA NS 21/687; Altheim to Höhne, 01.10.1937, BA, NS 21/687; Trautmann to RFSS, 15.10.1937, BA (ehem. BDC) Trautmann, Erika (15.04.1897); Höhne to Ullmann (Reichsführer-SS), 23.10.1937, BA, NS 21/687.

[54]Sievers to Trautmann, 01.11.1937, BA (ehem. BDC) Ahnenerbe: Trautmann, Erika: (15.04.1897).

[55]Sievers, "Aktennotiz," 16.11.1937, BA (ehem. BDC) Ahnenerbe: Altheim, Franz: (06.10.1898).

[56]Altheim to Professor [?], 16.03.1938, IfZ, MA 116/1.

[57]Personal communication, Dr. Dieter Metzler.

[58]Even important government institutes, such as the Deutsche Forschungsgemeinschaft, experienced grave difficulties in obtaining foreign exchange. See for example, DFG to Auswärtiges Amt. 05.02.1935, PAAA R65737, "Ausgrabungen im Irak" (1933–1937).

[59]Willi A. Boelcke, "Zur internationalen Goldpolitik des NS-Staates: Beitrag zu Deutschen Währungs und Außenwirtschaftspolitik 1933–1945," in Hitler, Deutschland und Mächte, ed. Manfred Funke (Düsseldorf: Droste, 1978), pp. 292–309.

[60]Trautmann, "Anmerkung," 30.06.1938, NARA, RG242, A3345 DS G112: Altheim, Franz (06.10.1898).

[61]Altheim and Trautmann to Ahnenerbe, 30.06.1938, NARA, RG242, A3345 DS G112, Altheim, Franz (06.10.1898).

[62]Ibid.

[63]Ibid.

[64]Ibid.

[65]Ibid.

[66]Ibid.

[67]David Price, "Cloak & Trowel," *Archaeology* (September / October 2003): 30–35.

[68]Personal communication, Dr. Bernhard Caemmerer. Dr. Caemmerer was a former student of Altheim.

[69]I am indebted to a 1934 article in *National Geographic* for many of the details of this scene. See Henrietta Allen Holmes, "The Spell of Romania," *National Geographic* (April 1934): 399–450.

[70]Altheim and Trautmann, "(Vertraulicher) Bericht: Rumänien" n.d., BA (ehem. BDC) Ahnenerbe: Altheim, Franz. (06.10.1898). This large report is divided into subsections by country.

[71]Ibid.

[72]Nicholas M. Nagy-Talavera, *The Greenshirts and the Others* (Stanford, California: Hoover Institute Press, 1970), pp. 251–254.

[73]Ibid, p. 247.

[74]Ibid.

[75]Ibid, pp. 276–277.

[76]Ibid, p. 277.

[77]Altheim and Trautmann, "(Vertraulicher) Bericht: Rumänien," n.d., BA (ehem. BDC) Ahnenerbe: Altheim, Franz (06.10.1898).

[78]Ibid.

[79]Ibid.

[80]Henrietta Allen Holmes, "The Spell of Romania," pp. 399–450.

[81]Altheim and Trautmann, "(Vertraulicher) Bericht: Rumänien," n.d., BA (ehem. BDC) Ahnenerbe: Altheim, Franz (06.10.1898).

[82]Ibid.

[83]Ibid.

[84]Altheim and Trautmann "(Vertraulicher) Bericht: Syrien," n.d. BA (ehem. BDC) Ahnenerbe: Altheim, Franz (06.10.1898).

[85]Ibid.

[86]*Enzyklopädie des Holocaust,* 2nd ed., ed. Israel Gutman (Piper: Munich, 1998), s.v. "Irak."

[87]Lukasz Hirszowicz, *The Third Reich and the Arab East* (London: Routledge & Kegan Paul, 1966), p. 18.

[88]*Enzyklopädie des Holocaust,* s.v. "Irak."

[89]Altheim and Trautmann, "(Vertraulicher) Bericht: Irak," n.d., BA (ehem. BDC) Ahnenerbe: Altheim, Franz (06.10.1898); See also Jordan, "Personalnachweis." BA (ehem. BDC) REM: Jordan, Julius (27.10.1877).

[90]For an example of this view, see Hans F.K. Günther, *Die Nordische Rasse bei den Indogermanen Asiens* (Munich: J.F. Lehmanns, 1934), pp. 124–125.

[91]Gertrude Bell to (her) father, 21.08.1921, RLUNT, Gertrude Bell Collection, online.

[92]Franz Altheim and Erika Trautmann, "(Vertraulicher) Bericht: Irak," n.d., BA (ehem. BDC) Ahnenerbe: Altheim, Franz (06.10.1898).

[93]Ibid.

[94]Ibid.

[95]Personal communication, Dr. Dieter Metzler. Moreover, Altheim's family made a point of listing this title among Altheim's more conventional honors—memberships in various European academies of science and the like—in his official death notice. See "Todesanzeige, Franz Altheim," *Frankfurter Allgemeine Zeitung,* 23.10.1976.

[96]Personal communication, Dr. Ruth Altheim-Stiehl.

[97]Franz Altheim and Erika Trautmann, "(Vertraulicher) Bericht: Irak," n.d., BA (ehem. BDC) Ahnenerbe: Altheim, Franz (06.10.1898).

[98]Franz Altheim and Erika Trautmann, "(Vertraulicher) Bericht: Unsere Vorschläge sind," n.d., BA (ehem. BDC) Ahnenerbe: Altheim, Franz (06.10.1898).

[99]Sievers to Himmler, 14.01.1939, NARA, RG242, T580/124/35; Chef des Sicherheitshauptamtes, Der Leiter der Zentralabteilung II 2 to Sievers, 23.05.1939, NARA, RG242, T580, R124/35; Sievers to Altheim, 13.05.1939, BA (ehem. BDC) Ahnenerbe: Altheim, Franz (06.10.1898).

[100]Chef des Sicherheitshauptamtes, Der Leiter der Zentralabteilung II 2 to Sievers, 23.05.1939, NARA, RG242, T580, R124/35; Sievers to Altheim. 13.05.1939. BA (ehem. BDC) Ahnenerbe: Altheim, Franz (06.10.1898).

[101]Altheim to Ahnenerbe, 21.06.1939, BA, NS 21/ 166.

10. CRO-MAGNON

[1]Himmler to Wüst, 05.09.1938, in Helmut Heiber, ed., *Reichsführer !. . . . Briefe an und von Himmler* (Stuttgart: Deutsche Verlags-Anstalt, 1968), pp. 59–60.

[2]Wüst, "Aktenvermerk: Betr. Die Rolle der Ferse," 05.11.1938, BA (ehem BDC) Wüst, Walther (07.05.1901); Brandt to Ahnenerbe, 10.08.1938, in Helmut Heiber, ed. *Reichsführer!* . . . , p. 58.

[3]Wüst to Sievers, 19.09.1938, BA, NS 21/225.

[4]Some of the Ahnenerbe researchers remarked after the war that they often laughed at Himmler's research requests and thought them ridiculous, but such comments may be more self-serving than truthful. Certainly, at the time, the Ahnenerbe staff responded to these requests in a serious way. See Wüst "Aktenvermerk: Betr.: Die Entwicklung des christlichen Kelches aus dem Gralskelch," 05.11.1938, BA (ehem. BDC) Ahnenerbe: Wüst, Walther (07.05.1901); Wüst, "Aktenvermerk: Herrscher-Abkunft-Sagen," 05.11.1938, BA (ehem BDC) Wüst, Walther (07.05.1901); Himmler to Ahnenerbe, 14.02.1940, in Helmut Heiber, ed. *Reichsführer!* . . . , pp. 71–72; Sievers to Otto Huth, 19.02.1940, BA, NS 21/225.

[5]This friendship is at once apparent in the intimate tone that Himmler adopts in

his letters to Wüst. For an example, see Himmler to Wüst, 10.12.1937, NARA, RG242, T580/207/725. Intriguingly, one of Wüst's friends at the University of Munich, Dr. Franz Dirlmeier, remarked upon the friendship between Wüst and Himmler after the war. See "Gedächtnisprotokoll. Unterredung Prof. Dr. Dirlmeier und Michael H. Kater," 09.07.1962, IfZ: ZS/A-25 vol. 1.

[6]"Gedächtnisprotokoll. Unterredung Dr. Ernst Schäfer und Michael H. Kater," 28.4.1964, IfZ: ZS/A-25, vol.2.

[7]Isabel Heinemann, "Rasse, Siedlung, deutsches Blut" (Göttingen: Wallstein, 2003), p. 95, footnote 129. See also Alexander Langsdorff and Hans Schleif, "Die Ausgrabungen der Schutzstaffeln," Germanien (January 1938), photograph 10, p. 10. The photograph shows SS men taking part in a dig. In addition, see Bettina Arnold, "The Past as Propaganda: Totalitarian Archaeology in Nazi Germany," Antiquity 64 (1990): 464–478.

[8]Reinhard Bollmus, Das Amt Rosenberg und seine Gegner (Stuttgart: Deutsche Verlags-Anstalt, 1970), pp. 332–334; Alexander Langsdorff and Hans Schleif, "Die Ausgrabungen der Schutzstaffeln," Germanien (January 1938): 6–11; Rolf Höhne, "Die Ausgrabungen der Schutzstaffeln," Germanien (1938): 224–230.

[9]Das Ahnenerbe (Offenbach am Main: Gebrüder Klingspor, n.d.), p. 14.

[10]Walther Gehl, Geschichte (Breslau: Ferdinand Hirt, 1940), p. 6. Gehl was an archaeologist well known to the SS and the Ahnenerbe. In 1936, he was one of the experts chosen to lead a large group of high-ranking SS officers on a prehistorical tour of Iceland.

[11]Bohmers, "Lebenslauf," 29.04.1937, BA (ehem. BDC) Ahnenerbe: Bohmers, Johan Christiaan Assien(16.01.1912).

[12]Personal communication, Prof. Dr. H.T. Waterbolk, a colleague of Bohmers after the war, at Groningen University; personal communication, Dr. Oebele Vries, a specialist in Frisian studies at Groningen University.

[13]Bohmers even seems to have tried to make his name sound Frisian. He was born Johan Christiaan Assien Böhmers, and at school he went by the name Han Böhmers—Han being a shortened form of Johan. Later, however, he dropped the umlaut from his surname and chose to be known as Assien. According to one of his former colleagues, Bohmers may have chose Assien because it was the diminutive of Asse, a very rare first name which might sound Frisian. Personal communication, Dr. H. T. Waterbolk.

[14]Bohmers, "Lebenslauf," 29.04.1937, BA (ehem. BDC) Ahnenerbe: Bohmers, Johan Christiaan Assien (16.01.1912).

[15]Bohmers to Sievers, 12.05.1940, BA (ehem. BDC) Ahnenerbe: Bohmers, Johan Christiaan Assien (16.01.1912). See also Sievers to Bohmers, 30.05.1938, BA, NS 21/60; personal communication, Dr. Oebele Vries.

[16]Bohmers, "Lebenslauf," 29.04.1937, BA (ehem. BDC) Ahnenerbe: Bohmers, Johan Christiaan Assien (16.01.1912).

[17]Personal communication, Prof. Dr. H.T. Waterbolk.

[18]Bohmers to Höhne, 29.04.1937, BA (ehem. BDC) Ahnenerbe: Bohmers, Johan Christiaan Assien (16.01.1912).

[19]Personal communication, Dr. Oebele Vries; Bohmers to Sievers, 12.05.1940, BA (ehem. BDC) Ahnenerbe: Bohmers, Johan Christiaan Assien (16.01.1912).

[20]Bohmers to Höhne, 29.04.1937, BA (ehem. BDC) Ahnenerbe: Bohmers, Johan Christiaan Assien (16.01.1912).

[21]Ibid., photograph of Johan Christiaan Assien Bohmers.

[22]Archaeologists refer to this site today as the Weinberg Caves of Mauern. Bohmers himself, however, referred to this site in the 1930s as the Mauern caves and, for the sake of simplicity, I have used this name.

[23]Herman Müller-Karpe, *Handbuch der Vorgeschichte* (Munich: C.H. Beck'sche Verslagsbuchhandlung, 1966), pp. 299–301.

[24]Hans F.K. Günther, *Rassenkunde des deutschen Volkes* (Munich: J.F. Lehmanns, 1924), pp. 257–258.

[25]Michael Kater, *Das "Ahnenerbe" der SS.* (Stuttgart: Deutsche Verlags-Anstalt, 1974), pp. 80.

[26]Donald Johansen and Blake Edgar, *From Lucy to Language* (New York: Simon & Schuster, 1996), p. 244.

[27]Erik Trinkhaus and Pat Shipman, *The Neandertals* (New York: Alfred A. Knopf, 1993), pp. 109–110.

[28]Julian Huxley and A.C. Haddon, *We Europeans* (London: Jonathan Cape, 1935), pp. 52–53.

[29]Hans F.K Günther, *Rassenkunde des deutschen Volkes*, p. 257.

[30]Ibid., p. 259.

[31]For an example of this, see Eugen Fischer, "Sind die alten Kanarier ausgestorben? Eine anthropologische Untersuchung auf den Kanarischen Inseln," *Zeitschrift für Ethnologie* 62 (1930): 258–273. Fischer, a prominent German anatomist, measured the heads of a sample group of modern and ancient Canary Islanders and compared the results to measurements of Cro-Magnon skulls. He observed from this that the Cro-Magnon looked very much like modern Canary Islanders. Then, since most of the Canary Islanders he examined had light or light brown hair, he concluded that the "Cro-Magnon race was blond when it arrived on the Canary Islands."

[32]Bohmers to Sievers, 18.07.1937 (sic; the date should read 18.07.1938, as indicated by the stamped date of the receiver on the letter), BA, NS 21/ 60.

[33]Assien Bohmers, "Die Mauerner Höhlen und ihre Bedeutung für die Einteilung der Altsteinzeit," in *Ahnenerbe Jahrestagungen. Bericht über die Kieler Tagung 1939*, ed. Herbert Jankuhn (Neumünster: Karl Wachholtz, 1944), p. 65, IfZ, DC 12.06.

[34]Michael Kater, *Das "Ahnenerbe"*, p. 80; Hermann Müller-Karpe, *Handbuch der Vorgeschichte*, pp. 299–301.

[35]Hermann Müller-Karpe, *Handbuch der Vorgeschichte*, p. 300.

[36]Rolf Höhne, "Die Ausgrabungen der Schutzstaffeln," p. 225.

[37]Hermann Müller-Karpe, *Handbuch der Vorgeschichte*, p. 300.

[38]Ibid.

[39]Ibid. See also Sievers to Himmler, 13.07.1938, BA, NS 21/60; Bohmers, "Stellungnahme zum Vorbericht," BA, NS 21/ 60.

[40]Bohmers, "Stellungnahme zum Vorbericht," BA, NS 21/ 60. See also Lothar Zotz, "Archaeological News: Germany," *American Journal of Archaeology* 53, no. 2 (1949): 175. For currently accepted dates on the beginning and end of the Last Interglacial, see "2.4 How Rapidly Did Climate Change in the Distant Past ?" in *Climate Change 2001: the Scientific Basis*, ed. J.T. Houghton et al (Cambridge: Cambridge University Press, 2001).

[41]Bohmers, "Stellungnahme zum Vorbericht," BA, NS 21/ 60. See Tables 1 and 2 specifically.

[42]Ibid.

[43]Bohmers to Sievers, 18.07.1937 (sic; the date should read 18.07.1938, as indicated by the stamped date on the letter), BA, NS 21/ 60. As subsequent research has shown, Bohmers was very mistaken in these views. Most Paleolithic archaeologists now agree that the Neandertal did not give rise to anatomically modern humans. Instead, modern humans migrated to Europe from elsewhere, most likely Africa, and replaced the Neandertal. See David van Reybrouck, "Boule's error: On the Social Context of Scientific Knowledge," *Antiquity* 76 (2002): 158–164. See also Donald Johanson and Blake Edgar, *From Lucy to Language*, p. 43.

[44]One surviving SS lesson plan boasts, for example, "Today, however, we know: all decisive cultural advancement and developments stemmed from our Nordic core area." Translated from SS Hauptamt, *Lehrplan für zwölfwöchige Schulung, Der Kampf um das Reich*, NARA, RG242, A3345 B, 122/573.

[45]In the fall of 1938, Bohmers went to Brussels to examine Rutot's collections. Bohmers to Abteilung Ausgrabungen des Ahnenerbes. 27.10.1938. BA, NS 21/60. Rutot believed in what he called "Tertiary man," a missing link between humans and apes who made simple flint tools some 30 million years ago. For more details, see Raf De Bont, "The Creation of Prehistoric Man: Aimé Rutot and the Eolith Controversy, 1900–1920," *Isis* 94 (2003): 604–630.

[46]Aimé Rutot, "Note sur l'âge," as translated and quoted in Raf De Bont, "The Creation of Prehistoric Man," p. 617.

[47]Bohmers, "Stellungnahme zum Vorbericht," BA, NS 21/60.

[48]Pers. Stab. RFSS to Sievers, 26.07.1938, BA, NS 21/60.

[49]Bohmers to Sievers, 18.07.1937 (sic; the date should read 18.07.1938, as indicated by the stamped date on the letter), BA, NS 21/60.

[50]Sievers to Bohmers, 30.05.1938, BA, NS 21/60.

[51]Douglas Chandler, "Belgium—Europe in Miniature," *National Geographic* (April 1938): 447–449.

[52]Bohmers to Abteilung Ausgrabungen des Ahnenerbes, 27.10.1938, BA, NS 21/60.

[53]Alan Houghton Broderick, *Father of Prehistory* (New York: Morrow, 1963), pp. 151–152.

[54]Bohmers to Abteilung Ausgrabungen des Ahnenerbes, 27.10.1938, BA, NS 21/60.

[55]Ibid.

[56]Abbé H. Breuil, *Four Hundred Centuries of Cave Art,* trans. Mary E. Boyle (Centre d'Études et Documentation Préhistoriques: Montignac, Dordogne, 1952), p. 16.

[57]"Grotte des Trois-Frères," in *L'Art des Cavernes: Atlas des Grottes Ornées Paléolithiques Françaises* (Ministère de la Culture: Imprimerie Nationale, Paris, 1984), pp. 400–409.

[58]Bohmers to Abteilung Ausgrabungen des Ahnenerbes, 07.11.1938, BA, NS 21/60.

[59]D. Peyrony, *Les Eyzies: Ses Musées d'Art préhistorique,* (Paris: Henri Laurens, 1931), pp. 9–11.

[60]Bohmers visited at least five of the caves: Les Combarelles, Font-de-Gaume, La Mouthe, Teyat, and Trois-Frères. Bohmers, "Memorandum," 29.01.1939, BA, NS 21/60.

[61]Abbé H. Breuil, *Four Hundred Centuries of Cave Art,* p. 95.

[62]Bohmers, "Memorandum," 29.01.1939, BA, NS 21/60.

[63]Ibid.

[64]Abbé H. Breuil, *Four Hundred Centuries of Cave Art,* p. 169.

[65]Sievers to Ullmann and Wolff, 18.02.1939, BA (ehem. BDC) Ahnenerbe: Bohmers, Johan Christiaan Assien (16.01.1912).

[66]Bohmers to Sievers, 12.03.1939, BA (ehem. BDC) Ahnenerbe: Bohmers, Johan Christiaan Assien (16.01.1912).

[67]Ibid. Recent scientific evidence shows that Himmler was correct in his idea that modern humans did not evolve directly from the Neandertal. But the Neandertal were by no means lumbering embarrassments in the human pedigree. They possessed brains larger than modern humans and were even known to have played flutes and created simple forms of art.

[68]Ibid.

[69]Ibid.

[70]Bohmers to Sievers, 26.12.1938, BA, NS 21/ 60.

[71]Ibid.

[72]Bohmers to Sievers, 12.03.1939, BA (ehem. BDC) Ahnenerbe: Bohmers, Johan Christiaan Assien (16.01.1912). Also Sievers to Ullmann and Wolff, 18.02.1939, BA (ehem. BDC) Ahnenerbe: Bohmers, Johan Christiaan Assien (16.01.1912).

[73]Bohmers, "Memorandum," 29.01.1939. BA, NS 21/60.

[74]Ibid.

[75]Ibid.

11. THE BLOSSOMING

[1]Gisela Lixfeld, "Das Ahnenerbe Heinrich Himmlers und die ideologisch-politische Funktion seiner Volkskunde," in *Völkische Wissenschaft*, ed. Wolfgang Jacobeit, Hannjost Lixfeld, and Olaf Bockhorn (Vienna: Böhlau, 1994), pp. 217–255.

[2]Robert Michael and Karin Doerr, *Nazi-Deutsch/Nazi German* (Westport, Conn.: Greenwood, 2002), s.v. "Bonzopolis." The character in question, Bonzo, was the creation of English cartoonist George Studdy and starred in some twenty-six animated films in the 1920s. As a small dog, Bonzo fled from the first whiff of danger and freely indulged in all the manly vices—drinking, gambling, and womanizing.

[3]Schulze, "Wertbericht über das Grundstück in Berlin-Dahlem," 26.10.1938, BA, NS 21/ 27a.

[4]Hermann Kaienburg, *Die Wirtschaft der SS* (Berlin: Metropol, 2003), p. 183.

[5]Schulze, "Wertbericht über das Grundstück in Berlin-Dahlem," 26.10.1938, BA, NS 21/ 27a.

[6]Hermann Kaienburg, *Die Wirtschaft der SS*, p. 183.

[7]*Enzyklopädie des Holocaust*, 2nd ed., ed. Israel Gutman, (Munich: Piper, 1998), s.v. "Arisierung."

[8]Schulze, "Wertbericht über das Grundstück in Berlin-Dahlem," 26.10.1938, BA, NS 21/ 27a.

[9]Ibid; Sievers's Diary 1941, 22.08.1941, BA, NS 21/127.

[10]One of the largest of these centers seems to have been the Research Site for Biology. See Greite, "Bericht der Forschungsstätte für Biologie," 11.07.1939, NARA, RG242, A3345 DS G119: Greite, Walther (13.06.1907).

[11]Hermann Kaienburg, *Die Wirtschaft der SS*, p. 184.

[12]Sievers to Wüst, 17.01.1940, BA (ehem. BDC) Ahnenerbe: Wüst, Walther (07.05.1901); Hermann Kaienburg, *Die Wirtschaft der SS*, p. 184; "Gedächtnisprotokoll Unterredung Prof. Dr. Walther Wüst und Michael H. Kater," 04.04.1963, IfZ, ZS/A-25, vol. 3.

[13]Susanne Heim, "Research for Autarky: The Contribution of Scientists to Nazi Rule in Germany," in *Geschichte der Kaiser-Wilhelm-Gesellschaft im Nationalsozialismus* (Berlin: Max-Planck-Gesellschaft zur Förderung der Wissenschaften, 2001), p. 8.

[14]Hermann Kaienburg, *Die Wirtschaft der SS*, p. 181.

[15]Ibid., p. 199.

[16]Ibid., p. 202.

[17]Ibid., p. 203.

[18]Ibid.

[19]Ibid., p. 494.

[20]So prevalent was this official form of theft that one modern German historian, Götz Aly, has argued that "economically, the Nazi state was a snowballing system of fraud." In Aly's view, Hitler managed to cut taxes and introduce expensive social

benefits in Nazi Germany only because of his criminal policies of robbing others of their land, real estate, bank accounts, and art treasures. Theft, suggests Aly, was the engine that drove the Nazi state. For further details, see Götz Aly, "I Am the People," *Sign and Sight*, 03.01.2004, http://www.signandsight.com/features/23.html; Jody K. Biehl, "How Germans Fell for the 'Feel-Good' Fueher," Spiegel Online, http://service.spiegel.de/cache/international/0,1518,347726,00.html.

[21]Hermann Kaienburg, *Die Wirtschaft der SS*, pp. 224–226.

[22]Ibid., p. 184.

[23]Irmelin Küttner, "Gutachtliche Äußerung zum Denkmalwert: Bauernhäuser mit Höfen und Gartenland, Mehrow, Lankreis Barnim," 29.09.1994, BLD; Kaufangebot der Eigenen Scholle an Frau Anna Bothe, 26.02.1937, Nr. 85 des Beurkundungsverzeichnisses des Kulturamtes Frankfurt (Oder) für 1937. As reproduced at http://www.mehrow.de/Geschichte/Bis_1945/Ende_des_Rittergutes.html.

[24]Ibid.

[25]"Bodenreform Mehrow," Magistrats der Stadt Hoppengarten/Orsteile Hönow, Mehrow, Eiche. Bodenfond Mehrow, 25.10.1945, as reproduced at http://www.mehrow.de/Geschichte/Nach_1945/Bodenreform.html.

[26]Hermann Kaienburg, *Die Wirtschaft der SS*, pp. 262–263.

[27]Ibid., p. 262.

[28]Robert R. Taylor, *The Word in Stone* (Berkeley, CA: Univ. of California Press, 1974), pp. 210–217.

[29]Hermann Kaienburg, *Die Wirtschaft der SS*, p. 262.

[30]Adolf Hitler, *Mein Kampf*, trans. Ralph Manheim (Boston: Houghton Mifflin, 1971), p. 410.

[31]Ibid.

[32]Irmelin Küttner, "Gutachtliche Äußerung zum Denkmalwert: Bauernhäuser mit Höfen und Gartenland, Mehrow, Landkreis Barnim." 29.09.1994. BLD Brandenburgisches Landesamt für Denkmalpflege. For more information about the antiquity of this housing style in Germany, see Hans Reichstein, *Die Fauna des Germanischen Dorfes Feddersen Wierde* (Stuttgart: Franz Steiner, 1991), pp. 7–9; Hans-Jürgen Häßler, ed., *Ur- und Frühgeschichte in Niedersachsen* (Stuttgart: Theiss, 1991), p. 245.

[33]Irmelin Küttner, "Gutachtliche Äußerung zum Denkmalwert:" Bauernhäuser mit Höfen und Gartenland, Mehrow, Landkreis Barnim." 29.09.1994. BLD.

[34]Hermann Kaienburg, *Die Wirtschaft der SS*, p. 262.

[35]Ibid.

[36]During the war, Himmler's officers argued continually about the size of the country estates they had been promised. See Felix Kersten, *The Memoirs of Doctor Felix Kersten*, ed. Herma Briffault (Garden City, N.Y.: Doubleday, 1947), pp. 49–50.

[37]Ibid.

[38]*Das Ahnenerbe* (Offenbach am Main: Gebrüder Klingspor, n.d.). This document lists some thirty different research sites. However, some of these departments seem to have existed only on paper, as future plans.

[39]Ibid.

[40]Walther Wüst, for example, was a regular contributor to both the *SS-Leitheft* and *Völkischer Beobachter*. See Victor and Victoria Trimondi, *Hitler, Buddha, Krishna* (Vienna: Ueberreuter, 2002), p. 59.

[41]The Ahnenerbe organized a scientific conference in Kiel in the summer of 1939 that was covered widely in the press. Alfons Paquet, "Planmäßige Vorgeschichts-forschung: Die Kieler Jahrestagung der Gemeinschaft 'Ahnenerbe,' " *Frankfurter Zeitung* (08.06.1939), IfZ, Presseauschnittsammlung; "Altsteinzeitliche Funde: Fort-setzung der Jahrestagung, 'Das Ahnenerbe,' " *Westdeutsche Beobachter* (02.06.1939), IfZ, Presseauschnittsammlung; "Wikinger beherrschten Osteuropa: Die Kieler Tagung des 'Ahnenerbes,' " *Berliner Börsen-Zeitung* (14.06.1939), IfZ, Presseauschnittsamm-lung.

[42]*Das Ahnenerbe*, p. 25.

[43]Ibid., pp. 32–33.

[44]Ibid., p. 33.

12. TO THE HIMALAYAS

[1]Victor and Victoria Trimondi, *Hitler, Buddha, Krishna* (Vienna: Ueberreuter, 2002), pp. 26–27.

[2]Richard Breitmann, *The Architect of Genocide* (New York: Alfred A. Knopf, 1991), p. 39.

[3]"The Activities of Dr. Ernst Schaefer, Tibet Explorer and Scientist with SS-sponsored Scientific Institutes," Headquarters United States Forces European The-ater, Military Intelligence Service Center, 12.02.1946, NARA, RG238, M1270/27. For further details on Oshima Hiroshi, see Ben Fischer, "The Japanese Ambassador Who Knew Too Much," *Center for the Study of Intelligence Bulletin* 9 (Spring, 1999): 6–9.

[4]Hans F.K. Günther, *The Racial Elements of European History*, trans. G.C. Wheeler (Port Washington, N.Y.: Kennikat Press, 1970), p. 133.

[5]The term Inner Asia refers to the interior of the Eurasian landmass. It encom-passes Uzbekistan, Turkmenistan, Tajikistan, Kyrgyzstan, Kazakhstan, the Republic of Mongolia, Xinjiang, Inner Mongolia, Tibet Autonomous Regions of the People's Republic of China, and neighboring parts of Afghanistan, Pakistan, Iran, China, and Siberia. I am indebted to the Research Institute for Inner Asian Studies, Indiana Uni-versity, Bloomington, for this definition. By contrast, the term Central Asia refers to a smaller portion of this region. Central Asia encompasses Uzbekistan, Turk-

menistan, Tajikistan, Kyrgyzstan, Xinjiang, eastern Iran, and Afghanistan. I have borrowed this definition from the department of Inner Asian and Altaic Studies at Harvard University.

⁶Hans F.K.Günther, *The Racial Elements of European History*, p. 132. For further details, see Hans. F.K. Günther, *Die Nordische Rasse bei den Indogermanen Asiens* (Munich: J.F. Lehmanns Verlag, 1934), pp. 194–196. The notion of an Aryan elite in Japan would have been deeply offensive to Japanese leaders of the day, however. They insisted upon the racial purity of the Japanese nation. For further details, see John W. Dower, *War without Mercy* (New York: Pantheon, 1986), pp. 9 and 215.

⁷Hans F.K. Günther, *The Racial Elements of European History*, p. 134. As one key piece of evidence, Günther notes that the invaders called themselves the Hari in Armenia, a Sanskrit word that he translates as "the Blonds." This translation is simply erroneous. The Sanskrit word "hari" describes a wide range of colors in the *Rig Veda*, including red, chestnut, and even green. For more details, see A. James Gregor, "Nordicism Revisited," *Phylon* 22, no. 4 (1961): 351–360.

⁸Hans. F.K Günther, *Die Nordische Rasse*, p. 57.

⁹Ibid., pp. 70–71. Günther never seems to have considered that other factors, such as dietary differences between upper and lower castes, might account for the disparities in physical height.

¹⁰Himmler to Wüst, 25.10.1937, NARA, RG242, T580/186/366.

¹¹"Gedächtnisprotokoll Unterredung Prof. Dr. Walther Wüst und Michael H. Kater," 04.04.1963, IfZ, ZS/A-25, vol.3.

¹²Elvira Weisenburger, "Der Rassepapst: Hans Friedrich Karl Günther, Professor für Rassenkunde," in *Die Führer der Provinz*, ed. Michael Kißener and Joachim Scholtyseck (Konstanz: Universitätsverlag Konstanz, 1997), p. 181.

¹³Deutsches Konsulat to Deutsches Gesandtschaft, 04.11.1935, PAAA, R 65588.

¹⁴Schäfer to Heissmeyer, 13.12.1935, BAK, R73/14198.

¹⁵Ibid.

¹⁶Deutsches Generalkonsulat to Geheimrat, 13.01.1936, PAAA: R65588.

¹⁷Schäfer to Heissmeyer, 13.12.1935, BAK, R73/14198.

¹⁸NSDAP, Auslands Amtsleiter to Dieckhoff, 17.02.1936, PAAA: R65588.

¹⁹SSO Ernst Schäfer, as quoted in Christopher Hale, *Himmler's Crusade* (Hoboken, N.J.: John Wiley & Sons, Inc., 2003), p. 114.

²⁰Sven Hedin, *Ohne Auftrag in Berlin* (Kiel: Arndt, 1991), pp. 139–140.

²¹"Unbekanntes Tibet," *Das Schwarze Korps*, 16.07.1936, BA, R135/80. This is the fifth installment in a series of articles the SS newspaper ran about Schäfer's second expedition to Inner Asia. The article appears without a byline, but there can be no doubt that Schäfer supplied the racial information for it. In the first installment, dated 18.06.1936, the newspaper's editors proudly noted that "we bring you the first part of a series for our readers using the original facts and picture material supplied for us by Ernst Schäfer." Some of Schäfer's surviving letters in public archives point-

edly refer to grave fields at Batang that he discovered. See, for example, Schäfer to Greite, 26.08.1936, BAK, R73/14198.

[22]"Unbekanntes Tibet," *Das Schwarze Korps*, 16.07.1936, BA, R135/80.

[23]"Führerstammkarte," NARA, RG242, A3343, SSO Schäfer, Ernst (14.03.1910).

[24]During one hearing after the war, Schäfer recounted a conversation he had had with Wüst during the war. "Wüst called me to him. He said to me, 'You studied ethnology, too, you are not just a biologist.' I said, 'I studied ethnology and I was very interested in it, in connection with Tibet.'" For more details see "Vernehmung des Ernst Schaefer vom 31.03.1947," Interrogation #1018-a. Mr. Lyon-Flick case. Question 18, NARA, RG238, M1019/62.

[25]"Unbekanntes Tibet," *Das Schwarze Korps*, 16.07.1936, BA, R135/80. For details on Schäfer's factual contributions to this article, see footnote 21.

[26]Ibid.

[27]Ibid.

[28]Ibid.

[29]In a private memoir that Schäfer wrote after the war for his family and friends, he claimed that Himmler asked him in this meeting whether he had encountered people with blond hair and blue eyes in Tibet and that he had denied this. This is almost certainly an evasion of the truth. Schäfer simply did not want to admit after 1945 that he had once had a strong personal interest in racial studies. For further details on Schäfer's memoir, see Rüdiger Sünner, *Schwarze Sonne* (Freiburg: Herder, 1999), p. 48.

[30]"The Activities of Dr. Ernst Schaefer, Tibet Explorer and Scientist with SS-sponsored Scientific Institutes," Headquarters United States Forces European Theater, Military Intelligence Service Center, 12.02.1946, NARA, RG238, M1270/27. In the American report, the city in Inner Asia is spelled "Urbe." It seems clear, however, that this is a transcription error by the American stenographers. In a later letter dated 26.05.1937, Himmler's chief of staff, Karl Wolff, notes that Schäfer's Tibet expedition has the aim of finding "the city of Obo in the Gobi Desert." Wolff to Höhne, 26.05.1937, BA, NS 21/687. It is possible that Himmler and Schäfer specifically meant Bayan Obo in the northwestern part of Outer Mongolia. The famous Swedish explorer Sven Hedin had visited the spot in 1927.

[31]"The Activities of Dr. Ernst Schaefer, Tibet Explorer and Scientist with SS-sponsored Scientific Institutes," Headquarters United States Forces European Theater, Military Intelligence Service Center, 12.02.1946, NARA, RG238, M1270/27.

[32]Galke "Aktennotiz," 06.08.1937, BA (ehem.BDC) Ahnenerbe: Schäfer, Ernst (14.03.1910); Wolff to Höhne, 26.05.1937, BA, NS 21/687.

[33]Rüdiger Sünner, *Schwarze Sonne*, p. 50. Sünner states that this story came from a book of memoirs written by an unnamed author. It is certain, however, that this author is Schäfer. During an interview I conducted with Mrs. Gabriele Winckler-Dechend, Wiligut's female assistant, she recalled a visit that Schäfer had paid to

Wiligut and clearly remembered their conversation about the telepathy of Tibetan lamas.

[34]Personal communication, Mrs. Gabriele Winckler-Dechend.

[35]"Unbekanntes Tibet," *Das Schwarze Korps*, 16.07.1936, BA, R135/80.

[36]Personal communication, Mr. Gerd Pucka.

[37]Personal communication, Mrs. Ursula Schäfer.

[38]"Unbekanntes Tibet," *Das Schwarze Korps*, 25.06.1936, BA, R135/80.

[39]Dolan Diary, Academy of Sciences Archives, vol. 1.95 (August 5–6, 1934), as quoted in Karl E. Meyer and Shareen Blair Brysac, *Tournament of Shadows* (Washington, D.C.: Counterpoint, 1999), p. 538.

[40]Personal communication, Mr. Gerd Pucka.

[41]"Führerstammkarte," NARA, RG242, A3343, SSO Schäfer, Ernst (14.03.1910).

[42]"Unbekanntes Tibet," *Das Schwarze Korps*, 25.06.1936, BA, R135/80.

[43]Schäfer to Himmler, 06.10.1936, NARA, RG242, T580/204/686.

[44]Ibid.

[45]Ibid.

[46]Sievers, "Aktennotiz," 04.10.1937, NARA, RG242, T580/143/167.

[47]Galke, "Aktennotiz," 06.08.1937, BA (ehem.BDC) Ahnenerbe: Schäfer, Ernst (14.03.1910).

[48]Ibid.

[49]The senior official was Dr. Rolf Höhne. For further details, see Bruno Beger, *Mit der deutschen Tibetexpedition Ernst Schäfer 1938/39 nach Lhasa* (Wiesbaden: Schwarz, 1998), pp. 5–6.

[50]Personal communication, Dr. Bruno Beger.

[51]Ibid.

[52]Ibid.

[53]Isabel Heinemann, *"Rasse, Siedlung, deutsches Blut"* (Göttingen: Wallstein, 2003), p. 18.

[54]Hofmann to Reichsführer-SS, 20.09.1940, NARA, RG242, A3343, SSO Beger, Bruno (27.04.1911).

[55]Personal communication, Dr. Bruno Beger. For details on this proposed Hawaii expedition, see Head of RuSHA to RFSS, 04.05.1938, BA, NS 21/361; Brandt to Ahnenerbe, 31.03.1939, BA, NS 21/361.

[56]RuSHA to RFSS, 04.05.1938, BA, NS 21/361.

[57]Personal communication, Dr. Bruno Beger.

[58]Personal communication, Mrs. Ursula Schäfer and Dr. Ulrich Gruber. German scholar Isrun Engelhardt recently discovered a written eyewitness account of this tragic accident that confirms the stories told by Mrs. Schäfer and Dr. Gruber. For details of this report, see Christopher Hale, *Himmler's Crusade*, p. 131.

[59]Personal communication, Dr. Ulrich Gruber.

[60]Gestapo to RFSS Pers. Stab, 04.03.1938, NARA, RG242, T580/204/686.

[61]Sievers's letters sometimes prompted great puzzlement from the German manufacturers. For an example of this, see Carl Walther Waffenfabrik to Galke, 12.01.1938, NARA, RG242, T580/204/686.

[62]Adolf Hitler, *Mein Kampf,* trans. Ralph Manheim (Boston: Houghton Mifflin, 1971), pp. 290–291; Ian Kershaw, *Hitler 1936–45: Nemesis* (London: Allen Lane, The Penguin Press, 2000), p. 26.

[63]John W. Dower, *War without Mercy,* p. 203.

[64]Ibid., pp. 272–275.

[65]"Die ersten Weißen in Jalung Podrang," *Niedersächische Tageszeitung Hannover,* 04.08.1939, BA, NS 21/633. See also Schäfer to Sievers, 27.12.1937, BA (ehem.BDC) Ahnenerbe: Schäfer, Ernst (14.03.1910).

[66]Karl E. Meyer and Shareen Blair Brysac, *Tournament of Shadows* (Washington, D.C.: Counterpoint, 1999), p. 514.

[67]Schäfer to Sievers, 27.12.1937, BA (ehem.BDC) Ahnenerbe: Schäfer, Ernst (14.03.1910).

[68]Sievers to Wolff, 23.01.1938, NARA, RG242, T580/204/686.

[69]Indeed, soon after the start of World War II, Himmler proceeded to place Schäfer in charge of a military expedition to Tibet. The purpose of the mission was to set up a base in Tibet from which to disrupt Indian railroads and telegraph and to carry out diversionary feints in order to keep British troops occupied. For further details, see Headquarters United States Forces European Theater Military Intelligence Service Center, "The Activities of Dr. Ernst Schaefer, Tibet Explorer and Scientist with SS-sponsored Scientific Institutes," Headquarters United States Forces European Theater, Military Intelligence Service Center, 12.02.1946, NARA, RG238, M1270/27.

[70]The Advertising Council of the German Economy donated 46,000 reichsmarks. The German Research Council contributed 30,000 reichsmarks, while the Eher Verlag paid 20,000 reichsmarks. The remaining 16,111 reichsmarks in revenue came from small donors, including the Foreign Ministry and two private donors. See Schäfer, "Gesamt-Abrechnung," 15.11.1940, NARA, RG242, T81/127/150165.

[71]RF-SS to Günther, 24.02.1938, BA-DH, ZM 1457 A5.

[72]Ullmann to Spitzi, 04.03.1938, BStU HA IX/11 FV 17/75 Bd. 17.

[73]This is the name that appears on the official letterhead of the expedition that Schäfer used in March 1938. See Schäfer to Beger, 01.03.1938, NARA, RG242, T81/130/163342.

[74]Bruno Beger, *Mit der deutschen Tibetexpedition,* p. 10.

[75]Deutsches Generalkonsulat, 11.06.1938, BA-DH, ZM 1457 A.5.

[76]Schäfer to Himmler, 05.06.1938, BA-DH, ZM 1457 A.5.

[77]Schäfer to Obersturmbannführer, 09.05.1938, BA-DH, ZM 1457 A.5.

[78]RF-SS to Sir Barry Domvile, 19.05.1938, BA-DH, ZM 1457 A.5.

[79]In addition, Domvile also passed word of Himmler's complaints on to his jour-

nalist son, Compton Domvile, who put a positive public spin on both the expedition and on Himmler himself in *The Anglo-German Review*. [Compton Domvile to RF-SS, translated 09.08.1938, NARA, RG242, A3343, SSO 067B: Schäfer, Ernst (14.03.1910).] Himmler, noted the younger Domvile, "has long taken a deep interest in the origin and prehistory of mankind. His studious and inquiring mind, which is held to be one of the keenest and most enlightened in Germany today, finds relaxation in archaeology, genealogy and kindred fields of research. Thus he holds the office of President of the Ahnenerbes (sic) organization, which is concerned with the study of the origins and early cultural development of the German people, and which, in pursuance of its researches into the early wanderings of man, has sponsored Dr. Schäfer's expedition." Compton Domvile, "Dr. Ernst Schäfer looks at India," *Anglo-German-Review*, 17.07.1938, NARA, RG242, A3343, SSO: Schäfer, Ernst (14.03.1910).

[80]Schäfer to Himmler, 05.06.1938, BA-DH, ZM 1457 A.5.

[81]H.A.F. Rumbold, India Office, to G.E. Hubbard, Political Intelligence Dept., FO, 13.01.43, as cited in Karl E. Meyer and Shareen Blair Brysac, *Tournament of Shadows*, p. 513.

[82]F. Spencer Chapman, *Lhasa: The Holy City* (London: Readers Union Ltd by arrangement with Chatto & Windus, 1940), p. 6.

[83]Ernst Schäfer, *Geheimnis Tibet: Erster Bericht der Deutschen Tibet-Expedition Ernst Schäfer 1938/39* (München: Bruckmann, 1943), pp. 57–58.

[84]Ernst Schäfer, *Geheimnis Tibet*, p. 56.

[85]Bruno Beger, *Mit der deutschen Tibetexpedition*, p. 14.

[86]"Aufstellung über die anthropologische Ausrüstung für SS-Untersturmführer Beger," NARA, RG242, T580/143/167.

13. TIBET

[1]For an example of Beger's measurements, see the data compiled for Kaiser Bahadur, the expedition's interpreter. "II, 2 Kaiser, Bahadur, Nepali, Gangtok (Dolmetscher)," NARA, RG242, T81/131/165016.

[2]Hans F.K. Günther, *The Racial Elements of European History*, trans. G.C. Wheeler (Port Washington, N.Y.: Kennikat Press, 1970), Map VIII, p. 105.

[3]Ibid., p. 10.

[4]Franz Boas, "Changes in Bodily Form of Descendants of Immigrants," 1910, as quoted in Clarence C. Gravlee, H. Russell Bernard, and William R. Leonard, "Heredity, Environment and Cranial Form: A Reanalysis of Boas's Immigrant Data," *American Anthropologist* 105, no. 1 (March 2003): 127. It is interesting to note that two American anthropologists have recently questioned the statistical significance of Boas's findings—the first major challenge to the study in ninety-two years. See Corey S. Sparks and Richard L. Jantz, "A Reassessment of Human Cranial Plasticity: Boas Revisited," *Proceedings of the National Academy of Science* 99, no. 23

(November 12, 2002): 14636–14639. Gravlee and his colleagues conclude, however, that "Boas got it right."

[5]Clarence C. Gravlee, H. Russell Bernard, and William R. Leonard, "Heredity, Environment and Cranial Form," pp. 125–138. As this paper points out, "Franz Boas's classic study, 'Changes in Bodily Form of Descendants of Immigrants,' is a landmark in the history of anthropology. More than any single study, it undermined racial typology in physical anthropology and helped turn the tide against early-20th-century scientist racism."

[6]Bruno Beger, "10.Bild.," NARA, RG242, T81/132/165403-165418. Although this document lacks both a title and a named author, these notes are clearly intended as a guide for a slide show presented by Beger. In this document, Beger briefly recalled the events of that day and the thoughts that went through his mind. "Suddenly a Tibetan with remarkably fine clothing arrives. My next thought is that I should measure him."

[7]Bruno Beger, "10.Bild.," NARA, RG242, T81/132/165403-165418.

[8]Ernst Schäfer, *Geheimnis Tibet: Erster Bericht der Deutschen Tibet-Expedition Ernst Schäfer 1938/39* (Munich: Bruckmann, 1943), p. 86.

[9]Ernst Schäfer, *Geheimnis Tibet*, pp. 86–90.

[10]The Tibetan Council of Ministers to Schäfer, Third Day of the Tenth Month of the Fire-Tiger Year, as quoted in Christopher Hale, *Himmler's Crusade* (Hoboken, N.J.: John Wiley & Sons, Inc., 2003), p. 200.

[11]SS-Reichskanzlei to Schäfer, 05.12.1938, NARA, RG242, A3343, SSO Schäfer, Ernst (14.03.1910).

[12]Bruno Beger, "14. Bild.," NARA, RG242, T81/132/165407.

[13]Bruno Beger, "25. Bild.," NARA, RG242, T81/132/165409.

[14]It is evident from the list of equipment that Beger took to Tibet that he used three substances—Negocoll, Hominit, and Celerit—to make casts and replica human heads. These three ingredients were all part of the Poller method of castmaking, favored by many physicians. "Aufstellung über die anthropologische Ausrüstung für SS-Untersturmführer Beger, NARA, RG242, T580/143/167; Alphons Poller, ed., *Das Pollersche Verfahren zum Abformen* (Berlin, Wien: Urban & Schwarzenberg, 1931).

[15]Alphons Poller, "Vorwort," *Das Pollersche Verfahren zum Abformen*, p. vi.

[16]"Aus dem 'Gesammt-Verlags-Katalog' des deutschen Buchhandels: Gebr. Schlagintweit, 'Sammlung ethnographischer Köpfe,'" BA, R135/58/151779. What is particularly interesting is that this price list came from a file containing many of Beger's personal papers on racial studies, including the racial measurements of his own wife, Hildegard.

[17]After the *Anschluss*, such exhibitions began opening all over Austria. See Klaus Taschwer, "Anthropologie ins Volk-Zur Austellungspolitik einer anwendbaren Wissenschaft bis 1945," in *Politik der Präsentation*, ed. Herbert Posch and Gottfried Fliedl

(Vienna: Turia and Kent, 1996) pp. 30–31; Gert Kerschbaumer, "Das 'Deutsche Haus der Natur' zu Salzburg," in *Politik der Präsentation*, p. 198.

[18]Beger, "Rassenkundliche Abformungen auf der Tibetexpedition Ernst Schäfer," 09.11.1942, NARA, RG242, T81/129/151690.

[19]Ibid.

[20]Ernst Schäfer, *Geheimnis Tibet*, p. 178.

[21]Bruno Beger, "36.Bild," NARA, RG242, T81/132/165410-165411.

[22]"Unterhaltung mit Labrang Kugnoe, dem geistlichen Oberhaupt von Gyantse," NARA, RG242, T81/128/151360.

[23]Schäfer, "Sachbericht," 23.01.1939, BA (ehem.BDC) Ahnenerbe: Schäfer, Ernst (14.03.1910). Intriguingly, this report reads as if it were intended for publication in *Das Schwarze Korps*, an important SS publication. Although it is unsigned, it was almost certainly written by Schäfer and is very similar in style to the "Unbekantes Tibet" articles that the zoologist wrote for *Das Schwarze Korps* in 1936.

[24]Ibid.

[25]Ibid.

[26]Bruno Beger, *Mit der deutschen Tibetexpedition Ernst Schäfer 1938/39 nach Lhasa* (Wiesbaden: Schwarz, 1998), p. 160.

[27]Ibid., p. 161.

[28]"SS-Männer-Pioniere-Wissenschaftler: Das Schwarze Korps sprach mit Dr. Ernst Schäfer," Mitte Januar, NARA, RG242, T81/132/165472-165477.

[29]Ibid.

[30]Ibid.

[31]"Desiderata der Tibetforschung," n.d., NARA RG242, T81/128/151362-151363.

[32]Handwritten English translation of a field-journal account, "Visit to Gyaldzong Place," NARA, RG242, T81/131/164706-07. This account is not signed, but it was found among a group of four other similar translated field-journal entries that all contained observations on anthropological matters. There seems to be no question that the original author was Bruno Beger. The original German version of this can be found in Beger's "Ethnologische Aufzeichnungen" folder at NARA, RG242, T81/128/151356-151357.

[33]Ernst Schäfer, *Geheimnis Tibet*, p. 178.

[34]Ibid., p. 180.

[35]Ibid., p. 182.

[36]Schäfer to Brandt, 25.06.1940, NARA, RG242, T84/257/6617401-6617415.

[37]"Empfang durch Himmler in München," *Berliner Börsenzeitung*, 05.08.1939, NARA, RG242, T580/143/167.

[38]Bruno Beger, "Rassen in Tibet," 19.02.1943, NARA, RG242, T81/128/151657-151669.

[39]Ibid.

[40]Schäfer to Brandt, 25.06.1940, NARA, RG242, T84/257/ 6617401-6617415.

[41]"The Activities of Dr. Ernst Schaefer, Tibet Explorer and Scientist with SS-sponsored Scientific Institutes," Headquarters United States Forces European Theater, Military Intelligence Service Center, 12.02.1946. NARA, RG238, M1270/27.

14. IN SIEVERS'S OFFICE

[1]David Clay Large, *Berlin* (New York: Basic Books, 2000), p. 316.

[2]Ibid., p. 317.

[3]Sievers to Ullmann, 11.08.1939, BA (ehem. BDC) 0.996.

[4]Edmund Kiss, *Das Sonnentor von Tihuanaku und Hörbigers Welteislehre* (Leipzig: Koehler & Amelang, 1937), p. 146; Donald Johanson and Blake Edgar, *From Lucy to Language* (New York: Simon & Schuster, 1996), p. 38.

[5]Kiss, "Personalangaben," 04.04.1944, NARA, RG242, A3343, SSO: Kiss, Edmund (10.12.1886); Abschrift (Ärztlicher Untersuchungsbogen), n.d., Kiss, Edmund. BA (ehem.BDC) Ahnenerbe: Kiss, Edmund (10.12.1886).

[6]Photograph of Edmund Kiss, "Dienstlaufbahn," n.d., NARA, RG242, A3343, SSO Kiss, Edmund (10.12.1886).

[7]Riepe, "Leumundszeugnis," 05.07.1948, HHA, Abt.520 KS-HL Nr.88, Spruchkammer Kassel; Rudolf Bury, "Eidesstattliche Erklärung," 06.07.1948, HHA, Abt.520 KS-HL Nr.88, Spruchkammer Kassel.

[8]Kiss, "SS-Stammrollenauszug," 31.01.1939, NARA, RG242, A3343, SSO: Kiss, Edmund (10.12.1886).

[9]Robert Bowen, preface to *Universal Ice*, (London: Bellhaven Press, 1993), p. viii.

[10]Indeed, Kiss and four other prominent believers in the World Ice Theory signed an official declaration, known as the Pyrmonter Protokoll, in 1936, stating that "Hans Hörbiger's World Ice Theory in its fundamental form is the intellectual gift of a genius that is important to all humanity in both practical and worldview terms. To Germans, it is a true Aryan gift of special importance." "Abschrift Pyrmonter Protokoll," 19.07.1936, NARA, RG242, T580/194/465. For further details on the popularity of the World Ice Theory in Nazi Germany, see Robert Bowen, *Universal Ice*, pp. 146–150.

[11]Robert Bowen, *Universal Ice*, pp. 59–60.

[12]Ibid., p. 76.

[13]Adolf Hitler, *Hitler's Table Talk 1941–1944*, ed. H.R.Trevor-Roper (London: Phoenix Press, 2000), no.125, p. 249.

[14]Indeed, Posnansky concluded that some twenty-seven degrees separated the current position of the sun from that when the temple was built. Adela Breton, "Proceedings of Americanists' Congress," *Man* 84 (1910). Subsequent archaeological excavation and radiocarbon dating—an absolute method of dating that came into use after the Second World War—has shown that Tiwanaku is no more than seventeen hundred years old. Alexei Vranich, "Tiwanaku Q&A," *Archaeology's Interactive Dig*, http://www.archaeology.org.

[15]Edmund Kiss, *Das Sonnentor von Tihuanaku*, pp. 130–132.

[16]Edmund Kiss, "Die Kordillerenkolonien der Atlantiden," *Schlüssel zum Welt-geschehen* 8/9 (1931): 259.

[17]Ibid., 261.

[18]Edmund Kiss, "Nordische Baukunst in Bolivien?" *Germanien* 5 (May 1933): 144.

[19]Edmund Kiss, *Das Sonnentor von Tihuanaku*, pp. 144–145.

[20]Edmund Kiss, *Das Sonnentor von Tihuanaku*, pp. 106–107.

[21]Rudolf von Elmayer-Vestenbrugg, "Versunkene Reiche," *Die H.J.*, 24.04.1937, BA, NS 21/714.

[22]Sievers to Koehler & Amelang, 09.12.1937, BA, NS 21/ 166.

[23]Wüst to Himmler, 07.03.1938, BA, NS 21/ 166.

[24]Himmler had taken a flight over Libya and, while soaring over mountains there, noticed what he believed to be the telltale white layers of fossil shorelines. He insisted that Kiss put aside his South American plans, take a leave of absence from his government post, and prepare immediately to depart for Libya. Kiss strongly suspected that the trip to Libya, an Italian possession at the time, would be futile. The North African coast was strewn with fossil records of rising and falling sea levels, but Kiss knew that these geological formations had nothing to do with the purported cataclysms of the World Ice Theory. Still, he agreed to take a look: he did not want to antagonize his powerful patron. He also agreed to include Sardinia on his itinerary, when someone in the Ahnenerbe pointed out an apparent similarity between stone towers there and the architecture at Tiwanaku.

Kiss set off by train for Rome on February 15 or 16, 1939. He had a cameraman and an assistant in tow. A short stint in a library there convinced him that a trip to Sardinia would be unnecessary. A few days later, he arrived in the Libyan capital of Tripoli. The Libyan governor, an ardent aviator and a friend of Charles Lindbergh, placed a plane, a pilot, and a truck at his disposal, and over the next two weeks, Kiss and his associates explored the Libyan countryside. The weather was not terribly conducive to aerial scouting—thick gray clouds heavy with rain hung over the coastal mountains, refusing to budge most days—but Kiss kept busy. He drew neat draftsmanlike maps of the mountains, carefully noting the coastal terraces and the notches of deep canyons and occasionally jotting down the word *Zeugen*, where he thought he found "pieces of evidence." He walked plains and canyons dotted with occasional palm trees, sketching profiles of the low mountains, and mulled over the origins of the Nalut Canyon not far from the Algerian border. The more he looked at it and at a number of similar gorges in the Libyan mountains, the more intrigued he became. "Strong water currents on the scale imagined by the World Ice Theory most likely formed them through erosion," he later wrote. "There is no other explanation."

For further information on this research trip, see Kiss, "Programm der Forschungsreise des SS-Hauptsturmführers Kiss," 30.01.1939, BA, NS 21/415; Sievers, "Abrechnung für die Tripolis-Reise von: SS-Hauptsturmführer Kiss, SS-Obersturmführer Bousset, SS-Anwärter Mohri" 16.02.1939, BA (ehem.BDC) Ahnenerbe, Kiss: Edmund (10.12.1886); Kiss to Sievers, 20.02.1939, BA NS, 21/415; "Niederschrift betreffend Vortrag des SS-Hauptsturmführers Kiss beim Reichsführer SS," 08.05.1939, BA (ehem. BDC) Ahnenerbe: Kiss, Edmund (10.12.1886); Kiss, "Ergebnisbericht der Forschungsreise des SS-Hauptsturmführers Kiss nach Tripolis," 15.05.1939, BA, NS 21/415; "Protokoll der öffentlichen Sitzung am 05.August 1948," HHA.Abt. 520 KS-HL Nr. 88. Spruchkammer Kassel.

[25]Sievers to Himmler, 18.08.1939, BA, NS 21/123.

[26]Kiss, "Vorläufiges Programm," 15.04.1939, BA, NS 21/171.

[27]Sievers to Himmler, 18.08.1939, BA, NS 21/123.

[28]Wüst to Menzel, 20.09.1938, BAK, R73/ 15896.

[29]Ibid. See also George G. Cameron, "Darius Carved History on Ageless Rock," *National Geographic* (December 1950): 825–844.

[30]Walther Wüst, "Ein indogermanisches Dokument," *SS-Leitheft* (July 1943), p. 5. NARA, RG242, A3345B, 124/ 646.

[31]Ibid.

[32]The portrait of Darius in the relief would have been of immense interest to racial researchers such as Günther, who relied on such images to illustrate popular picture books on race. Indeed, Günther had already included a rather crude sketch of Darius I, based on a sculpted relief, in his book *The Racial Elements of European History*, trans. G.C. Wheeler (Port Washington, N.Y.: Kennikat Press, 1970), pp. 145–147.

[33]R. Campbell Thompson, "The Rock of Behistun," in *Wonders of the Past*, ed. Sir J.A. Hammerton. Vol. II (New York: Wise and Co., 1937), p. 765. See also George G. Cameron, "Darius Carved History on Ageless Rock," p. 840.

[34]Wüst to Menzel, 20.09.1938, BAK, R73/ 15896.

[35]Sievers, "Aktenvermerk," 27.10.1938, BA (ehem.BDC) Ahnenerbe: Wüst, Walther (07.05.1901).

[36]P.L.O. Guy, "Balloon Photography and Archaeological Excavation," *Antiquity* 6 (June 1932):148–155. Others who followed this lead included the Polish excavators of Biskupin. See Jozef Kostrzewski, "Osada bagienna w Biskupinie, w pow. zininskim," *Przeglad Archeologiczny* 5 (1938): 121–140.

[37]R. Höhne, "Die Ausgrabungen der Schutzstaffeln," *Germanien* (1938): 224–230.

[38]Wüst to Menzel, 20.09.1938, BAK, R73/ 15896. Wüst does not mention the name of the Iranian student he had in mind, but it seems clear from other surviving pieces of correspondence that he planned to take Davoud Monchi-Zadeh. Monchi-Zadeh was working closely with him during this period. See Wüst: Aktenvermerk: Betr: Herrscher-Abkunft-Sagen, 05.11.1938, BA (ehem.BDC) Ahnenerbe: Wüst, Walther (07.05.1901).

[39]Wüst to Menzel, 20.09.1938, BAK, R73/ 15896.

[40]Wüst to Menzel, 20.09.1938, BAK, R73/ 15896.

[41]Huth, "Fragebogen für Mitglieder: Reichsverband Deutscher Schriftsteller," 05.12.1933, BA (ehem BDC) RKK: Huth, Otto Herbert (09.05.1906); Huth, "R.u.S.-Fragebogen." 30.01.1939. BA (ehem.BDC) RS: Huth, Otto Herbert (09.05.1906).

[42]Huth, "R.u.S.-Fragebogen," 30.01.1939, BA (ehem.BDC) RS: Huth, Otto Herbert (09.05.1906).

[43]Herman Wirth, *Der Aufgang der Menschheit* (Jena: Diederichs, 1934), pp. 105–109.

[44]Conrado Rodriguez-Martin, "The Guanche Mummies," in *Mummies, Disease and Ancient Cultures,* ed. Aiden Cockburn, Eve Cockburn, and Theodore A. Reyman, 2nd ed. (Cambridge: Cambridge University Press, 1998), p. 283.

[45]Otto Huth, "Die Gesittung der Kanarier als Schlüssel zum Ur-Indogermanentum," *Germanien* 2 (February 1937): 50.

[46]Earnest Hooton, *The Ancient Inhabitants of the Canary Islands* (Cambridge, MA: Peabody Museum of Harvard University, 1925), p. 44. In addition, as one modern mummy expert, Dr. Guido Lombardi, has pointed out to me, a form of malnutrition, known as kwashiorkor and caused by inadequate protein intake, can also lighten or redden the hair color of the living.

[47]Otto Huth, "Die Gesittung der Kanarier als Schlüssel zum Ur-Indogermanentum," p. 54.

[48]This is one of the expeditions shown on the Ahnenerbe map in the Klingspor volume.

[49]Huth to Wüst, 14.02.1939, BA (ehem.BDC) Ahnenerbe: Huth, Otto. (09.05.1906).

[50]Ibid.

[51]Ibid.

[52]Thomas Nußbaumer, *Alfred Quellmalz und seine Südtiroler Feldforschungen* (Innsbruck: Libreria Musicale Italiana, 2001), p. 163.

[53]It is akin to saying that the Pilgrim fathers of New England—who also spoke a Germanic language—were German forefathers and that the early remote New England colonies preserved a pure form of Germanic tradition.

[54]Paul Burkert to Notgemeinschaft (Arbeitsplan), 17.04.1935, BAK, R73/16788. In 1936, Burkert guided an SS study commission of scholars and very high-ranking SS officers—including Prinz Waldeck Pyrmont and Hermann Behrends, the first head of the SS Security Service—on a trip to Iceland and its ancient historic sites. The commission made a pilgrimage to Iceland's national shrine, Thingvellir, and to the home of Snorri Sturluson, the author of the *Prose Edda*. Without a doubt, the trip was intended to inspire the top echelon of the SS with a sense of their Nordic heritage and to prepare them for the terrible work that Himmler envisioned ahead of them. For more details, see Grundherr, "Abschrift zu Pol II 424," 19.06.1936, PAAA, Ges. Kopenhagen C 3 (1934/37) Band 1. For an account of the trip written by one of the scholars, see Otto Rahn, *Luzifers Hofgesind* (Leipzig: Schwarzhäupter, 1937). For

a brief scholarly account of the trip, see Thór Whitehead, *Íslandsaevintyri Himmlers 1935–1937*, (Reykjavík: Vaka-Helgafell, 1998).

[55]Sievers to Himmler. 13.01.1938, BA, NS 21/599. Also, Schweizer, "Zu Zoller-lassgesuch Schweizer vom 15.11.1938," n.d., BA (ehem BDC) Ahnenerbe: Schweizer, Bruno: (03.05.1897).

[56]Schweizer, "Zu Zollerlassgesuch Schweizer vom 15.11.1938," n.d., BA (ehem BDC) Ahnenerbe: Schweizer, Bruno: (03.05.1897).

[57]Schweizer, "Beiträge zum SS-Kalender," n.d., NARA, RG242, T580/199/570.

[58]Sievers to Gerlach, 18.04.1939, BA (ehem. BDC) Ahnenerbe: Schweizer, Bruno (03.05.1897).

[59]Sievers to Wüst, 21.04.1939, BA, NS 21/40.

[60]Indeed, so convinced was Himmler that Iceland was a repository of ancient Germanic lore that he sent a "study commission" of high-ranking SS officers to Iceland in 1936 to commune with the ancient past. See note 54.

[61]Hermann Kaienburg, *Die Wirtschaft der SS* (Berlin: Metropol, 2003), p. 262.

[62]As it turned out, the Scandinavian reporters had learned of plans for a smaller SS research trip to Iceland led by Dr. Kurt Tackenberg and Dr. Walter Gehl. Tackenberg and Gehl intended on excavating ancient Icelandic temples.

[63]"Übersetzung aus der dänischen Zeitung: 'Extrabladet,'" 11.03.1939, BA (ehem. BDC) Ahnenerbe: Schweizer, Bruno (03.05.1897).

[64]Himmler to Sievers, 22.03.1939, BA (ehem. BDC) Ahnenerbe: Schweizer, Bruno. (03.05.1897).

[65]SD-Hauptamt to Ahnenerbe, 20.05.1939, BA, NS 21/40.

[66]Sievers, "Besprechung Sievers-Schweizer-Gerlach," 21.04.1939, BA, NS 21/40.

[67]Sievers, "Betr. Islandreise," 26.05.1939, BA, NS 21/ 123.

[68]Sievers to president of Reichsgesundheitsamt, 25.01.1939. BA, NS 21/40.

[69]Hans F.K. Günther, *The Racial Elements of European History*, p. 74. Even the vile anti-Semitic film *Der Ewige Jude*, produced in 1940 at Josef Goebbels's request, publicly acknowledged the difficulty that many Nazis had in identifying Jews who did not wear forelocks or skullcaps. "Hair, beard, skullcap and caftan," noted the film's narrator, "make the Eastern Jew recognizable to all. If he appears without his trademarks, only the sharp-eyed can recognize his racial origins."

[70]For further details on the quest for racial diagnosis in Germany, see Ernst Klee, *Deutsche Medizin im Dritten Reich* (Frankfurt: S. Fischer, 2001), pp. 158–165.

[71]Isabel Heinemann, *"Rasse, Siedlung, deutsches Blut"* (Göttingen: Wallstein, 2003), p. 552. See also Himmler to Bormann, 22.05.1943, in *Reichsführer ! . . .* , ed. Helmut Heiber (Stuttgart: Deutsche Verlags-Anstalt, 1968), p. 213.

[72]Greite, "Lebenslauf," 28.11.37, NARA, RG242, SSO: Greite, Walther: (13.06.1907).

[73]*Enzyklopädie des Holocaust*, 2nd ed., ed. Israel Gutman (Munich: Piper, 1998), s.v. "Reichszentrale für Jüdische Auswanderung."

[74]Evelyne Polt-Heinz, review of *Die Rothschilds. Porträt einer Dynastie,* by Frederic Mortons, *Wiener Zeitung,* 03.12.2004.

[75]Greite, "Abschrift," 11.07.1939, NARA, RG242, A3345 DS G119. Greite, Walter (13.06.1907).

[76]"Gedächtnisprotokoll. Unterredung Frau Hella Sievers und Michael H. Kater," 26.04.1963, IfZ, ZS/A-25 vol. 2.

[77]Sievers, "Aktenvermerk für den Reichsführer SS Heinrich Himmler," 25.05.1939, BA (ehem.BDC) 0.996.

15. THIEVES

[1]Felix Kersten, *The Memoirs of Doctor Felix Kersten,* ed. Herma Briffault (Garden City, N.Y.: Doubleday, 1947), p. 44.

[2]Ian Kershaw, *Hitler 1936–1945: Nemesis* (London: Allen Lane, The Penguin Press, 2000), p. 223

[3]Ian Kershaw, *Hitler 1936–1945,* p. 226.

[4]Ian Kershaw, *Hitler 1936–1945,* pp. 240–241.

[5]Richard Rhodes, *Masters of Death* (New York: Alfred A. Knopf, 2002) pp. 3–4.

[6]Richard C. Lukas, *The Forgotten Holocaust* (Lexington: University Press of Kentucky, 1986), p. 3. As quoted in Richard Rhodes, *Masters of Death,* p. 6.

[7]Ian Kershaw, *Hitler 1936–1945,* p. 241.

[8]Isabel Heinemann, *"Rasse, Siedlung, deutsches Blut"* (Göttingen: Wallstein, 2003), p. 188.

[9]Peter Padfield, *Himmler: Reichsführer-SS* (London: Macmillan, 2001), p. 291.

[10]Sievers to Himmler, 04.09.1939, BA (ehem BDC) Ahnenerbe: Paulsen, Peter (08.10.1902).

[11]Ibid.

[12]Ibid.

[13]Rotraut Wolf, "Peter Paulsen Nachruf," *Fundberichte aus Baden Württemberg* 10, (1985): 727–728.

[14]Scheel, "Gutachten," 27.03.1938, BA (ehem. BDC) PK: Paulsen, Peter (08.10.1902).

[15]Paulsen, "R.u.S.–Fragebogen, Lebenslauf," 28.11.1936, BA (ehem BDC) RS: Paulsen, Peter (08.10.1902).

[16]Paulsen. "Lebenslauf," 28.11.1936, BA (ehem. BDC) PK: Paulsen, Peter (08.10.1902).

[17]Jörn Jacobs, "Peter Paulsen: Ein Wanderer zwischen zwei Welten," in *Prähistorie und Nationalsozialismus,* ed. Achim Leube and Morten Hegewisch (Heidelberg: Synchron, 2002), pp. 451–459.

[18]Paulsen, "R.u.S.-Fragebogen, Lebenslauf," 28.11.1936, BA (ehem BDC) RS: Paulsen, Peter (08.10.1902).

[19]Sievers to Himmler, 04.09.1939, BA (ehem BDC) Ahnenerbe: Paulsen, Peter (08.10.1902).

[20]Paulsen to Sievers, "Schutz-Maßnahmen für kulturgeschichtliche Denkmäler in Polen," 15.09.1939, BA (ehem. BDC) Ahnenerbe: Paulsen, Peter (08.10.1902). This letter is also quoted in "Document 2, Poland," in Hellmut Lehmann-Haupt, *Cultural Looting of the Ahnenerbe: Report prepared by Monuments, Fine Arts and Archives Section,* OMGUS, March 1, 1948, NARA, RG 260, M1926/R151. Paulsen's memo actually refers to the "protection of monuments in Poland." But as Lehmann-Haupt points out, the Ahnenerbe correspondence during the war is replete with ambiguity and the distortion of facts. Indeed, Lehmann-Haupt himself counted twenty-one different euphemisms for "looting" in the documents, from *sicherstellen* ("to secure"), to *aufnehmen* ("to register"). "Protection" should be added to this list.

[21]Petersen to Sievers, 18.09.1939, as reproduced in Andrzej Mezynski, *Kommando Paulsen* (Cologne: Dittrich-Verlag, 2000) pp. 26–33.

[22]Ullmann to Sievers, 21.09.1939, as reproduced in Andrzej Mezynski, *Kommando Paulsen*, p. 24.

[23]Franz Six, "Führerstammkarte," BA (ehem. BDC) SSO: Six, Franz Alfred (12.08.1909).

[24]Paulsen, "Dienstreisebericht von Peter Paulsen," 04.01.1940, as reproduced in Andrzej Mezynski, *Kommando Paulsen*, pp. 55–67.

[25]Arthur Burkhard, *The Cracow Altar of Veit Stoss* (Munich: Bruckmann, 1972), p. 26, footnote 21.

[26]Ibid., p. 23.

[27]Paulsen to Kaiser, 05.10.1939, as quoted in "Document 183, Poland," in Hellmut Lehmann-Haupt, *Cultural Looting of the Ahnenerbe: Report prepared by Monuments, Fine Arts and Archives Section,* OMGUS, March 1, 1948, NARA, RG260, M1926/R151.

[28]Ibid.

[29]Paulsen, "Aktenvermerk," 18.04.1941, as reproduced in Andrzej Mezynski, *Kommando Paulsen*, pp. 39–42.

[30]Sievers, Aktenvermerk, 14.10.1939, BA (ehem. BDC) SS HO-1324.

[31]The title of prince-bishop is a very old one, dating back to the Roman era. During that time, prince-bishops wielded great secular power in a city, in addition to their clerical authority.

[32]Paulsen, "Dienstreisebericht von Peter Paulsen," 04.01.1940, as reproduced in Andrzej Mezynski, *Kommando Paulsen*, pp. 55–67. The Black Madonna was painted sometime after the sixth century A.D. and was presented to the monks of Jasna Gora, near Czestochowa, in the fourteenth century. Miracles have reportedly befallen pilgrims traveling to see it, and in 1717 it was crowned "Queen of Poland" by the faithful.

[33]Liebel to Lammers, 16.11.1939, as noted in Jonathan Petropoulos, *Art as Politics*

in the Third Reich (Chapel Hill & London: University of North Carolina Press, 1996), p. 108.

[34]Ibid., p. 109.

[35]Adolf Hitler, *Hitler's Table Talk 1941–1944*, ed. H.R.Trevor-Roper (London: Phoenix Press, 2000), no.77, pp. 146–149.

[36]In 1942, members of the Polish resistance in Switzerland discovered that the altar had been hidden either in or near Nuremberg. They relayed the information to American authorities, who began searching for the altar after they entered Nuremberg in 1945. The American army retrieved it and sent it back to Kraków in 1946. After eleven years of restoration, it was reinstalled in St. Mary's Church. For further details, see Ferdinand and Delia Kuhn, "Poland Opens Her Doors," *National Geographic* (September 1958): 357–398.

[37]Sievers, Aktenvermerk, 14.10.1939, BA (ehem. BDC) SS HO-1324.

[38]Sievers, Aktenvermerk, 14.10.1939, BA (ehem. BDC) SS HO-1324.

[39]Paulsen. "Dienstreisebericht von Peter Paulsen," 04.01.1940, as reproduced in Andrzej Mezynski, *Kommando Paulsen*, pp. 55–67.

[40]Sievers, "Aktenvermerk," 14.10.1939, BA (ehem. BDC) SS HO-1324.

[41]Gustav Scheel to Martin Bormann, 12.11.1942, as translated and quoted in Jonathan Petropoulos, *Art as Politics in the Third Reich*, p. 105.

[42]Paulsen, "Dienstreisebericht von Peter Paulsen," 04.01.1940, as reproduced in Andrzej Mezynski, *Kommando Paulsen*, pp. 55–67.

[43]Schleif to Sievers, 06.01.1940, BA (ehem.BDC) Ahnenerbe: Schleif, Hans (23.02.1902).

[44]Paulsen, "Dienstreisebericht von Peter Paulsen," 04.01.1940, as reproduced in Andrzej Mezynski, *Kommando Paulsen*, pp. 55–67.

[45]Schleif to Sievers, 06.01.1940, BA (ehem.BDC) Ahnenerbe: Schleif, Hans (23.02.1902).

[46]Petersen to Sievers, 01.11.1940, BA (ehem. BDC) Ahnenerbe: Petersen, Ernst (28.04.1905).

[47]Peter Bogucki, "Konrad Jazdzewski (1908–1985) European Prehistorian," *The Polish Review* 31, no. 1 (1986): 73–77. See also Konrad Jazdzewski, *Poland* (London: Thames and Hudson, 1965).

[48]Petersen to Sievers, 01.11.1940, BA (ehem. BDC) Ahnenerbe: Petersen, Ernst (28.04.1905). Petersen even reported in this letter a complaint of Jazdzewski that there was "no civility in these 'men of the SS.' "

[49]Petersen to Sievers, 01.11.1940, BA (ehem. BDC) Ahnenerbe: Petersen, Ernst (28.04.1905).

[50]Jazdzewski, however, returned after Paulsen's detachment had plundered the collection, and remained there throughout the war, trying to rebuild the museum. For further details, see Petersen to Sievers, 01.11.1940, BA (ehem. BDC) Ahnenerbe: Petersen, Ernst (28.04.1905). As Jazdzewski himself noted in a book he published on

Polish archaeology after the war, "The Nazi invasion of Poland was a catastrophe without precedent for all branches of the nation's cultural life and archaeology was no exception. One quarter of Polish archaeologists perished on the battlefield or in concentration camps. Others, threatened with arrest and almost certain death, had to live in hiding (as did the doyen of Polish archaeologists, Professor Józef Kostrzewski in Poznán); the few remaining ones were forbidden to engage in scientific work." See Konrad Jazdzewski, *Poland*, pp. 18–19.

[51]Paulsen, "Dienstreisebericht von Peter Paulsen," 04.01.1940, as reproduced in Andrzej Mezynski, *Kommando Paulsen*, pp. 55–67.

[52]Eduard Tratz, "Abschrift: Verzeichnis der in Polen beschlagnahmten naturwissenschaftlichen Gegenstände," n.d., BA (ehem. BDC) Ahnenerbe: Tratz, Eduard (25.09.1888). This list was originally included in a letter from Sievers to Six dated 18.12.1939. "Document 203, Poland," in Hellmut Lehmann-Haupt, *Cultural Looting of the Ahnenerbe: Report prepared by Monuments, Fine Arts and Archives Section*, OMGUS, March 1, 1948, NARA, RG 260, M1926/R151.

[53]Eduard Tratz, "Über die Aufgaben der naturwissenschaftlichen Museen im allgemeinen und über Arbeiten im 'Haus der Natur' in Salzburg im besonderen," *Der Biologe* 8 (1939), as cited in Gottfried Fliedl, *Das Haus der Natur in Salzburg als Institut des SS-Ahnenerbes*, http://homepage.univie.ac.at/gottfried.fliedl/mouseion/hausdernatur.html.

[54]Tratz, "Abschrift: Verzeichnis der in Polen beschlagnahmten naturwissenschaftlichen Gegenstände," n.d., BA (ehem. BDC) Ahnenerbe: Tratz, Eduard (25.09.1888).

[55]Gert Kerschbaumer, "Der Deutsche Haus der Natur zu Salzburg," in *Politik der Präsentation*, ed. Herbert Posch and Gottfried Fliedl (Vienna: Turia and Kant, 1996), pp. 180–212.

[56]Tratz, "Abschrift: Verzeichnis der in Polen beschlagnahmten naturwissenschaftlichen Gegenstände," n.d., BA (ehem. BDC) Ahnenerbe: Tratz, Eduard (25.09.1888).

[57]Ibid.

[58]Werner Schroeder, "Die Bibliotheken des RSHA: Aufbau und Verbleib" (lecture, Weimar, November 9, 2003). Online transcript: http://www.initiativefortbildung.de/pdf/provenienz_schroeder.pdf.

[59]Andrej Angrick, *Besatzungspolitik und Massenmord* (Hamburg: Hamburger Edition, 2003), pp. 326–331. Worren Green, "The Fate of the Crimean Jewish Communities," *Jewish Social Science* 46, no. 2 (Spring 1984): 169–176.

[60]Paulsen, "Dienstreisebericht von Peter Paulsen," 04.01.1940, as reproduced in Andrzej Mezynski, *Kommando Paulsen*, pp. 55–67.

[61]Sievers to Apfelstädt, 03.04.1942, "Document 214, Poland," in Hellmut Lehmann-Haupt, *Cultural Looting of the Ahnenerbe: Report prepared by Monuments, Fine Arts and Archives Section*, OMGUS, March 1, 1948, NARA, RG260, M1926/R151.

[62]Paulsen, "Dienstreisebericht von Peter Paulsen," 04.01.1940, as reproduced in Andrzej Mezynski, *Kommando Paulsen*, pp. 55–67.

[63]Felix Kersten, *The Kersten Memoirs, 1940–1945* (New York: Macmillan, 1957), p. 23.

[64]Indeed, Frank eventually adopted such a regal lifestyle in Warsaw that Hitler himself sardonically referred to the former lawyer as King Stanislaus V. For further details, see Jonathan Petropoulos, *Art as Politics in the Third Reich*, p. 226.

[65]Sievers, "Aktenvermerk," 24.07.1939, BA (ehem. BDC) Ahnenerbe: Schleif, Hans (23.02.1902).

[66]Schleif to Sievers, 24.01.1940, in "Document 205, Poland," in Hellmut Lehmann-Haupt, *Cultural Looting of the Ahnenerbe: Report prepared by Monuments, Fine Arts and Archives Section*, OMGUS, March 1, 1948, NARA, RG260, M1926/R151.

[67]Paulsen, "Dienstreisebericht von Peter Paulsen," 04.01.1940, as reproduced in Andrzej Mezynski, *Kommando Paulsen*, pp. 55–67.

[68]Sievers, "Aktenvermerk," 20.05.1940, BA (ehem. BDC) Ahnenerbe: Paulsen, Peter (08.10.1902).

[69]Sievers found Paulsen a teaching job in the Germanische Leitstelle and later in the Junker school in Bad Tölz. For more details, see Mezynski, *Kommando Paulsen*, p. 78.

[70]As quoted in Stanislaw Strzetelski, *Where the Storm Broke* (New York: Roy Slavonic Publications, 1942), p. 110.

[71]Schleif to Sievers, 24.01.1940, "Document 205, Poland," in Hellmut Lehmann-Haupt, *Cultural Looting of the Ahnenerbe: Report prepared by Monuments, Fine Arts and Archives Section*, OMGUS, March 1, 1948, NARA, RG260, M1926/R151.

[72]Michael H. Kater, *Das "Ahnenerbe" der SS 1935–1945* (Stuttgart: Deutsche Verlags-Anstalt, 1974), pp. 150–153.

[73]Dettenberg to Kommans, 14.09.1940, BA (ehem. BDC) Ahnenerbe: Schleif, Hans (23.02.1902).

[74]Ibid.

[75]Dettenberg, "Report on activities of Generaltreuhänder in the occupied Eastern Territories," 28.03.1941, "Document 33, Poland," in Hellmut Lehmann-Haupt, *Cultural Looting of the Ahnenerbe: Report prepared by Monuments, Fine Arts and Archives Section*, OMGUS, March 1, 1948, NARA, RG260, M1926/R151.

[76]Ibid.

[77]Michael H. Kater, *Das "Ahnenerbe" der SS 1935–1945*, p. 153.

16. THE TREASURE OF KERCH

[1]As quoted in Steven Lehrer, *Hitler Sites* (Jefferson, North Carolina: McFarland & Company, Inc., 2002), p. 182.

[2]Ian Kershaw, *Hitler 1936–1945* (London: Allen Lane: The Penguin Press, 2000), p. 389.

[3]Ibid., p. 393 and p. 579.

[4]Ibid., p. 397.

[5]H.R. Trevor-Roper, "The Mind of Adolf Hitler," introductory essay in Adolf Hitler, *Hitler's Table Talk, 1941–1944* ed. H.R. Trevor-Roper (New York: Oxford University Press, 1998), p. xvii; Ian Kershaw, *Hitler 1936–1945*, p. 397.

[6]Adolf Hitler, *Hitler's Table Talk, 1941–1944* ed. H.R. Trevor-Roper (London: Phoenix Press, 2000), pp. 13, 25, 48, 290.

[7]Ibid., no. 245, p. 548.

[8]Ibid., no. 248, pp. 621–622.

[9]Ibid., no. 11, p. 16.

[10]Ibid., no. 20, p. 35.

[11]Chris Bishop and Chris McNab, *Campaigns of World War II Day by Day* (Hauppauge, NY: Barron's, 2003), p. 68.

[12]Alan Clark, *Barbarossa* (London: Weidenfeld & Nicolson, 1965), p. 55.

[13]"Tätigkeitsbericht der Ortskommandantur I/853 von Simferopol vom 14.11.1941," as reproduced in *Verbrechen der Wehrmacht,* ed. Hamburger Institut für Sozialforschung (Hamburg: Hamburger Edition, 2002), p. 176.

[14]"Nachkriegsaussagen-Simferopol: Jean B., ehemaliger Angehöriger der Geheimen Feldpolizei 647. 14.03.1969," in *Verbrechen der Wehrmacht*, p. 177.

[15]Ibid.

[16]Ibid.

[17]Ibid.

[18]"Simferopol," *Verbrechen der Wehrmacht*, p. 175.

[19]*Enzyklopädie des Holocaust,* 2nd ed., ed. Israel Gutman (Piper: Munich, 1998), s.v. "Krim."

[20]"Affidavit of Otto Ohlendorf, 5.11.1945," *Trials of War Criminals before the Nuernberg Military Tribunals,* Vol. 4, Case 9: The Einszatzgruppen Case (Washington, D.C: U.S. Government Printing Office, 1949–1950) pp. 205–207. The volumes from the Trials of War Criminals before the Nuremberg Military Tribunal are in the process of being placed online at http://www.mazal.org.

[21]*Enzyklopädie des Holocaust,* s.v. "Krim."

[22]Ian Kershaw, *Hitler 1936–1945*, p. 455.

[23]Himmler, "SS-Befehl," 24.02.1943, BA, NS 19/281.

[24]Herwig Wolfram, *History of the Goths* (Berkeley: University of California Press, 1990) pp. 36–39.

[25]It should be noted that the term "Germanic" here refers simply to the name of a broad subgroup of languages that includes English, Frisian, German, Yiddish, Icelandic, Norwegian, Danish, and Swedish. Just because someone speaks a Germanic language, it does not follow that they are of German ancestry.

[26]Herwig Wolfram, *History of the Goths*, p. 44.

[27]Personal communication, Dr. Alexander Gertsen.

[28]As one of the world's experts on the Goths points out, "it is often necessary, however, to remind Central Europeans of the plain fact that a history of the Goths is not part of a history of the German people and certainly not part of 'the history of the Germans in foreign countries.'" For more details, see Herwig Wolfram, *History of the Goths*, pp. 1–2.

[29]For a good example of this, see Wilhelm Wolfslast, *Die Germanische Völkerwanderung* (Stuttgart: Robert Lutz Nachfolger, 1941), pp. 24–25.

[30]Alexander Alexandrovich Vasiliev, *The Goths in the Crimea* (Cambridge, MA: The Mediaeval Academy of America, 1936), p. 51.

[31]Personal communication, Dr. Alexander Gertsen, University of Simferopol.

[32]Personal communications, Dr. Thomas S. Burns and Dr. David Braund.

[33]"Goten in der Krim und Wikinger in Nowgorod," *SS-Leitheft*, Kriegsausgabe 4a (1941), NARA, RG242, A3345, 124/ 69; "Das Germanenreich am Schwarzen Meer/Ein Gespräch unter dem Himmel der Krim," *SS-Leitheft*, Kriegsausgabe 6b (1941), NARA, RG242, A3345, 124/ 161–164; "Von der Ostsee bis zum Schwarzen Meer," *Germanische Leithefte* 3/4 (1942), NARA, RG242, A3345, 124/1198–1199.

[34]"Das Germanenreich am Schwarzen Meer/Ein Gespräch unter dem Himmel der Krim," *SS-Leitheft*, Kriegsausgabe 6b (1941), NARA, RG242, A3345, 124/ 161–164.

[35]Chris Bishop and Chris McNab, *Campaigns of World War II*, pp. 66–67.

[36]Mechtild Rössler and Sabine Schleiermacher, "Der 'Generalplan Ost' und die 'Modernität' der Großraumordnung," *Der 'Generalplan Ost,'* ed. Mechtild Rössler and Sabine Schleiermacher (Berlin: Akademie-Verlag, 1993), p. 8; Czelsaw Madajczyk, "Vom 'Generalplan Ost' zum 'Generalsiedlungsplan,'" in *Der 'Generalplan Ost,'* pp. 13–14.

[37]Czelsaw Madajczyk, "Vom 'Generalplan Ost' zum 'Generalsiedlungsplan,' in *Der 'Generalplan Ost,'*" pp. 13–14.

[38]Ibid p. 14.

[39]Isabel Heinemann, *"Rasse, Siedlung, deutsches Blut,"* (Göttingen: Wallstein, 2003), p. 368–372.

[40]Czelsaw Madajczyk, "Vom 'Generalplan Ost' zum 'Generalsiedlungsplan,'" in *Der 'Generalplan Ost,'* p. 16

[41]Felix Kersten, *The Kersten Memoirs 1940–1945* (New York: Macmillan, 1957), p. 133.

[42]Ibid., p. 134.

[43]Ibid., p. 133. Also Himmler, "Allgemeine Anordnung Nr.20/VI/42 über die Gestaltung der Landschaft in den eingegliederten Ostgebieten vom 21.12.1942," as reproduced in *Der 'Generalplan Ost'* pp. 142–144.

[44]Himmler, "Allgemeine Anordnung Nr.20/VI/42 über die Gestaltung der Landschaft in den eingegliederten Ostgebieten vom 21.12.1942," in *Der 'Generalplan Ost,'* pp. 136–147.

[45]Felix Kersten, *The Kersten Memoirs 1940–1945*, p. 137.

[46]Ibid.

[47]Uwe Hoßfeld, "Im Spannungsfeld von 'Deutscher Biologie,' Lyssenkoismus und evolutions-ideologischer Axolotl-Forschung," *Lomonossow: Sonderheft "Der Agrarbiologe Lyssenko—ein Exempel für die Ideologisierung der Wissenschaft* 3 (1999), http://www.lomonossow.de/1999_03/.

[48]Sabine Schleiermacher, "Begleitende Forschung zum 'Generalplan Ost,' " in *Der 'Generalplan Ost,'* pp. 339–340; "The Activities of Dr. Ernst Schaefer, Tibet Explorer and Scientist with SS-sponsored Scientific Institutes," Headquarters United States Forces European Theater, Military Intelligence Service Center, 12.02.1946, NARA, RG238, M1270/27.

[49]Letter from Ludolf von Alvensleben to Dr. Rudolf Brandt, 17.07.1942, as quoted in Andrej Angrick, *Besatzungspolitik und Massenmord* (Hamburg: Hamburger Edition, 2003) pp. 530–531.

[50]Sievers, "Bescheinigung," 21.07.1942, BA (ehem. BDC) Ahnenerbe: Jankuhn, Herbert (08.08.1905).

[51]Walther Gehl, "Bild 145. Gotische Kinderkrone von Kertsch," *Geschichte*. (Breslau: Ferdinand Hirt, 1940), p. 36 of plate insert; "Von der Ostsee bis zum Schwarzen Meer," *Germanische Leithefte* 3/4 (1942), NARA, RG242, A3345, 124/1198–1199; Alfred Frauenfeld, *Die Krim* (Aufbaustab für den Generalbezirk Krim, n.d.) p. 41.

[52]Jankuhn to Sievers, 27.05.1941, BA (ehem. BDC) Ahnenerbe: Jankuhn, Herbert (08.08.1905). Jankuhn was an immensely intelligent and perceptive man, and he had somehow learned of the impending mobilization against the Soviet Union almost a month before Operation Barbarossa began. As soon as he caught word of the forthcoming invasion, he pointed out the importance to Sievers of studying the "The Gothic empire in Russia." See also Jankuhn, "Aktenvermerk," 04.06.1941, BA (ehem. BDC) Ahnenerbe: Jankuhn, Herbert (08.08.1905).

[53]Heiko Steuer, "Herbert Jankuhn und seine Darstellungen zur Germanen- und Wikingerzeit," in *Eine hervorragende nationale Wissenschaft* (Berlin: Walter de Gruyter 2001), p. 417.

[54]*Nachruf. Dr.phil. Herbert Jankuhn*, Georg-August Universität Göttingen, Oktober 1990.

[55]Henning Hassman and Detlef Jantzen, " 'Die Deutsche Vorgeschichte—eine nationale Wissenschaft,' Das Kieler Museum vorgeschichtlicher Altertümer im Dritten Reich," *Offa* 51 (1994): 9–23.

[56]Höhne to Jankuhn, 26.11.1937, BA (ehem. BDC) Ahnenerbe: Jankuhn, Herbert (08.08.1905).

[57]Wüst to Jankuhn, 30.04.1938, BA (ehem. BDC) Ahnenerbe: Jankuhn, Herbert (08.08.1905).

[58]Henning Hassmann, "Archaeology in the Third Reich," *Archaeology, Ideology and Society*, ed. Heinrich Härke (Frankfurt am Main: Peter Lang, 2000), p. 85.

[59]Jankuhn to Ahnenerbe, Oslo, 20.06.1940, BA (ehem. BDC) Ahnenerbe: Jankuhn, Herbert: (08.08.1905).

[60]Jankuhn to Sievers, 16.10.1940, BA (ehem. BDC) Ahnenerbe: Jankuhn, Herbert: (08.08.1905). See also Jankuhn, "Bericht über meinen Aufenthalt in der Bretagne," 24.01.1941, BA (ehem. BDC) Ahnenerbe: Jankuhn, Herbert (08.08.1905).

[61]Jankuhn to Sievers, 04.09.1940, BA (ehem. BDC) Ahnenerbe: Jankuhn, Herbert (08.08.1905).

[62]Peter Padfield, *Himmler* (London: Macmillan, 2001), p. 382.

[63]Jankuhn to Seefeld, 25.06.1942, BA (ehem. BDC) Ahnenerbe: Seefeld, Wolf von (19.06.1912).

[64]Jankuhn, "Bericht über die Tätigkeit des Sonderkommandos Jankuhn bei der SS-Division Wiking, für die Zeit vom 20. Juli bis 1. Dezember 1942," BA (ehem. BDC) Ahnenerbe: Jankuhn, Herbert (08.08.1905).

[65]Ibid.

[66]Andrej Angrick, *Besatzungspolitik und Massenmord*, pp. 581–582.

[67]Jankuhn to Sievers, 06.09.1942, BA (ehem.BDC) Ahnenerbe: Jankuhn, Herbert (08.08.1905).

[68]Jankuhn to Ahnenerbe, 16.08.1942, BA (ehem. BDC) Ahnenerbe: Jankuhn (08.08.1905).

[69]Andrej Angrick, *Besatzungspolitik und Massenmord*, pp. 581–582.

[70]Ibid.

[71]Daniel Jonah Goldhagen, *Hitler's Willing Executioners* (New York: Alfred A. Knopf, 1996), pp. 200–201.

[72]Beumelburg, "Die Goten auf der Krim," 14.07.1942, BA, NS 19/ 2212. Beumelburg was a major in the Luftwaffe and it seems almost certain that it was von Alvensleben, who forwarded this report to Himmler. See Sievers to Jankuhn, 07.09.1942, BA (ehem. BDC) Ahnenerbe: Jankuhn (08.08.05).

[73]Beumelburg, "Die Goten auf der Krim," 14.07.1942, BA, NS 19/ 2212.

[74]The exact dates of the Maikop massacres are currently unknown because few documents regarding the massacres have survived. Most scholars agree that they occurred in late summer and fall of 1942. However, it is clear that the *Einsatzgruppe* squads were present in Maikop when Jankuhn was there, because some assisted him in packing up the museum collection. For further details, see Jankuhn to Amt Ahnenerbe, 29.08.1942, BA (ehem. BDC) Ahnenerbe: Jankuhn, Herbert (08.08.1905).

[75]Andrej Angrick, *Besatzungspolitik und Massenmord*, pp. 581–585.

[76]In reality, the helmet was made in Rome between the third and first century B.C. See Christian Hufen, "Gotenforschung und Denkmalpflege: Herbert Jankuhn und die Kommandounternehmen des 'Ahnenerbe' der SS," in *'Betr.: Sicherstellung' NS-Kunstraub in der Sowjetunion*, ed. Wolfgang Eichwede and Ulrike Hartung (Bremen: Temmen, 1998) p. 84, footnote p. 35.

[77]Jankuhn to Amt Ahnenerbe, Anlage 2 & 3 zum Bericht vom, 29.08.1942, "Liste der beim EK 11 in Maikop sichgergestellten Funde aus dem Museum in Maikop," BA (ehem. BDC) Ahnenerbe: Jankuhn, Herbert (08.08.1905).

[78]"Affidavit of Karl Rudolf Werner Braune, 8 July 1947," *Trials of War Criminals before the Nuernberg Military Tribunals*, Vol. 4, Case 9: The Einsatzgruppen Case (Washington, D.C: U.S. Government Printing Office, 1949–1950) pp. 214–226.

[79]Andrej Angrick, *Besatzungspolitik und Massenmord*, pp. 445–446.

[80]Jankuhn to Amt Ahnenerbe, 29.08.1942, BA (ehem. BDC) Ahnenerbe: Jankuhn, Herbert: (08.08.1905).

[81]Jankuhn, "Bericht über die Tätigkeit des Sonderkommandos Jankuhn bei der SS-Division Wiking, für die Zeit vom 20. Juli bis 1. Dezember 1942," BA (ehem. BDC) Ahnenerbe: Jankuhn, Herbert (08.08.1905).

[82]Andrej Angrick, *Besatzungspolitik und Massenmord*, pp. 445–446.

[83]I. I. Vdovichenko, *Antique Painted Vases from the Collections of the Crimean Museums* (Simferopol: SONAT, 2003), p. 55; Jankuhn, "Bericht über die Tätigkeit des Sonderkommandos Jankuhn bei der SS-Division Wiking, für die Zeit vom 20. Juli bis 1. Dezember 1942," BA (ehem. BDC) Ahnenerbe: Jankuhn, Herbert (08.08.1905).

[84]Jankuhn, "Bericht über die Tätigkeit des Sonderkommandos Jankuhn bei der SS-Division Wiking, für die Zeit vom 20. Juli bis 1. Dezember 1942," BA (ehem. BDC) Ahnenerbe: Jankuhn, Herbert (08.08.1905).

[85]Jankuhn, "Liste der in Armawir durch das Sonderkommando Jankuhn sichergestellten Funde aus der Krim," 02.09.1942, BA (ehem. BDC) Ahnenerbe: Jankuhn, Herbert (08.08.1905).

[86]Jankuhn, "Bericht über die Tätigkeit des Sonderkommandos Jankuhn bei der SS-Division Wiking, für die Zeit vom 20. Juli bis 1. Dezember 1942," BA (ehem. BDC) Ahnenerbe: Jankuhn, Herbert (08.08.1905). The current whereabouts of these objects is unclear; Christian Hufen, "Gotenforschung und Denkmalpflege," pp. 92–95.

17. LORDS OF THE MANOR

[1]Peter Witte et al., eds. *Der Dienstkalender Heinrich Himmlers 1941/42.* (Hamburg: Christians Verlag, 1999), pp. 527–528. For a brief description of Himmler's Hegewald compound, see Wendy Lower, "A New Ordering of Space and Race: Nazi Colonial Dreams in Zhytomyr, Ukraine, 1941–1944," *German Studies Review* 25, no. 2 (May 2002): 227–254.

[2]Peter Padfield, *Himmler* (London: Cassell & Co. 2001), p. 204.

[3]Stephan and Norbert Lebert, *My Father's Keeper* (Boston: Little, Brown and Company, 2001) p. 113.

[4]Ibid.

[5]Peter Padfield, *Himmler*, p. 366.

[6]Ibid., p. 279.

[7]"The Activities of Dr. Ernst Schaefer, Tibet Explorer and Scientist with SS-sponsored Scientific Institutes," Headquarters United States Forces European Theater, Military Intelligence Service Center, 12.02.1946, NARA, RG238, M1270/27.

[8]Wendy Lower, "A New Ordering of Space and Race: Nazi Colonial Dreams in Zhytomyr, Ukraine, 1941–1944," *German Studies Review* 25, no.2 (May 2002): 227.

[9]Adolf Hitler, *Hitler's Table Talk 1941–1944*, ed. H.R. Trevor-Roper (London: Phoenix Press, 2000), no. 245, p. 548.

[10]Michael Kater, *Das 'Ahnenerbe' der SS 1935–1945* (Stuttgart: Deutsche Verlags-Anstalt, 1974), p. 164.

[11]Alfred Frauenfeld, *Die Krim* (Aufbaustab für den Generalbezirk Krim, n.d.), p. 39. Frauenfeld was also the author of the proposal concerning the colonization of the Crimea. His views on the Crimea were of great interest to Hitler. Adolf Hitler, *Hitler's Table Talk 1941–1944*, ed. H.R.Trevor-Roper (London: Phoenix Press, 2000), no. 245, p. 548.

[12]Thomas Nußbaumer, *Alfred Quellmalz und seine Südtiroler Feldforschungen 1940–1942* (Innsbruck: Studien Verlag, 2001), p. 93.

[13]Adolf Hitler, *Hitler's Table Talk 1941–1944*, no. 245, p. 548.

[14]It is unclear precisely when Himmler first disclosed these plans to Hitler. However, the SS leader told his personal physician on July 16 that Hitler had just given him verbal approval for the *Wehrbauern* settlements. Felix Kersten, *The Kersten Memoirs, 1940–1945* (New York: Macmillan, 1957), pp. 132–134.

It is possible that Himmler began to press for a decision on the Master Plan East during a dinner he had with Hitler on July 8, soon after the fall of Sevastopol. As two prominent German historians have noted, the purpose of the dinner meeting was very likely to discuss the "Germanization" of the Crimea. Moreover, three days earlier, Wolf-Karl Wolff, Himmler's liaison officer at Hitler's headquarters, had informed him that an important meeting would take place on July 6 and advised him that the SS should attend. The purpose of the meeting was to discuss "the position of the collection camps, the resettlement, the racial recordings, the necessary security needed for resettlement, the defense of the camps, the liquidation by the *Einsatzkommandos* and all the problems associated with these questions." For further details, see Jochen von Lang, *Der Adjutant* (Munich: Herbig Verlag, 1985), p. 177. See also Peter Witte et al., eds., *Der Dienstkalender Heinrich Himmlers 1941/42*, pp. 480–481.

[15]Felix Kersten, *The Kersten Memoirs*, p. 132.

[16]Ibid., pp. 132–133.

[17]Ibid., p. 228.

[18]Ibid., pp. 239–240.

[19]Ibid., p. 239.

[20]Ibid., p. 240.

[21]Felix Kersten, *The Memoirs of Doctor Felix Kersten*, ed. Herma Briffault (Garden City, N.Y.: Doubleday, 1947), pp. 49–50.

[22]Alfred Frauenfeld, *Die Krim*, pp. 78–79.

[23]Kersten, "Betrifft: Geplante Fahrt des Reichsführers-SS zur Besichtigung der gotischen Bergfestungen und Höhlenstädte auf der Krim," 04.10.1942, BA (ehem.BDC) Ahnenerbe: Jankuhn, Herbert (08.08.1905).

[24]Kersten, "Lebenslauf," Berlin, 15.10.1944, BA (ehem. BDC) REM: Kersten. Karl. (08.08.1909).

[25]Jankuhn, "Gutachten über Dr.Karl Kersten," n.d., BA (ehem. BDC) REM: Kersten, Karl (08.08.1909); Jes Martens, "Die Nordische Archäologie und das 'Dritte Reich,'" in *Prähistorie und Nationalsozialismus*, ed. Achim Leube and Morten Hegewisch (Heidelberg: Synchron, 2002), p. 609.

[26]Kersten, "Betrifft: Geplante Fahrt des Reichsführers-SS zur Besichtigung der gotischen Bergfestungen und Höhlenstädte auf der Krim," 04.10.1942, BA (ehem.BDC) Ahnenerbe: Jankuhn, Herbert (08.08.1905).

[27]Kersten, "Betrifft: Bergfestung Tschufut-Kale bei Bachtschissaraj," 02.10.1942, BA (ehem.BDC): Jankuhn, Herbert (08.08.1905).

[28]Ibid.

[29]Kersten, "Betrifft: Bergstadt Tepe-Kermen bei Schury auf der Krim," 04.10.1942, BA (ehem.BDC) Ahnenerbe: Jankuhn, Herbert (08.08.1905).

[30]Alexander Alexandrovich Vasiliev, *The Goths in the Crimea* (Cambridge, MA: The Mediaeval Academy of America, 1936) p. 51.

[31]Kersten, "Betrifft: Gotische Bergfestung Eski-Kermen bei Tscherkess-Kermen," 04.10.1942, BA (ehem.BDC) Ahnenerbe: Jankuhn, Herbert (08.08.1905).

[32]Natalia Dobrynina, "Report Summarizing Russian Scholarly and Scientific Literature on the History of four Crimean Cave Cities," p. 7.

[33]Kersten, "Betrifft: Gotische Bergfestung Eski-Kermen bei Tscherkess-Kermen," 04.10.1942, BA (ehem.BDC) Ahnenerbe: Jankuhn, Herbert (08.08.1905).

[34]Kersten, "Betrifft: Geplante Fahrt des Reichsführers-SS zur Besichtigung der gotischen Bergfestungen und Höhlenstädte auf der Krim," 04.10.1942, BA (ehem.BDC) Ahnenerbe: Jankuhn, Herbert (08.08.1905).

[35]Peter Witte et al., eds. *Der Dienstkalender Heinrich Himmlers 1941/42*, pp. 600.

[36]Adolf Hitler, *Hitler's Table Talk, 1941–1944*, no. 285, pp. 621–622.

[37]Andrej Angrick, *Besatzungspolitik und Massenmord* (Hamburg: Hamburger Edition, 2003), pp. 540–544.

[38]Peter Witte et al., eds., *Der Dienstkalender Heinrich Himmlers 1941/42*, p. 601. See also Kersten, "Betrifft: Museum in Bachtchissaraj," 04.10.1942, BA (ehem.BDC): Jankuhn, Herbert: (08.08.1905).

[39]Peter Witte et al., eds., *Der Dienstkalender Heinrich Himmlers 1941/42*, p. 601.

[40]Andrej Angrick, *Besatzungspolitik und Massenmord*, pp. 540–544.

18. SEARCHING FOR THE STAR OF DAVID

[1]*Encyclopaedia Britannica* (Chicago: Britannica Inc, 1968), s.v. "Union of the Soviet Socialist Republics."

[2]Otmar Freiherr von Verschuer, "Rassenbiologie der Juden," *Forschungen zur Judenfrage* 3 (Hamburg: Hanseatische Verlaganstalt, 1938): 137–151.

[3]Ibid., pp. 140–141.

[4]German racial scientists went to enormous lengths in their attempt to develop reliable methods of accurately detecting an individual's race. They searched for measurable differences in everything from skull seams, calf muscles, heart rates, and earwax to body odor, fingerprints, and blood groups. They were unable to find any quick, accurate way of identifying even supposedly "pure" groups, such as the Nordic or Mediterranean races. So Jews, who were purportedly mixtures of many different races, presented an immensely difficult "diagnostic" problem. For further details on the quest for racial diagnosis in Germany, see Ernst Klee, *Deutsche Medizin im Dritten Reich* (Frankfurt: S. Fischer, 2001) pp. 158–165.

[5]Andrej Angrick, *Besatzungspolitik und Massenmord* (Hamburg: Hamburger Edition, 2003), p. 326.

[6]*The Jewish Encyclopedia* (New York and London: Funk and Wagnalls, 1902), s.v. "Krimchaks."

[7]Neal Ascherson, *Black Sea* (London: Jonathan Cape, 1995), p. 23; Andrej Angrick, *Besatzungspolitik und Massenmord*, p. 330.

[8]As cited in Ian Kershaw, *Hitler 1936–1945* (London: Allen Lane, The Penguin Press, 2000), p. 470.

[9]Andrej Angrick, *Besatzungspolitik und Massenmord*, pp. 326–331.

[10]Andrej Angrick, *Besatzungspolitik und Massenmord*, p. 330; Warren Green, "The Fate of the Crimean Jewish Communities: Ashkenazim, Krimchaks, and Karaites," *Jewish Social Science* 46, no. 2 (Spring 1984): 169–176.

[11]*The Jewish Encyclopedia*, s.v. "Caucasus."

[12]Ibid.

[13]Michael Kater, *Das "Ahnenerbe" der SS 1935–1945* (Stuttgart: Deutsche Verlags-Anstalt, 1974) pp. 231–232.

[14]Sievers to Himmler, 10.02.1942, NARA, RG242, A3345 DS G119: Greite, Walter (13.06.1907).

[15]Mollison to Stuck, 18.08.194(?), NARA, RG242, A3345 DS G119: Greite, Walter (13.06.1907).

[16]Sievers, "Tagebuch: 10.12.1941," BA, NS 21/127. The entry reads: "SS O'stuf Dr. Beger: discussion of a proposal to get Jewish skulls for anthropological examination. Cooperation with Hirt. Strassburg. Cooperation with RuSHA [Race and Settlement office]. Report on the meeting Beger-Schäfer at Munich. Permission to employ a (female) assistant."

[17] Isabel Heinemann, *"Rasse, Siedlung, deutsches Blut"* (Göttingen: Wallstein, 2003), pp. 75–76, 544–559; Beger, 'Führerstammkarte,' NARA, RG242, A3345, SSO Beger, Bruno (27.04.1911).

[18] Isabel Heinemann, *"Rasse, Siedlung, deutsches Blut,"* p. 610

[19] Sievers, "Tagebuch: 10.12.1941," BA, NS 21/127. In the statement that Beger gave during his trial after the war, he stated that "part of my assignment was to find as many varieties of Jewishness as possible." "IV. Die Einlassung der Angeschuldigten und die Beweiswürdigung 1.) Der Angeschuldigte Dr.Beger," p. 80. Frankfurter Schwurgericht. Strafverfahren gegen Bruno Beger, Hans Fleischhacker, Wolf-Dietrich Wolff. 27.10.1970-06.04.1971. IfZ, Gf 03.32.

[20] Josef Wastl, as quoted in Klaus Taschwer, "'Anthropologie ins Volk,'—Zur Ausstellungspolitik einer anwendbaren Wissenschaft bis 1945," in *Politik der Präsentation,* ed. Herbert Posch and Gottfried Fliedl (Vienna: Turia und Kant, 1996), p. 248.

[21] Bernhard Purin, "Die museale Darstellung jüdischer Geschichte und Kultur in Österreich zwischen Aufklärung und Rassismus," in *Politik der Präsentation,* pp. 33–34. Wastl did eventually obtain permission in 1942 to dig up bodies from one of Vienna's largest Jewish cemeteries, the Währinger Friedhof. Indeed, he and his colleagues exhumed some 220 skeletons from the cemetery for their studies.

[22] Götz Aly, "The Posen Diaries of the Anatomist Herman Voss," in *Cleansing the Fatherland,* ed. Götz Aly, Peter Chroust, and Christian Pross (Baltimore: The Johns Hopkins University Press, 1994) p. 141. Posen is the German name for the city of Poznan in western Poland.

[23] Ibid., p. 144.

[24] Amos Elon, "Death for Sale: Masks, an Attempt about Shoah," *New York Review of Books,* November 20, 1997.

[25] Personal communication, Dr. Guido Lombardi. See also Tony Waldron, *Counting the Dead* (Chichester, New York: John Wiley & Sons, 1994) pp. 24–25.

[26] Straßburg, or as it is often spelled in English, Strassburg, is the German name for Strasbourg.

[27] "I. Die Angeschuldigten: Lebensläufe und politische Werdegang. 1) Dr. Bruno Beger," pp. 953–957, Frankfurter Schwurgericht. Strafverfahren gegen Bruno Beger, Hans Fleischhacker, Wolf-Dietrich Wolff. 27.10.1970–06.04.1971, StA Mchn, Stanw 34.878/91. See also Beger "Führerstammkarte," NARA, RG242, A3343, SSO Beger, Bruno (27.04.1911); Hirt "Führerstammkarte," NARA, RG242, A3343, SSO Hirt, August (29.04.1898). In a letter that Hirt wrote to Beger in the fall of 1942, the anatomist addressed the younger man as "comrade Beger" and "du." This suggests that the two men knew each other quite well. Hirt to Beger, 05.09.1942, BA, R135/49.

[28] Hirt, "Lebenslauf," BA (ehem. BDC) PK, Hirt, August (29.04.1898); Hans-Joachim Lang, *Die Namen der Nummern* (Hamburg: Hoffmann und Campe, 2004), p. 210.

[29]Frederick H. Kasten, "Unethical Nazi Medicine in Annexed Alsace-Lorraine: The Strange Case of Nazi Anatomist Professor Dr. August Hirt," *Historians and Archivists: Essays in Modern German History and Archival Policy.* ed. George O. Kent (Fairfax, Virginia: George Mason University Press, 1991) p. 178.

[30]Ibid.

[31]Hirt, "Gesundheitsbogen," BA (ehem. BDC) PK Hirt. August (29.04.1898). "Gedächtnisprotokoll Unterredung Prof. Dr. Walther Wüst und Michael H. Kater." 22.04.1963. IfZ, ZS/A-25, vol.3.

[32]As Frederick H. Kasten has observed, it is very difficult to determine when Hirt became a strident anti-Semite. It is possible, however, that he became infected with this bigotry in his youth and merely concealed it while working with his Jewish colleague, Dr. Phillipp Ellinger. Ellinger was an important medical researcher at the time, and Hirt may have swallowed his prejudice in order to advance his career. See Frederick H. Kasten, "Unethical Nazi Medicine in Annexed Alsace-Lorraine," p. 178.

[33]For further details on the plans for the university, see Frederick H. Kasten, "Unethical Nazi Medicine in Annexed Alsace-Lorraine," pp. 180–182.

[34]Ernest Lachman, "Anatomist of Infamy: August Hirt," *Bull.Hist.Med* 51 (1977): 594–602.

[35]Frederick H. Kasten, "Unethical Nazi Medicine in Annexed Alsace-Lorraine," p. 190. Mustard gas goes by a variety of names, including yperite, sulphur mustard, and LOST. The latter name is the one most frequently found in the archival documents concerning Hirt's experiments.

[36]Roy C. Ellis and Donald I. Perry, "The ABC of Safe Practices for the Biological Sciences Laboratory: An Easy to Use Reference Manual for Laboratory Personnel," 2001, http://www.hoslink.com/Ellis/INDEX.html.

[37]Frederick H. Kasten, "Unethical Nazi Medicine in Annexed Alsace-Lorraine," p. 191.

[38]Sievers to Hirth (sic), 03.01.1942, BA (ehem.BDC) Ahnenerbe: Hirt, August (29.04.1898).

[39]Ian Kershaw, *Hitler 1889–1936* (London: Allen Lane, The Penguin Press, 1999), p. 97.

[40]Brandt to Sievers, 29.12.1941, BA (ehem.BDC) WI Hirt August (29.04.1898).

[41]It is impossible to determine with certainty who among these three men first came up the idea of obtaining the heads from executed commissars. There are no surviving documents to shed light on this question. All three men, moreover, were in a position to have heard something about the commissar policy. Sievers was an extremely well-informed SS officer. Hirt was a former military physician with excellent army contacts. And Beger had done a month-long stint at the eastern front in 1941. Indeed, he had traveled as a propagandist with the Viking Division as it pushed east from Ternopol toward Dnepropetrovsk in the Ukraine. Beger, "SS Kriegs-

berichterkompanie Bogen," 07.11.1941, NARA, A3356, RG242, German Army Officer Personnel File T201-1.1. Beger, Bruno (27.04.1911).

[42]Richard Breitmann, *The Architect of Genocide* (New York: Alfred A. Knopf 1991), p. 149

[43]Ibid. Immediately after the invasion of the Soviet Union, for example, the SS education office published photos of Russian officials whom it described as Jewish political commissars. "Sonnenwende-Schicksalswende," *Germanische Leitheft* 1, no. 2 (1941): 11. BA, NSD, 41/78.

[44]"Subject: Securing Skulls of Jewish-Bolshevik Commissars for the purpose of scientific research at the Strassburg Reich University," February 1942, NARA, Records of the U.S. Nuremberg War Crimes Trials: United States of America v. Karl Brandt et.al. (Case 1), Nov. 21, 1946–Aug. 20, 1947, RG238, M887/16/Jewish Skeleton Collection. Historians have debated over the authorship of this memo. For the best summation of these arguments, see Michael H. Kater, *Das Ahnenerbe*, pp. 245–248. I find it particularly interesting that Sievers's personal secretary, Dr. Gisela Schmitz, explained to investigators after the war that Beger was the author of all but the last section of the memo, which Hirt wrote. Nevertheless, Beger's handwriting does not match that in the document. See "Anklage der Generalstaatsanwalt Frankurt gegen Beger, Fleischhacker, Wolff" Frankfurter Schwurgericht. Strafverfahren gegen Bruno Beger, Hans Fleischhacker, Wolf-Dietrich Wolff. 27.10.1970-06.04.1971, pp. 984–985. StA Mchn, Stanw 34.878/91; Schmitz, "Eidesstattliche Erklärung Dr. Schmitz," 27.03.1947. Frankfurter Schwurgericht. Strafverfahren gegen Bruno Beger, Hans Fleischhacker, Wolf-Dietrich Wolff. 27.10.1970-06.04.1971. StA Mchn, Stanw 34.878/11.

[45]Richard Breitmann, *The Architect of Genocide*, p. 229. As Breitmann and many other historians have pointed out, the Final Solution began well before the Wannsee Conference. Himmler's *Einsatzgruppen* had already embarked upon a massive slaughter of Jews during the invasion of the Soviet Union. Moreover, the first death camp, complete with gas chambers, went into operation at Chelmno in Poland on December 8, 1941, and the SS staff had already designed other such facilities.

[46]Himmler to Bormann, 22.05.1943, in *Reichsführer!...*, ed. Helmut Heiber (Stuttgart: Deutsche Verlags-Anstalt, 1968), p. 213.

[47]Willi Frischauer, *Himmler: The Evil Genius of the Third Reich* (London: Odhams Press, 1953), p. 127.

[48]Ibid.

[49]Brandt to Sievers, 27.02.1942, NARA, RG242, T175/103/2625109.

[50]Peter Witte et al., eds. *Der Dienstkalender Heinrich Himmlers 1941/42* (Hamburg: Christians Verlag, 1999), pp. 390–391.

[51]Michael Kater, *Das "Ahnenerbe,"* pp. 231–233.

[52]Robert E. Conot, *Justice at Nuremberg* (New York: Harper & Row, 1983), p. 292.

[53]Ibid., p. 286.

[54]Peter Witte et al., eds., *Der Dienstkalender Heinrich Himmlers 1941/42*, pp. 390–391.

[55]Kater to Vogt, 05.01.1970, StA Mchn, Stanw. 34.878/5; Wüst, "Prof. Dr.Hirt," 06.02.1944, StA Mchn, Stanw. 34.878/75; Wüst to Sievers, 16.03.1944, StA Mchn, Stanw. 34.878/75. See also Michael H. Kater, *Das "Ahnenerbe,"* p. 256.

[56]Sievers to Hirt, end of August or beginning of September, 1942, as quoted in "Urteil. V. Die Beteiligung des Angeklagten Dr. Beger," p. 23. Frankfurter Schwurgericht. Strafverfahren gegen Bruno Beger, Hans Fleischhacker, Wolf-Dietrich Wolff. 27.10.1970-06.04.1971. IfZ, Gf 03.32.

[57]Beger to Brandt, 13.04.1943, as translated and quoted in Benno Müller-Hill, *Murderous Science* (Oxford: Oxford University Press, 1988), p. 52.

[58]Historians often refer to this camp today as Natzweiler-Struthof. However, the vast majority of surviving German documents concerning the Jewish Skeleton Collection refer to it as Konzentrationlager Natzweiler, or concentration camp Natzweiler. I have thus chosen to refer to it simply as Natzweiler throughout the text.

[59]Josef Kramer, Statement 26.07.1945, NARA, Records of the U.S. Nuremberg War Crimes Trials: United States of America v. Karl Brandt et.al. (Case 1), Nov. 21, 1946–Aug. 20, 1947, RG238, M887/16/Jewish Skeleton Collection.

[60]Wolff to Hirt, 22.04.1943, BA, NS 21/906.

[61]Wolff to Firma Franz Bergmann u. Paul Altmann K.G., 04.11.1942, BA, NS 21/905.

[62]Doreen Moser et al., Marine Mammal Skeletal Preparation and Articulation, Poster Presentation at the Biennial Conference on the Biology of Marine Mammals, November 28–December 3, 2002, Vancouver, B.C.

[63]Wolff to Firma Franz Bergmann u. Paul Altmann K.G., 04.11.1942, BA, NS 21/905.

[64]Dr. Miklós Nyiszli, *Auschwitz: A Doctor's Eyewitness Account* (Geneva: Ferni Pub. House, 1979), p. 204. During his imprisonment at Auschwitz, Nyiszli was forced to work for Dr. Josef Mengele, a physician and racial biologist. At one point, Mengele ordered Nyiszli to perform anatomical measures on two Jewish prisoners, a father and a son. The two prisoners were then taken away and murdered. Mengele then instructed Nyiszli to render their bodies down to skeletons for the anthropological museum in Berlin. In his book, Nyiszli described the chemicals commonly used at the time for chemical maceration, and I have borrowed upon his description for this passage.

[65]Doreen Moser et al., Marine Mammal Skeletal Preparation and Articulation, Poster Presentation at the Biennial Conference on the Biology of Marine Mammals, November 28–December 3, 2002, Vancouver, B.C.

[66]Wolff to Firma Franz Bergmann u. Paul Altmann K.G., 04.11.1942, BA, NS 21/905.

[67]Wolff to Hirt, 08.10.1942, BA, NS 21/905.

[68]Beger to Sievers, 03.10.1942, BA, NS 21/905.

[69]Rudolf Vrba, *I Cannot Forgive* (Vancouver: Regent College Publishing, 1997), p. 123.

[70]Danuta Czech, *Auschwitz Chronicle 1939–1945* (New York: Henry Holt and Co., 1990), pp. 202–247.

[71]Brandt to Reichssicherheitshauptamt, 10.08.1942, BA, NS19/3638.

[72]Members of the proposed team included Dr. Davoud Monchi-Zadeh, the Iranian student who was to accompany Wüst on his proposed Bisitun expedition; Dr. Viktor Christian, Dean of the Philosophy Faculty at the University of Vienna and an authority on Near Eastern studies; Dr. Alfred Rust, an archaeologist; Dr. Karl Vogt, a racial scientist interested in the Transcaucasus region; and Wolfram Sievers. See Sievers, Vermerk. Betr: Einsatz des 'Ahnenerbes' im vorderen Orient, Iran und indoiranischen Raum, 12.02.1942, LOC Manuscript Division. Captured German Documents. Section 19, ODN 511, Reel 46.

[73]Schäfer to Sievers, 17.08.1942, BA, NS 21/42; "Personalaufstellung für das geplante Unternehmen," 18.08.1942, NARA, RG242, T81/131, 164290-164292; Sievers to Reichssicherheitshauptamt, 20.08.1942, NARA, RG242, T175/34.

[74]"Personalaufstellung für das geplante Unternehmen," 18.08.1942, NARA, RG242, T81/131, 164290-164292.

[75]"The Activities of Dr. Ernst Schaefer, Tibet Explorer and Scientist with SS-sponsored Scientific Institutes," Headquarters United States Forces European Theater, Military Intelligence Service Center, 12.02.1946, NARA, RG238, M1270/27.

[76]"Vernehmung Ernst Schäfer, 08.12.1970," p. 4, Frankfurter Schwurgericht. Strafverfahren gegen Bruno Beger, Hans Fleischhacker, Wolf-Dietrich Wolff. 27.10.1970-06.04.1971, StA Mchn, Stanw 34.878/18. Exactly when Schäfer proposed the Caucasus mission is unclear, but in August 1942, he wrote to Sievers, noting that "as I was already able to inform you months ago, the total research of the Caucasus area has great importance and weight for science and for worldview." Schäfer to Sievers, 17.08.1942, BA, NS 21/42.

[77]Sievers to Reichssicherheitshauptamt, 20.08.1942, NARA, RG242, T175/34. See also Sonderkommando K documents found in file NARA, RG242, T81/131/164293-164311, which includes documents titled, "Fahrzeugaufstellung," "Persönliche Ausrüstung," and "Allgemeine Ausrüstung."

[78]Brandt to Sievers, 24.08.1942, NARA, RG242, T175/34.

[79]SS-Führungshauptamt to Reichsführer-SS, 29.01.1943, NARA, RG242, T175/R124; "Betr.Eskorte" n.d. NARA, RG242, T81/131/164293-164294.

[80] "The Activities of Dr. Ernst Schaefer, Tibet Explorer and Scientist with SS-sponsored Scientific Institutes," Headquarters United States Forces European Theater, Military Intelligence Service Center, 12.02.1946, NARA, RG238, M1270/27.

[81]Ibid; "Kurzgefasste Darstellung der im Rahmen des Sonderkommando 'K' geplanten Rassenkundlichen Untersuchungen," n.d., BA, R135/44, 164287–164288.

"Gedächtnisprotokoll Unterredung Dr. Ernst Schäfer und Michael H. Kater," 28.04.1964, IfZ, ZS/A-25, vol. 2. Moreover, as Beger himself noted in a letter to the Security Service, about Special Command K, "For the realization of the above mentioned undertaking, which has political goals but is of a purely military character, it is necessary to conduct scientific work partially as a cover-up, but also partially to clarify specific ethnic questions." Beger to Kern. 06.10.1942. BA, R135/50.

[82]"Kurzgefasste Darstellung der im Rahmen des Sonderkommando 'K' geplanten Rassenkundlichen Untersuchungen,"n.d., BA, R135/44, 164287-164288.

[83]"Vernehmung Ernst Schäfer, 08.12.1970," p. 5, Frankfurter Schwurgericht. Strafverfahren gegen Bruno Beger, Hans Fleischhacker, Wolf-Dietrich Wolff. 27.10.1970-06.04.1971, StA Mchn, Stanw 34.878/18. See also "Gedächtnisprotokoll Unterredung Ernst Schäfer und Michael H. Kater," 28.4.1964, IfZ, ZS/ A-25. vol.2.

[84]"Vernehmung Ernst Schäfer. 08.12.1970," p. 5, Frankfurter Schwurgericht. Strafverfahren gegen Bruno Beger, Hans Fleischhacker, Wolf-Dietrich Wolff. 27.10.1970-06.04.1971, StA Mchn, Stanw 34.878/18.

[85]*The Universal Jewish Encyclopedia* (New York: The Universal Jewish Encyclopedia Inc, 1939–43), s.v. "Mountain Jews."

[86]Ibid.

[87]Ibid.

[88] "Personalaufstellung für das geplante Unternehmen," 18.08.1942, NARA, RG242, T81/131, 164290-164292; "Kurzgefasste Darstellung der im Rahmen des Sonderkommando 'K' geplanten rassenkundlichen Untersuchungen," n.d., BA, R135/44, 164287-164288.

[89]Beger, "Auffassung der Hilfsmannschaft," NARA, RG242, T81/131/164289.

[90]Isabel Heinemann, *"Rasse, Siedlung, deutsches Blut,"* pp. 615–616.

[91]Ibid., p. 632.

[92]Ibid., pp. 615–616, 632.

[93]Ibid., pp. 234–238.

[94]Ibid., pp. 615–616, 632. A program for one of these courses, taught in Prague in 1942, has survived, and demonstrates the high opinion that RuSHA held of both Rübel and Fleischhacker as racial specialists. The two men teach many of the sessions, from Racial Diagnosis using Photographs, Part I, II, III, and IV, to the Soul of European Races, Part I and II. "Programmfolge für den Eignungsprüferlehrgang in Prag v. 20.7.-8.8.42," BA, NS 2/89.

[95]Trojan to Reichsdozentenführung, 26.01.1943, BA (ehem. BDC) PK: Trojan, Rudolf (26.02.1917). See also Ernst Klee, *Deutsche Medizin im Dritten Reich.* p. 165.

[96]Beger, "Auffassung der Hilfsmannschaft," NARA, RG242, T81/131/164289.

[97]Ibid.

[98]"Kurzgefasste Darstellung der im Rahmen des Sonderkommando 'K' geplanten rassenkundlichen Untersuchungen,"n.d., BA, R135/44, 164287-164288.

[99]Fleischhacker to Beger, 31.10.1942, BA, R135/44.

[100]Himmler, moreover, favored this deception. In 1940, he asked the head of the Race and Settlement Department, Otto Hofmann, to develop a deceptive question-naire to send to school doctors in Czechoslovakia. In addition to a number of stan-dard medical questions, Himmler instructed Hofmann to include questions that would elicit racial biological characteristics of Czech schoolchildren, such as height, weight, and eye, skin, and hair color. This practice seems to have been endemic in RuSHA. On the advice of the department, for example, Reinhard Heydrich ordered such camouflaged investigations to be carried out on adults in Czechoslovakia in 1942. For further details, see Himmler to Otto Hofmann, 09.10.1940, and Hofmann to Himmler, 24.10.1940, in *Die Deutschen in der Tschechoslowakei 1933–1947* ed. Václav Král (Prague: Nakladatelstvi Ceskioslovenske Akademie Ved, 1964), p. 424. See also Isabel Heinemann, *"Rasse, Siedlung, deutsches Blut,"* pp. 159–160; Richard Breitmann, *The Architect of Genocide*, p. 94.

[101]As Beger noted in his plan for a racial project in Norway, "It might be useful to disguise the racial investigations as medical research on physical fitness." Beger to RFSS via RuSHA, 30.06.1941, NARA, R135/ 52/162779-162784.

[102]Otmar Freiherr von Verschuer, "Rassenbiologie der Juden," p. 141.

[103]"Kurzgefasste Darstellung der im Rahmen des Sonderkommando 'K' ge-planten rassenkundlichen Untersuchungen," n.d., BA, R135/44, 164287-164288.

[104]Endres to Beger, 27.09.1942, BA, R135/48/164005-164006.

[105]"Kurzgefasste Darstellung der im Rahmen des Sonderkommando 'K' ge-planten rassenkundlichen Untersuchungen," n.d., BA, R135/44, 164287-164288.

[106]"Aufstellung der Ausrüstung für Sonderkommando K," n.d., BA, R135/44.

[107]Scholtz to Chef des RuSHA, 18.07.1941, BA, NS 2/79. Bl 118-120, as quoted in Isabel Heinemann, *"Rasse, Siedlung, deutsches Blut,"* p. 532.

[108]Isabel Heinemann, *"Rasse, Siedlung, deutsches Blut,"* p. 531–32.

[109]Ian Kershaw, *Hitler 1936–1945* (London: Allen Lane, The Penguin Press, 2000), p. 534.

[110]Ibid., p. 547.

[111]Ibid., p. 550.

[112]RFSS to Schäfer, 05.02.1943, BA NS 19/2681.

19. THE SKELETON COLLECTION

[1]Hans-Joachim Lang, *Die Namen der Nummern* (Hamburg: Hoffmann und Campe, 2004), p. 278. After painstaking research, German journalist Hans-Joachim Lang has recently identified the names and compiled brief biographies of all eighty-six vic-tims of the Jewish Skeleton Collection conspiracy.

[2]The story of the Hospital of the Jewish Community is a particularly remarkable one. During the war, the hospital was under the authority of the Reich Main Secu-rity Office, which took over several of its buildings for a soldiers' hospital and a

prison camp for Jews. But the Nazi authorities permitted the Jewish hospital to continue providing medical services to Jews, and, astonishingly, the hospital survived until the end of the war, thanks largely to the tenacity of its director, Dr. Walter Lustig. Daniel B. Silver, *Refuge in Hell: How Berlin's Jewish Hospital Outlasted the Nazis* (Boston: Houghton Mifflin, 2004).

[3]"Aussage Reineck und Toch." Anklage der Generalstaatsanswaltschaft Frankfurt gegen Beger, Fleischhacker, Wolff, pp. 1006–1008, Frankfurter Schwurgericht. Strafverfahren gegen Bruno Beger, Hans Fleischhacker, Wolf-Dietrich Wolff. 27.10.1970-06.04.1971, StA Mchn, Stanw 34.878/91. Other former Auschwitz prisoners remembered details of this event a little differently. The Polish doctor Wladyslaw Fejkiel, for example, recalled seeing groups of prisoners assembled near Block 24. I have based my description on the testimony of witnesses Reineck and Toch, which agrees on several key points.

[4]Robert Jay Lifton and Amy Hackett, "Nazi Doctors," in *Anatomy of the Auschwitz Death Camp*, ed. Yisrael Gutman and Michael Berenbaum (Published in Association with the United States Holocaust Memorial Museum. Bloomington: Indiana University Press, 1994), p. 305.

[5]Hans-Joachim Lang, *Die Namen der Nummern*, p. 115.

[6]"Aussage Reineck und Toch." Anklage der Generalstaatsanwaltschaft Frankfurt gegen Beger, Fleischhacker, Wolff, pp. 1006–1008, Frankfurter Schwurgericht. Strafverfahren gegen Bruno Beger, Hans Fleischhacker, Wolf-Dietrich Wolff. 27.10.1970-06.04.1971, StA Mchn, Stanw 34.878/91.

[7]One Holocaust survivor, Gershon Evan, underwent similar racial measurements after being arrested by the Gestapo in 1939 in Vienna. Like the prisoners of Auschwitz, Evan was measured by an anthropologist. He later wrote a poignant account of his experience. Unlike the Auschwitz prisoners, Evan was alerted ahead of time by other prisoners as to what was going to happen. "Had I not known what to expect," he explained in his account, "the instruments would have given me the creeps." For further details, see Gershon Evan, *Winds of Life* (Riverside, California: Ariadne Press, 2000), pp. 153–154.

[8]"Aussage Toch." Anklage der Generalstaatsanwaltschaft Frankfurt gegen Beger, Fleischhacker, Wolff, pp. 1007–1008, Frankfurter Schwurgericht. Strafverfahren gegen Bruno Beger, Hans Fleischhacker, Wolf-Dietrich-Wolff. 27.10.1970-06.04.1971, StA Mchn, Stanw 34.878/91.

[9]Although Beger claimed after the war to have arrived in Auschwitz on June 11, 1943, he seems to have arrived several days earlier. A report written by Wolff states that Beger left for Auschwitz on June 6. The train trip from Berlin to Auschwitz took only one day. Moreover, Beger had already selected prisoners and begun some measurements by June 11. Wolff, "Vermerk," 11.06.1943, BA NS21/ 907; Schäfer to Beger, 24.06.1943, BA, R135/45/151544.

[10]"Aussage Gabel," Anklage der Generalstaatsanwaltschaft Frankfurt gegen

Beger, Fleischhacker, Wolff, pp. 1003–1004. Frankfurter Schwurgericht. Strafverfahren gegen Bruno Beger, Hans Fleischhacker, Wolf-Dietrich Wolff. 27.10.1970-06.04.1971, StA Mchn, Stanw 34.878/91.

[11]Fleischhacker, "R.u.S.-Fragebogen: Lichtbilder," BA (ehem. BDC) RS: Fleischhacker, Hans (10.03.1912).

[12]"Diary of Johann Paul Kremer. April 30th, 1942," in KL Auschwitz Seen by the SS, ed. Jadwig Bezwinska (New York: Howard Fertig, 1984), p. 213.

[13]Ibid., pp. 213–220.

[14]Beger to Fleischhacker, 16.06.1943, NS 21/907.

[15] "Aussage Gabel." Anklage der Generalstaatsanwaltschaft Frankfurt gegen Beger, Fleischhacker, Wolff," pp. 1003–1004. Frankfurter Schwurgericht. Strafverfahren gegen Bruno Beger, Hans Fleischhacker, Wolf-Dietrich Wolff. 27.10.1970-06.04.1971, StA Mchn, Stanw 34.878/91.

[16] "Diary of Johann Paul Kremer. Sept. 5th, 1942," in KL Auschwitz Seen by the SS p. 216.

[17]Ibid.

[18]Personal communication, Dr. Bruno Beger.

[19]Yisrael Gutman, "Auschwitz—An Overview," in Anatomy of the Auschwitz Death Camp, p. 10 See also "Reminiscences of Pery Broad," in KL Auschwitz Seen by the SS, p. 139.

[20]Rudolf Vrba, I Cannot Forgive (Vancouver: Regent College Publishing, 1997), p. 77.

[21]Yisrael Gutman, "Auschwitz—An Overview," in Anatomy of the Auschwitz Death Camp, pp. 20–21.

[22]Miklós Nyiszli, Auschwitz: A Doctor's Eyewitness Account (Geneva: Ferni Pub. House 1979), p. 36.

[23]Henrypierre, "Affidavit," 17.11.1946, Office of U.S. Chief of Counsel. Translation of Document No. NO-880. StA Mchn. Stanw 34878/14.

[24]"Urteil. IV. Die Einlassung der Angeschuldigten und die Beweiswürdigung," p. 80, Frankfurter Schwurgericht. Strafverfahren gegen Bruno Beger, Hans Fleischhacker, Wolf-Dietrich Wolff. 27.10.1970-06.04.1971. IfZ, Gf 03.32.

[25]Beger to Schäfer, 24.06.1943, NARA T 81/ 128/151545.

[26]Ibid.

[27]"Discovery of human testicles in the Strasbourg Institute," 25.05.1945, Document N0-521. NARA, Records of the U.S. Nuremberg War Crimes Trials: United States of America v. Karl Brant et al. (Case 1), Nov. 21, 1946–Aug. 20, 1947, RG238 Nuernberg, Organization Series 516–517. Box 11, Entry 174. NM-70.

[28]"Aussage Toch." Anklage der Generalstaatsanwaltschaft Frankfurt gegen Beger, Fleischhacker, Wolff," pp. 1007–1008, Frankfurter Schwurgericht. Strafverfahren gegen Bruno Beger, Hans Fleischhacker, Wolf-Dietrich Wolff. 27.10.1970-06.04.1971, StA Mchn, Stanw 34.878/91.

[29]"Aussage Wörl." Anklage der Generalstaatsanwaltschaft Frankfurt gegen Beger, Fleischhacker, Wolff," pp. 1005, Frankfurter Schwurgericht. Strafverfahren gegen Bruno Beger, Hans Fleischhacker, Wolf-Dietrich Wolff. 27.10.1970-06.04.1971, StA Mchn, Stanw 34.878/91. For details about the conditions inside Block 10, see Robert Jay Lifton and Amy Hackett, "Nazi Doctors," in *Anatomy of the Auschwitz Death Camp*, pp. 301–316.

[30]Robert Jay Lifton and Amy Hackett, "Nazi Doctors," in *Anatomy of the Auschwitz Death Camp*, p. 304.

[31]It is possible, as Hans-Joachim Lang suggests, that the selected men and women were drawn from prisoners confined in the barracks used for medical experiments. See Hans-Joachim Lang, *Die Namen der Nummern*, p. 114.

[32]Robert Jay Lifton and Amy Hackett, "Nazi Doctors," in *Anatomy of the Auschwitz Death Camp*, p. 305.

[33]"Aussage Gabel." Anklage der Generalstaatsanwaltschaft Frankfurt gegen Beger, Fleischhacker, Wolff, pp. 1003–1004, Frankfurter Schwurgericht. Strafverfahren gegen Bruno Beger, Hans Fleischhacker, Wolf-Dietrich Wolff. 27.10.1970-06.04.1971 Staatsarchiv München. Stanw 34.878/91. pp. 1003–1004; "Urteil. V. Die Einlassung der Angeschuldigten und die Beweiswürdigung," p. 79, Frankfurter Schwurgericht. Strafverfahren gegen Bruno Beger, Hans Fleischhacker, Wolf-Dietrich Wolff. 27.10.1970-06.04.1971 IfZ, Gf 03.32. p. 79.

[34]"Aussage Gabel." Anklage der Generalstaatsanwaltschaft Frankfurt gegen Beger, Fleischhacker, Wolff," pp. 1003–1004, Frankfurter Schwurgericht. Strafverfahren gegen Bruno Beger, Hans Fleischhacker, Wolf-Dietrich Wolff. 27.10.1970-06.04.1971, StA Mchn, Stanw 34.878/91.

[35]Otmar Freiherr von Verschuer, "Rassenbiologie der Juden," *Forschungen zur Judenfrage* 3 (Hamburg: Hanseatische Verlaganstalt, 1938): 137–151.

[36]See notes 16 and 17 in Chapter 13.

[37]"Aussage Gabel." Anklage der Generalstaatsanwaltschaft Frankfurt gegen Beger, Fleischhacker, Wolff," pp. 1003–1004, Frankfurter Schwurgericht. Strafverfahren gegen Bruno Beger, Hans Fleischhacker, Wolf-Dietrich Wolff. 27.10.1970-06.04.1971, StA Mchn, Stanw 34.878/91.

[38] "Urteil. V. Die Einlassung der Angeschuldigten und die Beweiswürdigung," p. 81, Frankfurter Schwurgericht. Strafverfahren gegen Bruno Beger, Hans Fleischhacker, Wolf-Dietrich Wolff. 27.10.1970-06.04.1971. IfZ, Gf 03.32.

[39]Ibid.

[40]Beger to Fleischhacker, 16.06.1943, NS 21/907; Sievers to Eichmann, 21.06.1943, NARA T-175/103 2625099.

[41]Sievers to Eichmann, 21.06.1943, NARA, RG242, T175/ 103/2625099. In all likelihood, Beger lacked all the inoculations necessary for immunity to typhus. Physicians at Auschwitz gave the SS staff members three innoculations, spaced out over three weeks, to ensure that they possessed sufficient immunity. Beger was not

at Auschwitz long enough, however, to receive all three. See "Diary of Johann Paul Kremer, Sept. 5th, 1942," in *KL Auschwitz Seen by the SS*, pp. 214–219.

[42]Beger to Sievers, 09.07.1943, BA, NS 21/907.

[43]Danuta Czech, *Auschwitz Chronicle* (New York: Henry Holt and Co., 1990) p. 191.

[44]Hans Joachim Lang, *Die Namen der Nummern*, p. 161.

[45]Sievers to Eichmann, 21.06.1943, NARA, RG242, T175/103/2625099.

[46]Wolff to Hirt, 07.07.1943, BA, NS 21/907.

[47]Wolff to Beger, "Telegramm," 30.07.1943, NARA, RG242, T580/R153/241.

[48]Prisoners were housed in Hotel Struthof before construction was finished on the camp facilities at Natzweiler. Thus the camp was sometimes called Struthof.

[49]Frederick H. Kasten, "Unethical Nazi Medicine in Annexed Alsace-Lorraine," *Historians and Archivists: Essays in Modern German History and Archival Policy*, ed. George O. Kent (Fairfax, Virginia: George Mason University Press, 1991), p. 175.

[50]Ibid., p. 192.

[51]Sievers to Brandt, 27.04.1943, BA (ehem. BDC) WI, Hirt, August (29.04.1898).

[52]Ibid.

[53]Josef Kramer, "Statement," 26.07.1945, NARA, Records of the U.S. Nuremberg War Crimes Trials: United States of America v. Karl Brant et.al. (Case 1), Nov. 21, 1946–Aug. 20, 1947, RG238, M887/16/Jewish Skeleton Collection.

[54]Wolff to Hirt, Approx. date 20.04.1943, BA, NS 21/906.

[55]Ibid. In this letter, Wolff explains that the machines should be ready in six weeks' time.

[56]"Urteil. IV. Die Einlassung der Angeschuldigten und die Beweiswürdigung," p. 19, Frankfurter Schwurgericht. Strafverfahren gegen Bruno Beger, Hans Fleischhacker, Wolf-Dietrich Wolff. 27.10.1970-06.04.1971. IfZ, Gf 03.32.

[57]Beger, "Reisekostenabrechnung," 06.10.1943, BA, NS 21/506.

[58]Hans-Joachim Lang, *Die Namen der Nummern*, p. 172.

[59]Isabel Heinemann, *"Rasse, Siedlung, deutsches Blut,"* (Göttingen: Wallstein, 2003) pp. 544–547; *Trials of War Criminals Before the Nuernberg Military Tribunals* (Washington: U.S. Government Printing Office, 1950), Vol. 1, pp. 695–696.

[60]Robert Jay Lifton and Amy Hackett, "Nazi Doctors," in *Anatomy of the Auschwitz Death Camp*, pp. 306–308.

[61]"Discovery of human testicles in the Strasbourg Institute," 25.05.1945, Document N0-521 NARA, Records of the U.S. Nuremberg War Crimes Trials: United States of America v. Karl Brandt et al. (Case 1), Nov. 21, 1946–Aug. 20, 1947, RG238 Nuernberg, Organization Series 516-517. Box 11, Entry 174. NM-70. In 1958, a team of British biologists demonstrated that mouse spermatozoa treated with Trypaflavine were less motile. Indeed the treatment "lowered the fertilization rate." Hirt had experimented with this dye for years, and may well have suspected that it would damage human fertility. For further details on the 1958 British work, see R. G.

Edwards, "The Experimental Induction of Gynogenesis in the Mouse. III. Treatment of Sperm with Trypaflavine, Toluidine Blue, or Nitrogen Mustard," *Proceedings of the Royal Society of London. Series B, Biological Sciences* 149, no. 934 (July 1, 1958): 117–129.

[62]"Discovery of human testicles in the Strasbourg Institute," 25.05.1945, Document N0-521. NARA, Records of the U.S. Nuremberg War Crimes Trials: United States of America v. Karl Brandt et al. (Case 1), Nov. 21, 1946–Aug. 20, 1947, RG238 Nuernberg, Organization Series 516-517. Box 11, Entry 174. NM-70.

[63]Ibid.

[64]Ibid.

[65]Hans-Joachim Lang, *Die Namen der Nummern*, p. 173.

[66]Olga Lengyel, *Five Chimneys: The Story of Auschwitz* (London: Mayflower, Granada Publ., 1972) p. 143.

[67]There is some doubt whether the women were executed in one group or in two. I believe that the balance of evidence suggests that they were murdered in two groups, the first on April 11 and the second on April 13.

[68]Josef Kramer, Statement, 26.07.1945, NARA, Records of the U.S. Nuremberg War Crimes Trials: United States of America v. Karl Brandt et al. (Case 1), Nov. 21, 1946–Aug. 20, 1947, RG238, M887/16/Jewish Skeleton Collection.

[69]Ibid.

[70]Ibid.

[71]Ibid.

[72]Ibid. A British military court sentenced Kramer to death for his war crimes in November 1945. He was executed on December 13, 1945.

[73]Henrypierre, "Affidavit," 17.11.1946, Office of U.S. Chief of Counsel. Translation of Document No. NO-880. StA Mchn. Stanw 34878/14. Hirt, however, was mistaken about the number. He ultimately received eighty-six corpses—twenty-nine females and fifty-seven males.

[74]Hans-Joachim Lang, *Die Namen der Nummern*, p. 176.

[75]Henrypierre, "Affidavit," 17.11.46, Office of U.S. Chief of Counsel. Translation of Document No. NO-880. StA Mchn. Stanw 34878/14.

[76]Ibid.

[77]Ibid.

[78]Ibid.

[79]Waltraud Schwab, "Identifizierung nach 60 Jahren," *Die Tageszeitung*, September 23, 2003, p. 7.

[80]Henrypierre, "Affidavit." 17.11.1946, Office of U.S. Chief of Counsel. Translation of Document No. NO-880. StA Mchn. Stanw 34878/14.

20. REFUGE

[1]Ian Kershaw, *Hitler 1936–1945* (London: Allen Lane, The Penguin Press, 2000), p. 596.

[2]Ian Kershaw, *Hitler 1936–1945*, p. 597; W.G. Sebald, "A Natural History of Destruction," *The New Yorker*, November 4, 2002, p. 66–77.

[3]Anordnung Himmler, 29.07.1943, BAK, NS 21/265.

[4]Thomas Greif, *Der SS-Standort Waischenfeld 1934–1945* (Erlangen: Palm & Enke, 2000) pp. 45–50.

[5]As historian Michael Kater has pointed out, Schäfer often insisted after the war that this institute was not part of the Ahnenerbe, but the documentary evidence demonstrates that this allegation is untrue. Michael H. Kater, *Das "Ahnenerbe" der SS 1935–1945* (Stuttgart: Deutsche Verlags-Anstalt, 1974), p. 213.

[6]Thomas Greif, *Der SS-Standort Waischenfeld 1934–1945*, p. 19, pp. 54–58.

[7]Ibid., pp. 18–19.

[8]Wolff to Ludwig Müller. 02.11.1943. BA, NS 21/67.

[9]"Report of 10 October 1942, on cooling experiments on human beings," *Trials of War Criminals before the Nuernberg Military Tribunals*, Vol. 1 Case 1: The Medical Case (Washington, D.C: U.S. Government Printing Office, 1949–1950), pp. 226–243.

[10]Rascher was greatly disturbed to find that one of the female prostitutes showed "unobjectionably Nordic racial characteristics: blond hair, blue eyes, corresponding head and body structure, 21¾ years of age." As he noted in a memorandum, "I questioned the girl, why she had volunteered for the brothel. I received the answer: 'To get out of the concentration camp, for we were promised that all those who would volunteer for the brothel for half a year would then be released from the concentration camp.'" Impressed by her "Nordic" appearance, Rascher refused to use her in the experiments. "Memorandum of Rascher on women used for rewarming in freezing experiments, 5 November 1942," *Trials of War Criminals before the Nuernberg Military Tribunals*, Vol. 1, Case 1: The Medical Case (Washington, D.C: U.S. Government Printing Office, 1949–1950), p. 245.

[11]"Letter from Rascher to Himmler, 17 February 1943, and Summary of Experiments for Rewarming of Chilled Human Beings by Animal Warmth, 12 February 1943," *Trials of War Criminals before the Nuernberg Military Tribunals*, Vol. 1, Case 1: The Medical Case (Washington, D.C: U.S. Government Printing Office, 1949–1950), pp. 249–251.

[12]Ibid.

[13]"Freezing Experiments," *Trials of War Criminals before the Nuernberg Military Tribunals*, Vol. 2, Case 2: The Milch Case (Washington, D.C: U.S. Government Printing Office, 1949–1950), pp. 847–848.

[14]"Extract from the Closing Brief against Defendant Sievers: Freezing Experiments," *Trials of War Criminals before the Nuernberg Military Tribunals*. Vol. 1, Case 1:

The Medical Case (Washington, D.C: U.S. Government Printing Office, 1949–1950), p. 200.

[15]Ernst Klee, *Auschwitz, die NS-Medizin und ihre Opfer* (Frankfurt am Main: S. Fischer, 1997), p. 351.

[16]Fritz Rascher,"Affidavit 31 December 1946," *Trials of War Criminals before the Nuernberg Military Tribunals*, Vol. 1, Case 1: The Medical Case (Washington, D.C: U.S. Government Printing Office, 1949–1950), pp. 670.

[17]Ibid.

[18]Ibid.

[19]"Letter from Haagen to Hirt, 15 November 1943," *Trials of War Criminals before the Nuernberg Military Tribunals*, Vol. 1, Case 1: The Medical Case (Washington, D.C: U.S. Government Printing Office, 1949–1950), pp. 578–579.

[20]Ernst Klee, *Auschwitz, die NS-Medizin und ihre Opfer*, pp. 378–380. Later in 1944, Sievers also obtained the services of Dr. Wilhelm Beiglböck, who conducted experiments at Dachau on making seawater drinkable. As part of these experiments, Beiglböck forced his subjects to drink chemically treated seawater, resulting in serious medical injuries. "Sea Water experiments," *Trials of War Criminals before the Nuernberg Military Tribunals*, Vol. 1, Case 1: The Medical Case (Washington, D.C: U.S. Government Printing Office, 1949–1950), pp. 418–493.

[21]Ernst Klee, *Auschwitz, die NS-Medizin und ihre Opfer*, pp. 389–391. As Klee points out, some of the best evidence concerning these tests comes from one of the prisoners, Rudolf Guttenberger, a young Gypsy. My description of the experiment is drawn from Guttenberger's recollection of the events.

[22]Ibid.

[23]Ibid., p. 381.

[24]Schäfer, "Zur Filmvorführung in Mittersill," n.d., BA, R135/31. In this speech, Schäfer outlines some of the rumors about Mittersill castle circulating among the local inhabitants.

[25]Historical Exhibition, Writers' Room, 24.04.2002, Schloss Mitersill.

[26]Statement of Dr. Ernst Schäfer, 27.11.1945, BA (ehem.BDC) 0.916.

[27]Michael H. Kater, *Das "Ahnenerbe" der SS*. p. 213. In January 1943, Schäfer's department was renamed the Sven Hedin Reich Institute for Inner Asian Studies in order to win support from the University of Munich and the Reichs Ministry of Education. It remained, however, a department of the Ahnenerbe.

[28]"Urlaubschein Nr. Verlängerung. Der Umsiedler Evert, Maria," 10.05.1944, NARA, RG242, T81/132/166106; "Aufstellung der am 24. März 1944 in Schloß Mittersill eingetroffenen 15 Bibelforscherinnen," NARA, RG242, T81/132/166107; Forschungsstätte für Innerasien to Reichsgesundheitsführer, 23.06.1944, NARA, RG 242, T81/132/166157; "Forderungsnachweis: 1.10-31.10.1944," 01.11.1944, NARA, RG242, T81/132/166138; "Forderungsnachweis: 01.08.-31.08.44," 01.08.1944," BA, R135/12, 166145. These documents record the use of concentration-camp prisoners

at Mittersill, which became a subcamp of Mauthausen. It should be noted that many of the prisoners that Himmler sent to Mittersill were Jehovah's Witnesses: no Jewish prisoners worked at the castle. Perhaps this explains why, as Mauthausen historian Andreas Baumgartner has pointed out, "the conditions at Mittersill were probably slightly better than elsewhere." For further details on this see Andreas Baumgartner, *Die vergessenen Frauen von Mauthausen* (Vienna: Verlag Österreich, 1997), pp. 133–139.

[29]Beger, "Vermerk: Betr. Ethnologische Tibetgegenstände," 27.09.1943, NARA T81/130/162887.

[30]Frederick H. Kasten, "Unethical Nazi Medicine in Annexed Alsace-Lorraine: The Strange Case of Nazi Anatomist Professor Dr. August Hirt," in *Historians and Archivists*, ed. George O. Kent (Fairfax, Virginia: George Mason University Press, 1997), p. 188.

[31]Indeed, Beger seems to have worked on the anthropological research data from the murdered prisoners throughout the fall of 1943 and into the spring of 1944. "Urteil. V. Die Beteiligung des Angeklagten Dr. Beger," pp. 34–35, Frankfurter Schwurgericht. Strafverfahren gegen Bruno Beger, Hans Fleischhacker, Wolf-Dietrich Wolff. 27.10.1970-06.04.. IfZ, Gf 03.32.. See also Wolff to Beger, 03.11.1943, BA, 135/49/163551; Beger to Wolff, 06.11.1943, BA, R135/52/162956.

[32]Trojan to Beger, 23.06.1944, NARA, RG242, T81/131/164370. In this letter, Rudolf Trojan, one of the racial experts at Mittersill, asks Beger, "What is supposed to happen with the Jewish skulls? They are just lying around here and taking up space. What was originally planned for these? I think it would be best if you send them to Strassburg and they should deal with them." It is indeed possible that these skulls came from Strassburg. When Allied investigators arrived at the anatomical institute in Strassburg in late 1944, they discovered only sixteen complete bodies from those murdered at Natzweiler. In addition to these, they discovered a number of other de-fleshed bodies missing their heads. The investigators concluded that the heads had been incinerated in the crematorium of Strassburg, in order to prevent their identification. But it is possible that Hirt's staff managed to deflesh these skulls earlier by other methods of maceration, such as boiling in water. They could then have been sent to Beger. For further details, see "Photographs of treatment inflicted upon political and racial deportees and others detained at the Struthof camp (Laboratory for medical experiments)," translation of document No-483. NARA, Records of the U.S. Nuremberg War Crimes Trials: 'United States of America v. Karl Brandt et al.' (Case 1), Nov. 21, 1946–Aug. 20, 1947, RG238, M887/16/Jewish Skeleton Collection.

[33]Trojan to Beger, 08.08.1944, NARA, RG238, T81/131/164382-164385.

[34]Beger to Oberste Stelle für Kriegsgefangene Torgau, 10.02.1944, BA, R135/51/162463-162464.

[35]Beger, "Entwurf! Waffen SS. Wehrwissenschaftliche Forschungen des 'Ahnenerbes,'" BA, R135/52/162729; Sievers to Beger, 23.11.1943, NARA, RG242, T81/132/166160.

[36]Sievers, "Vermerk," 24.03.1945, BA, NS 21/329.

[37]Personal communication, Mrs. Ursula Schäfer.

[38]Sievers to Grau, 07.12.1944, BA, NS 21/ 329.

[39]Personal communication, Mrs. Ursula Schäfer.

[40]United States Fleet Headquarters of the Commander in Chief, "Amphibious Operations: Invasion of Northern France Western Task Force, June 1944," COM-INCH Pub 006, October 1944.

[41]Sievers to Brandt, 05.09.1944, NARA, RG242, T175/103/2625096.

[42]Berg, "Note re: skeleton collection in the Strassburg anatomical institute: dated 15.10.1944," translation of document NO-091. NARA, Records of the U.S. Nuremberg War Crimes Trials: 'United States of America v. Karl Brandt et al.' (Case 1), Nov. 21, 1946–Aug. 20, 1947, RG238, M887/16/Jewish Skeleton Collection.

[43]Sievers to Hirt, 18.10.1944, BA, NS 21/908.

[44]Henrypierre, "Affidavit," 17.11.1946, Office of U.S. Chief of Counsel. Translation of Document No. NO-880. StA Mchn. Stanw 34878/14.

[45]Berg, "Aktenvermerk für Brandt," 26.10.1944, BA, NS 19/1582. Also Sievers to Brandt, 07.12.1944, BA, NS 21/908.

[46]"Discovery of human testicles in the Strasbourg Institute," 25.05.1945, translation of Document NO-521. NARA, Records of the U.S. Nuremberg War Crimes Trials: 'United States of America v. Karl Brandt et al. (Case 1),' Nov. 21, 1946–Aug. 20, 1947, RG238 Nuernberg, Organization Series 516–517. Box 11, Entry 174. NM-70.

[47]Thomas Greif, Der SS-Standort Waischenfeld: 1934–1945, pp. 80–82.

[48]Sievers to Hirt, 20.01.1945, BA, NS 21/909.

[49]Hirt to Sievers, 19.10.1944, BA, NS 21/908.

[50]Sievers to Six, 28.03.1945, BA, NS 21/ 909.

[51]Hirt, "Stellungnahme zu der Veröffentlichung der 'Daily Mail' vom 03.01.1945," 25.01.1945, BA, NS 19/2281.

[52]Ibid.

[53]Sievers to Six, 16.02.1945. BA, NS 21/909.

[54]Wolff to Trojan, 19.02.1945, BA, NS 21/909; Wolff to Beger, 19.02.1945. BA, NS 21/909.

[55]Thomas Greif, Der SS-Standort Waischenfeld, p. 83.

21. THOR'S HAMMER

[1]Albert Speer, Infiltration (New York: Macmillan, 1981) p. 146.

[2]As quoted in Richard Rhodes, The Making of the Atomic Bomb (New York: Touchstone, 1988) p. 403.

[3]Richard Rhodes, The Making of the Atomic Bomb, pp. 404–405.

[4]Ibid.

[5]Ibid.

[6]German historian Rainer Karlsch has suggested that the German government did indeed succeed in building a "hybrid tactical nuclear weapon," but his contentions have been met with skepticism. See Rainer Karlsch, *Hitlers Bombe* (Munich: Deutsche Verlags-Anstalt, 2005).

[7]Proposal by Elektro-Mechanische Apparatebaugesellschaft, 28.10.1944, as quoted in Albert Speer, *Infiltration*, p. 146.

[8]Felix Kersten, *The Memoirs of Doctor Felix Kersten*, ed. Herma Briffault (Garden City, N.Y.: Doubleday, 1947), pp. 252–253.

[9]Office VI to Dr. R. Brandt, 08.01.1945, as translated and quoted in Albert Speer, *Infiltration*, p. 146.

[10]Oseberg to Brandt, 07.02.1945, as translated and quoted in Albert Speer, *Infiltration*, p. 147.

[11]Antony Beever, *The Fall of Berlin 1945* (New York: Viking, 2002), pp. 28–32, 67.

[12]Ian Kershaw, *Hitler 1936–1945* (London: Allen Lane, The Penguin Press, 2000) p. 818.

[13]Ibid., pp. 817–819.

[14]Ibid., pp. 716–717.

[15]Stephen Cook and Stuart Russell, *Heinrich Himmler's Camelot* (Andrews, N.C.: Kressmann-Backmeyer, 1999), p. 204.

[16]Ibid., p. 204–205.

[17]Ibid., p. 206.

[18]"Wewelsburg Kr.Paderborn.-Museumeinrichtung 1935–1944," Nach Erinnerungen von Wilhelm Jordan, 29.12. 1979, KWA 70/1/2/14.

[19]It is currently unknown exactly how much wartime plunder made its way to Wewelsburg. But some inkling can be gained from a few surviving documentary sources. Paulsen. "Dienstreisebericht von Peter Paulsen," 04.01.1940, as reproduced in Andrzej Mezynski, *Kommando Paulsen* (Cologne: Dittrich-Verlag, 2000) pp. 55–67; Jordan, "Verzeichnis von vorgeschichtlichen Funden aus dem Museum für Vorgeschichte in Dnjepropetrowsk," 31.12.1943, BA, NS 19/ 3638.

[20]Jochen von Lang, *Der Adjutant* (Munich: Herbig Verlag, 1985), p. 162.

[21]Stuart Russell and Jost W. Schneider, *Heinrich Himmlers Burg* (Essen: Heitz & Höffkes, 1989), pp. 180–181.

[22]Stephen Cook and Stuart Russell, *Heinrich Himmler's Camelot*, p. 212.

[23]Ian Kershaw, *Hitler 1936–1945*, pp. 816–817.

[24]Ibid., p. 819.

[25]Peter Padfield, *Himmler* (London: Cassel & Co, 2001), p. 608. See also Willi Frischauer, *Himmler, the Evil Genius of the Third Reich* (London: Odhams Press, 1953) p. 256.

[26]"Churchill Sure Himmler Faces a Glowing Future," *The New York Times*, May 17, 1945.

[27]One of Himmler's companions was SS-Hauptsturmführer Macher, the man Himmler had sent to destroy Wewelsburg.

[28]"Search for Himmler Widens," *The New York Times*, May 15, 1945.

[29]Peter Padfield, *Himmler*, p. 610.

[30]"Himmler, Caught, Ends Life by Poison: Arch Criminal Dies," *The New York Times*, May 25, 1945.

[31]Peter Padfield, *Himmler*, p. 610.

[32]Peter Padfield, *Himmler*, p. 611.

[33]"A Grave on the Heath," *Time*, June 26, 1945. For a picture of one of these plaster casts, see "Time's Winged Chariot, Heinrich Himmler," *Time*, June 24, 1945.

[34]Willi Frischauer, *Himmler*, p. 258.

22. NUREMBERG

[1]Martha Gellhorn, *The Face of War* (New York: Simon and Schuster, 1959) p. 213.

[2]Ibid.

[3]"Protocol of Proceedings," *Foreign Relations of the United States Diplomatic Papers, The Conference of Berlin (The Potsdam Conference)*, Vol. II (Washington: U.S. Government Printing Office, 1960), p. 1483.

[4]Thomas Greif, *Der SS-Standort Waischenfeld 1934–1945* (Erlangen: Palm & Enke, 2000) p. 82.

[5]In fact, there are two accounts of how the files were hidden. One, provided by Sievers's assistant, Wolff, noted that they were placed behind the rubble of a blast. A second, provided by the Ahnenerbe's cave expert, Dr. Hans Brand, suggested that they were placed behind a false wall. I think Wolff's version is somewhat more credible, as Brand and his coworkers noted after the war that they had not taken part in this action. As Sievers's trusted assistant at Waischenfeld, Wolff would more likely know what had happened to the files. For further details see Thomas Greif, *Der SS-Standort Waischenfeld*, p. 82.

[6]Thomas Greif, *Der SS-Standort Waischenfeld*, pp. 85–86.

[7]Major Frank Wallace to Murray Bernays, as quoted in Robert E. Conot, *Justice at Nuremberg* (New York: Harper & Row, 1947), p. 38.

[8]Michael H. Kater, *Das "Ahnenerbe" der SS 1935–1945* (Stuttgart: Deutsche Verlags-Anstalt, 1974), pp. 239–243. Some doubt now exists about whether all the children were stolen. See Ernst Klee, *Auschwitz, die NS-Medizin und ihre Opfer* (Frankfurt am Main: S. Fischer, 1997), p. 353.

[9]Michael H. Kater, *Das "Ahnenerbe,"* pp. 243.

[10]Local officials arranged for Hirt's body to buried, and it was not until the mid-1960s that Israeli investigators closed the books on the case. According to medical historian Frederick Kasten, the Israeli secret service contacted officials in the Black Forest region and had them exhume the body of the man who committed suicide

there in the early summer of 1945. An Israeli pathologist conclusively identified the bones as those of Dr. August Hirt.

[11]"Indictment," *Trials of War Criminals before the Nuernberg Military Tribunals*, Vol. 1, Case 1: The Medical Case (Washington, D.C: U.S. Government Printing Office, 1949–1950), pp. 8–10.

[12]"Opening Statement of the Prosecution by Brigadier General Telford Taylor, 9 December 1945," *Trials of War Criminals before the Nuernberg Military Tribunals*, Vol. 1, Case 1: The Medical Case (Washington, D.C: U.S. Government Printing Office, 1949–1950), pp. 27–29.

[13]"Final Statement of Defendant Sievers," *Trials of War Criminals before the Nuernberg Military Tribunals*, Vol. 2, Case 1: The Medical Case. (Washington, D.C: U.S. Government Printing Office, 1949–1950), pp. 157–159.

[14]Gedächtnisprotokoll Unterredung Frau Hella Sievers–Michael H. Kater, 26–27.4.1963, IfZ. ZS/A-25. vol. 2.

[15]Ibid.

[16]For years, prominent historians such as Michael Kater cast considerable doubt on the existence of this resistance group. Recently, however, a German sociologist, Ina Schmidt, has conducted interviews with several close associates of Hielscher, who claim that this group did indeed exist and that Sievers was a member. Ina Schmidt, "Der Herr des Feuers" (dissertation, Hamburger Universität für Wirtschaft und Politik, 2002), p. 229, pp. 235–239.

[17]One poem that Sievers kept among his Ahnenerbe papers is entitled "Poem dedicated to Wolfram Sievers to remember the [ ] nights from 21–24.01.1933." The author of the poem is identified as "F.H." One stanza reads "We burn like flames/and taste every pleasure/and fall together/and fight breast to breast." BA (ehem BDC) RS Sievers, Wolfram (10.07.1905).

[18]Michael H. Kater, *Das "Ahnenerbe,"* p. 313.

[19]Ina Schmidt, "Der Herr des Feuers," pp. 228–229.

[20]Ibid., pp. 235–236.

[21]Ibid., pp. 228–229.

[22]"Extract from the Closing Statement of the Prosecution," *Trials of War Criminals before the Nuernberg Military Tribunals*, Vol. 2, Case 1: The Medical Case (Washington, D.C: U.S. Government Printing Office, 1949–1950), p.4–5.

23. SECRETS

[1]Anna Funder, *Stasiland* (London: Granta Books, 2003), p. 284, footnote for p. 119.

[2]*The Encyclopedia of the Third Reich*, ed. Christian Zentner and Friedemann Bedürftig (New York: Da Capo Press, 1997), s.v. "Questionnaire."

[3]Adalbert Rückerl, *NS-Verbrechen vor Gericht* (Heidelberg: C.F. Müller, 1984), pp. 117–120.

[4]C.M. Clark, "West Germany Confronts the Nazi Past," *The European Legacy* 4, no. 1 (1999): 114–115.

[5]Adalbert Rückerl, *NS-Verbrechen vor Gericht*, pp. 117–120.

[6]C.M Clark, "West Germany Confronts the Nazi Past," p. 116.

[7]Ibid.

[8]Ibid., p. 117; *The Encyclopedia of the Third Reich*, ed. Christian Zentner and Friedemann Bedürftig (New York: Da Capo Press, 1997), s.v. "Persil Certificate"; Adalbert Rückerl, *NS-Verbrechen vor Gericht*, pp. 117–120.

[9]C.M Clark, "West Germany Confronts the Nazi Past," p. 116.

[10]Wirth to Zaharia, 18.07.1946, StA Mchn, SpkA K 2015 vol.2.

[11]Ibid.

[12]Ibid.

[13]Sievers to Wüst, 04.01.1940, BA, NS 21 /46.

[14]Ingo Wiwjorra, "Herman Wirth—Ein gescheiterter Ideologe zwischen 'Ahnenerbe' und Atlantis," *Historische Rassismusforschung*, eds. Barbara Danckwortt et al (Hamburg: Argument, 1995). p. 108.

[15]Personal communication, Dr. Luitgard Löw.

[16]Ibid.

[17]Ibid.

[18]Riksantikvarieämbetet to Wirth Roeper Bosch, 20.03.1965, ATA Stockholm.

[19]Riksantikvarieämbetet to Wirth Roeper Bosch, 20.03.1965, ATA Stockholm.

[20]Personal communication, Mr. Paul Rohkst.

[21]Peter Adam, "Schenkel der Göttlichen," *Der Spiegel* 34, No. 40 (29.09.1980).

[22]Ibid.

[23]Personal communication, Mr. Paul Rohkst.

[24]After the opening of the museum exhibit of Wirth's casts, the government of upper Austria established a commission to investigate the role of Wirth and his colleague Ernst Burgstaller during the Nazi regime. Personal communication, Dr. Luitgard Löw.

[25]Franz Mandl, "Das Erbe der Ahnen: Ernst Burgstaller/Herman Wirth und die österreichische Felsbildforschung," *Archäologie und Felsbildforschung: Mitteilungen der ANISA* 19/20, no.1/2 (1999).

[26]Personal communication, Mr. Juhani von Grönhagen. Yrjö von Grönhagen does indeed seem to have worked in some capacity for the Finnish foreign ministry during the war. But the exact nature of this work is unclear. For further details, see Grönhagen, "RKK Fragebogen," 07.03.1942, BA (ehem. BDC) RKK: Grönhagen, Yrjö von (03.11.1911); Herta von Grönhagen to Mag. Puntila, 10.11.1940, ARK, A 3860.

[27]Grönhagen, "RKK Fragebogen," 07.03.1942, BA (ehem. BDC) RKK: Grönhagen, Yrjö von (03.11.1911).

[28]Volckmar, "Erklärung," 01.07.1946, ARK, A 3860.

[29]British Security Service, "To whom it may concern," 28.06.1946, ARK, A 3860.

[30]"Einige Angaben über das Geschlecht Grönhagen," unpublished document in collection of Mr. Juhani von Grönhagen.

[31]While researching this book, I and my colleague Charlotte Stenberg managed to track down many of these lost recordings in an archive in Berlin, bringing to scholarly attention again the voices of long-dead singers.

[32]Personal communication, Dr. Juha Pentikäinen.

[33]Ibid.

[34]Ibid.

[35]Personal communication, Mr. Juhani von Grönhagen.

[36]Personal communication, Dr. Ruth Altheim-Stiehl.

[37]Ibid.

[38]Altheim and Trautmann, "(Vertraulicher) Bericht über eine im Sommer und Herbst 1938 unternommen Forschungsreise in Schweden, Rumänien, Syrien und Irak," approximate date 05.12.1938, BA (ehem. BDC) Ahnenerbe: Altheim, Franz. 06.10.1898; Altheim and Trautmann. "(Vertraulicher) Bericht" n.d. BA (ehem. BDC) Ahnenerbe: Altheim, Franz (06.10.1898); Sievers to RFSS, 15.01.1941, NARA, RG242,T175/48/2585099; Altheim and Trautmann, "Notwendigkeit einer deutschen Initiative im arabischen Raum Vorderasiens," 14.01.1941, NARA, RG242, T175/48/2585100.

[39]Sievers to RFSS, 15.01.1941, NARA, RG242, T175/48/2585099. This letter included an attachment entitled "Notwendigkeit einer deutschen Initiative im arabischen Raum Vorderasiens," 14.01.1941, NARA, RG242, T175/48/2585100.

[40]"Sonderbericht über Iran," 11.06.1941, BA, NS 21/2414. The original covering letter for this report appears to have been lost, thus making indisputable identification of the author extremely difficult. However, a number of clues point to Altheim and Trautmann. The report was found in a file belonging to Himmler's personal staff. This same file contained the secret Iraq report that Altheim and Trautmann submitted five months earlier. In addition, the Iran report was typed in a similar format to the Iraq report. Also, it contained information similar to that which Altheim and Trautmann gleaned during their 1938 trip to Iraq. For example, this Iran report notes that "in the lower classes, one can often come across the opinion that the Führer is the thirteenth imam who will deliver all Islamic Iranians from earthly suffering."

[41]Personal communication, Dr. Ruth Altheim-Stiehl.

[42]Ibid. Surviving letters show that Altheim did indeed try to obtain the release of the young Hungarian woman—Grazia Kerényi, the daughter of the prominent Hungarian historian Karl Kerényi, who had been a close friend of Altheim before the war—from an Austrian concentration camp in 1944. For further details, see Völker Losemann, "Die 'Krise der Alten Welt' und der Gegenwart," Völker Imperium Romanum, ed. Peter Kneissl and Völker Losemann (Stuttgart: Franz Steiner, 1998).

[43]Personal communication, Dr. Ruth Altheim-Stiehl.

[44]Indeed, by 1944, Altheim had taken to describing Trautmann as an "elderly" lady. She was forty-seven years old. See Altheim to Neugebauer, 16.08.1944, DAI Nachlass Neugebauer, Karl Anton.

[45]Personal communication, Dr. Ruth Altheim-Stiehl.

[46]Personal communication, Dr. Bernhard Caemmerer.

[47]Personal communication, Dr. Dieter Metzler.

[48]Eberhard Merkel, "Bibliographie Franz Altheim," *Beiträge zur Alten Geschichte und deren Nachleben*, ed. Ruth Stiehl and Hans Erich Stier (Berlin: Walter de Gruyter & Co., 1970), pp. 390–426.

[49]"Am 17.Oktober 1976 verstarb im 79.Lebensjahr," *Der Tagesspiegel*, Berlin, October 24, 1976.

[50]Personal communication, Dr. Bernhard Caemmerer.

[51]". . . unsere verstorbene Base Erika Trautmann," *Familienblatt Nehring* 9 (1968): 54.

[52]Personal communication, Dr. Dieter Metzler.

[53]Even Bohmers's former colleagues at the Ahnenerbe found this claim utterly unbelievable. "The idea that Bohmers was part of the resistance is absurd," wrote Lothar Zotz, a fellow Ahnenerbe archaeologist, after the war. "Gedächtnisprotokoll. Unterredung Zotz/Freund und Michael H. Kater," 18.03.1963, IfZ, ZS/A-25 vol. 3. p. 787.

[54]Personal communication, Dr. Oebele Vries.

[55]Personal communication, Dr. H.T. Waterbolk.

[56]Bohmers to Sievers, 11.11.1940, BA (ehem.BDC) Ahnenerbe: Bohmers, John Christiaan Assien (16.1.1912).

[57]Personal communication, Dr. H.T. Waterbolk.

[58]Personal communication, Dr. Oebele Vries.

[59]It is clear from surviving pieces of correspondence that Bohmers did indeed meet Himmler, but it seems highly unlikely that he ever addressed him to his face as Heini.

[60]"Zeuge Dr. Schäfer," 07.11.1962, StA Mchn, Stanw 34878/18. See also Wüst to Kater, 07.06.1964, IfZ, ZS/A-25. vol. 3.

[61]Personal communication, Mrs. Ursula Schäfer. Also,"Zeuge Dr. Schäfer," 07.11.1962, StA Mchn, Stanw 34878/18.

[62]"Urteil Ernst Schäfer," 13.06.1949, StA Mchn, Spruchkammer-Akt 1573.

[63]Personal communication, Mrs. Ursula Schäfer.

[64]"Der Anti-Disney," *Der Spiegel*, No. 13 (1959): 60–61.

[65]Ibid.

[66]Exhibit on Mittersill's history in Writers Room, 24.04.2002, Schloss Mittersill.

[67]Andrew Lycett, *Ian Fleming* (London: Weidenfeld & Nicolson, 1995), pp. 294–295.

[68]Ibid., p. 295.

[69]Victor and Victoria Trimondi, *Hitler, Buddha, Krishna* (Vienna: Ueberreuter, 2002), p. 111.

[70]"Entlassungsschein," 19.06.1947, HHA, Wiesbaden. Abt. 520 KS-HL Nr. 88, Spruchkammer Kassel; Kellner to Spruchkammer Kassel. 29.06.1948, HHA. Abt. 520 KS-HL Nr. 88 Spruchkammer Kassel.

[71]Kellner to Spruchkammer Kassel, 14.07.1948, HHA, Wiesbaden. Abt. 520 KS-HL Nr. 88 Spruchkammer Kassel.

[72]"Protokoll der öffentlichen Sitzung am: 5 August, 1948," HHA, Wiesbaden. Abt. 520 KS-HL Nr. 88, Spruchkammer Kassel.

[73]Ibid.

[74]Ibid.

[75]"Spruch," 05.08.1948, HHA, Wiesbaden. Abt. 520 KS-HL Nr. 88, Spruchkammer Kassel.

[76]Robert E. Lester, ed., *Art Looting and Nazi Germany* (Bethesda, MD: LexisNexis, 2002), p. v.

[77]Rotraut Wolf, "Nachruf Peter Paulsen 1902–1985," *Fundberichte aus Baden-Württemberg* 10 (1985): 727–728.

[78]Ibid.

[79]Gert Kerschbaumer, "Das Deutsche Haus der Natur zu Salzburg," in *Politik der Präsentation,* ed. Herbert Posch, and Gottfried Fliedl (Vienna: Turia and Kant, 1996), p. 182.

[80]Ibid., pp. 182–183.

[81]Gottfried Fliedl, *Das Haus der Natur in Salzburg als Institut des SS-Ahnenerbes,* http://homepage.univie.ac.at/gottfried.fliedl/mouseion/hausdernatur.html.

[82]Personal communication, Dr. Dieter Jankuhn.

[83]Ibid.

[84]Ibid.

[85]Heiko Steuer, "Herbert Jankuhn und seine Darstellungen zur Germanen- und Wikingerzeit," in *Eine hervorragend nationale Wissenschaft,* ed. Heiko Steuer, (Berlin: Walter de Gruyter, 2001), p. 425.

[86]Personal communication, Dr. Dieter Jankuhn.

[87]"Nachruf. Dr.phil. Herbert Jankuhn," Georg-August Universität Göttingen, October 1990.

[88]Personal communication, Dr. Anders Hagen.

[89]Anders became so worried that the courageous resistance of Norwegian archaeologists against Jankuhn and the Ahnenerbe would be forgotten that he wrote an article on the subject in 1985 and published it in a prominent Norwegian archaeological journal, *Viking.* Jankuhn had retired from teaching by then, but he remained active in European archaeology, working as a consultant for the Commission for Antiquities Studies in Middle and Northern Europe in Göttingen. Soon after the article

was published, one of Hagen's students translated the article into German and passed it around among archaeological students in Germany. Most, noted Hagen, had never heard of the Ahnenerbe and had little idea that one of the leading figures in German archaeology had played such an important role in it. "They were shocked," recalled Hagen. "They hadn't heard about this."

[90]"Gedächtnisprotokoll. Unterredung Dr. Herbert Jankuhn und Michael H. Kater," 14.05.1963, IfZ, ZS/A-25, vol. 1.

[91]Personal communication, Dr. Dieter Jankuhn.

[92]*Völkischer Beobachter*, 06.07.1941, as quoted in Victor and Victoria Trimondi, *Hitler, Buddha, Krishna*, p. 39.

[93]Wüst to Sievers, 13.11.1944, BA (ehem. BDC) Ahnenerbe: Wüst, Walther (07.05.1901); Wüst. "Garantieschein." 01.11.44. BA, R51/ 10146.

[94]That the institute was an integral part of the Ahnenerbe can be seen from a letter that Wüst wrote to Himmler's personal staff on February 6, 1944. In this letter, Wüst notes that Hirt's experiments "are conducted under the Institute for Military Scientific Research in the Amt Ahnenerbe." StA Mchn. Stanw 34878/75.

[95]"Results of Detailed Interrogation, Prof. Dr. Walther Wüst," AIC NO. 1760. n.d. StA Mchn. SpKa Karton 2015, Wüst Walther.

[96]Minister für politische Befreiung, Bayern. "Entwurf." 18.03.1953, BHA, MSo 1921; Wüst to Hauptkammer Munich, 27.06.1952, StA Mchn. SpkA K 2015.

[97]Prof. Gustav Freytag to Wüst, 05.08.1954, BStM, Ana 625: Correspondence.

[98]Personal communication, Dr. Helmut Humbach.

[99]Adalbert Rückerl, *NS-Verbrechen vor Gericht* (Heidelberg: C.F. Müller, 1984) pp. 139–167.

[100]Zentrale Stelle Ludwigsburg, "Vermerk über Sievers Tagebücher," 18.05.1967, StA Mchn. Stanw 34878/1.

[101]Max Vogt, "Aktenvermerk," 24.11.1969, StA Mchn, Stanw 34878/5. Recent efforts by several researchers to locate Wüst's missing records have also met with failure.

[102]Staatsanwalt München I, 23.03.1972, StA Mchn, Stanw 34878/10; Staatsanwalt München I, 13.06.1972, StA Mchn, Stanw 34878/10.

[103]Wüst to Pers. Stab., 06.02.1944, StA Mchn, Stanw 34878/75.

[104]"I. Die Angeschuldigten: Lebenslauf, Dr. Bruno Beger," p. 6, Frankfurter Schwurgericht. Strafverfahren gegen Bruno Beger, Hans Fleischhacker, Wolf-Dietrich Wolff. 27.10.1970-06.04.1971. IfZ, Gf 03.32; Grau to Sievers, 19.12.44, BA, NS 21/39; Beger to Sievers, ? 02.45, BA, NS 21/39.

[105]"Urteil. II. Der Angeklagte Dr. Beger," p. 10, Frankfurter Schwurgericht. Strafverfahren gegen Bruno Beger, Hans Fleischhacker, Wolf-Dietrich Wolff. 27.10.1970-06.04.1971. IfZ, Gf 03.32.

[106]Ibid.

[107]The publishing house was run by Margarete Landé and therein lies a very com-

plicated story. Landé was a German woman of Jewish ancestry who was raised as a Christian and who entered into a close and rather enigmatic relationship with Clauss. Clauss was a handsome, charismatic man who attracted much adulation from his female students. During the war, Clauss was accused of breaking the Nazi racial laws by employing Landé. Beger intervened on behalf of Clauss, helping to get him off. Clauss then proceeded to hide Landé at his hunting lodge, where Beger finally discovered her. But Beger kept the information secret, quite possibly out of his deep friendship for Clauss. For further details, see Peter Weingart, *Doppel-Leben* (Frankfurt am Main.: Campus, 1995).

[108]"I. Die Angeschuldigten: Lebenslauf, Dr. Bruno Beger," p. 6, Frankfurter Schwurgericht. Strafverfahren gegen Bruno Beger, Hans Fleischhacker, Wolf-Dietrich Wolff. 27.10.1970-06.04.1971. IfZ, Gf 03.32.

[109]Personal communication, Bruno Beger. See also Vareschi, "Dienstvertrag," 15.06.1943, NARA, T81/130/163316; "Urteil I. Die Angeschuldigten: Lebenslauf und politischer Werdegang," Frankfurter Schwurgericht. Strafverfahren gegen Bruno Beger, Hans Fleischhacker, Wolf-Dietrich Wolff. 27.10.1970-06.04.1971; "Zeuge Prof. Clauss 30.10.1962," Voruntersuchungsache gegen Dr. Beger, StA Mchn, Stanw. 34878/13.

[110]Bruno Beger, *Mit der deutschen Tibetexpedition Ernst Schäfer 1938/39 nach Lhasa* (Wiesbaden: Schwarz, 1998), p. 5.

[111]Frankfurter Schwurgericht. Strafverfahren gegen Bruno Beger, Hans Fleischhacker, Wolf-Dietrich Wolff. 27.10.1970-06.04.1971. HHA, 4 Ks 1/70.

[112]"Urteil. V. Die Beteiligung des Angeklagten Dr. Beger," pp. 33–34. Frankfurter Schwurgericht. Strafverfahren gegen Bruno Beger, Hans Fleischhacker, Wolf-Dietrich Wolff. 27.10.1970-06.04.1971. IfZ, Gf 03.32.

[113]Max Kaase, "Demokratische Einstellungen in der Bundesrepublik Deutschland," *Sozialwissenschaftliches Jahrbuch für Politik*, vol. 2, ed. Rudolf Wildenmann (Munich: Olzog, 1971), Vol. 2, p. 325.

[114]"Strafverfahren gegen Dr. Beger, Dr. Fleischhacker und Wolff wegen Beihilfe zum Mord," IfZ, ZS/A-25, vol.1.pp. 20–21.

[115]Ibid; "Im Namen des Volkes: Strafsache gegen Beger und Wolff," Frankfurter Schwurgericht. Strafverfahren gegen Bruno Beger, Hans Fleischhacker, Wolf-Dietrich Wolff. 27.10.1970-06.04.1971. StA Mchn, Stanw 34878/93. See also "SS-Anthropologe Beger zu drei Jahren Freiheitsstrafe verurteilt," *Frankfurter Allgemeine Zeitung*, April 7, 1971, p. 8.

[116]"Strafverfahren gegen Dr. Beger, Dr. Fleischhacker und Wolff wegen Beihilfe zum Mord," IfZ, ZS/A-25, vol.1.pp. 20.–21; "Im Namen des Volkes: Strafsache gegen Beger und Wolff," Frankfurter Schwurgericht. Strafverfahren gegen Bruno Beger, Hans Fleischhacker, Wolf-Dietrich Wolff. 27.10.1970-06.04.1971. StA Mchn, Stanw 34878/93.

[117]"Urteil gegen Beger," Frankfurter Schwurgericht. Strafverfahren gegen Bruno

Beger, Hans Fleischhacker, Wolf-Dietrich Wolff. 27.10.1970-06.04.1971. StA Mchn, Stanw 34878/93.

[118]The four prisoners listed in a letter of commendation that Sievers sent at Beger's request to the commander of the Auschwitz camp were Ludwig Wörl, Kasimir Kott, Josef Weber, and Adolf Laatsch. Ludwig Wörl was born in Munich on February 28, 1906. He was arrested for his political activities as a Communist and imprisoned first in Dachau and deported to Auschwitz on August 20, 1942. There he became a *Blockältester,* or senior block prisoner in charge of a barrack. Josef Weber, also known as Józef Weber, was born in Poland on July 27, 1920. He, too, was arrested as a political prisoner and transported to Auschwitz on May 2, 1941. Kasimir Kott, also known as Kazimierz Kot, was born in Poland on February 11, 1914. He arrived in Auschwitz on May 22, 1942, as a political prisoner. Adolf Laatsch was born on April 18, 1892, in the German city of Witten. He was sent first to Dachau, and then to Auschwitz on April 1, 1943. There he quickly became a *Blockältester* responsible for one of the camp barracks. Sievers to Commander of KZ Auschwitz, 22.07.1943, BA, NS 21/907; Jerzy Wróblewski, Panstwowe Muzeum Auschwitz-Birkenau, to Charlotte Stenberg, 24.03.2005, collection of the author; personal communication, Professor Rudolf Vrba. Professor Vrba is not only one of the few prisoners who succeeded in escaping from Auschwitz, but he also coauthored *The Vrba Wetzler Report* in 1944, which described the Auschwitz killing machine, supplied estimates of the numbers of Jews murdered, and warned that an additional eight hundred thousand Hungarian Jews were in danger of extermination at Auschwitz. Professor Vrba knew one of the prisoners, Ludwig Wörl, personally,

[119]"Urteil gegen Beger," Frankfurter Schwurgericht. Strafverfahren gegen Bruno Beger, Hans Fleischhacker, Wolf-Dietrich Wolff. 27.10.1970-06.04.1971. StA Mchn, Stanw 34878/93; "Im Namen des Volkes: Strafsache gegen Beger und Wolff," Frankfurter Schwurgerichts. Strafverfahren gegen Bruno Beger, Hans Fleischhacker, Wolf-Dietrich Wolff. 27.10.1970-06.04.1971. StA Mchn. Stanw34878/93.

[120]Beger to author, 03.10.2004, author's collection.

24. SHADOWS OF HISTORY

[1]Ashley Montagu, *Statement on Race* (New York: Oxford University Press, 1972) p. 143.

[2]Ibid., p. 146.

[3]Ibid., p. 146.

[4]Some of the most famous inner emigrants include poet and novelist Ricarda Huch, sculptor and playwright Ernst Barlach, composer Carl Orff, and the artist Emil Nolde. Personal communication, Dr. Peter Stenberg.

GUIDE TO THE MOST IMPORTANT PERSONALITIES

Dr. Franz Altheim (1898–1976) An urbane bohemian with a wry sense of humor, Altheim was an expert on the origins of Roman religion and the history of the Latin language. While teaching at the University of Frankfurt, he met Erika Trautmann, who became his mistress and who helped introduce him to Himmler. With financial assistance from the SS leader, the couple traveled to Italy and Dalmatia in 1937, searching for evidence of the Nordic origins of Roman civilization. A year later, with funding from the Ahnenerbe, the pair journeyed across Eastern Europe and the Middle East, arriving in Iraq. En route, they gathered intelligence on Iraqi pipelines and tribal leaders for the SS Security Service.

Dr. Bruno Beger (1911–) A member of a prominent Heidelberg family, Beger was an expert in *Rassenkunde*, or racial studies, a growth industry in Nazi Germany. At the invitation of Ernst Schäfer, he joined the SS expedition to Tibet, serving as its anthropologist and racial expert. During the war, he was assigned as a racial expert to Schäfer's Special Command K in the Caucasus. His mission was to racially diagnose tribal groups in the region—information that could be used by the SS command to slate Jewish groups for extermination. In 1942, Beger took part in a major war crime known as the Jewish Skeleton Collection.

Dr. Assien Bohmers (1912–1988) Born in the Netherlands, Bohmers was a renegade archaeologist fascinated by the German Nazi party and by Frisian politics. In 1937, he conducted excavations for the SS at the Cro-Magnon site of Mauern in Germany, claiming to have uncovered the origins of the Aryan race. His findings fascinated Himmler, and in 1938 Bohmers joined the staff of the Ahnenerbe. That fall, he journeyed across the Netherlands, Belgium, England, and France, searching for the earliest traces of Aryan art and culture.

Dr. Fritz Bose (1906–1975) A musicologist who theorized that the world's richly varied musical styles were a reflection of innate racial traits rather than cultural differences, Bose became head of the Berlin Acoustics Institute in 1934, after his Jewish mentor was forced out. In 1936, Himmler sent Bose to Finland with Yrjö von Grönhagen to record the ancient songs of Finnish witches and sorcerers. After the war, he taught for a time at the Technical University in Berlin.

Richard Walther Darré (1895–1953) An agriculturalist specializing in animal breeding, Darré believed that Germany had to take serious measures to restore the purity of the Nordic race. He advocated exterminating the handicapped and using scientific knowledge to assist Germans in selecting their mates in order to produce superior human stock. In 1932, he became the head of the Race and Settlement Office of the SS, better known as RuSHA, a position he held until 1938. RuSHA examiners were responsible for the racial purity of the SS. In 1933, he became the minister of food and agriculture in Germany.

Yrjö von Grönhagen (1911–2003) A handsome young Finnish nobleman who loathed communism, Grönhagen met Himmler while trekking on foot from Paris to Helsinki. His knowledge of ancient Finnish myths and folklore so charmed Himmler that he put Grönhagen to work in the Ahnenerbe. In 1936, Himmler sent the young Finn and German musicologist Fritz Bose on a research trip to the Finnish province of Karelia. There they recorded ancient magical spells preserved by elderly wizards and recorded the primeval songs of the region.

Dr. Hans F. K. Günther (1891–1968) A German philologist and anthropologist who became one of the Third Reich's foremost exponents of racial studies, Günther wrote a series of popular books on race that proved enormously influential in Nazi circles. Adolf Hitler greatly admired his work, and in 1930

Hitler attended Günther's inaugural lecture at the University of Jena. After the Nazi seizure of power, Günther took up an important post at the University of Berlin.

Heinrich Himmler (1900–1945) The son of a prominent Bavarian schoolmaster, Himmler joined the Nazi party in 1923 and soon rose to the position of Reichsführer-SS (1929–1945). Ruthless, intelligent, and utterly dependable, he became one of Hitler's favorite henchmen, holding the posts of chief of the German Police (1936), minister of the interior (1943), commander in chief of the Replacement Army (1944), and commander in chief of the Rhine (1944) and Vistula armies (1945). As diligently as he performed Hitler's bidding, however, Himmler always found time to oversee one of his fondest creations, the Ahnenerbe, even serving as its first superintendent (1935–1938). He committed suicide at a British interrogation center.

Dr. August Hirt (1898–1945) A talented anatomist who suffered a severe facial injury in the First World War, Hirt specialized in the study of the human nervous system. In 1933, while teaching at the University of Heidelberg, he joined the SS and there he met a young Bruno Beger. During the war, Hirt became a department head in the Ahnenerbe's Institute for Military Scientific Research and conducted a series of notorious medical experiments on concentration-camp prisoners. In 1942 and 1943, he planned and directed a major war crime known as the Jewish Skeleton Collection.

Adolf Hitler (1889–1945) An Austrian drifter and artist turned politician, Hitler became Reich chancellor in January 1933 and Germany's head of state a year later. He assumed the role of minister of defense in February 1938, and took on the responsibility of supreme commander in the field in December 1941. He committed suicide in a bunker beneath the Reich Chancellery on April 30, 1945.

Dr. Herbert Jankuhn (1905–1990) One of the brightest and most respected archaeologists in Germany, Jankuhn joined the Ahnenerbe in 1937, becoming the head of its department of archaeology three years later. In 1942, Jankuhn led a small unit of archaeologists to southern Russia and the Caucasus to search for the treasure of the Goths and find proof of a classical-era German empire along the Black Sea.

Dr. Karl Kersten (1909–1992) An expert on the northern European Bronze Age, Kersten was an old friend of Herbert Jankuhn. In 1942, he joined Jankuhn's archaeological unit, and spent the early fall of that year surveying archaeological sites in the Crimea and preparing an itinerary for Himmler's tour to the region.

Edmund Kiss (1886–1960) A much-decorated soldier in the First World War and an architect and writer by profession, Kiss claimed to have found the ruins of primordial Aryan palaces and temples in South America. In 1939, he planned to lead a team of twenty—the Ahnenerbe's largest and most expensive expedition—to Tiwanaku in the Bolivian Andes.

Dr. Peter Paulsen (1902–1985) An archaeologist with an international reputation as a Viking expert, Paulsen headed a unit of scholars dispatched to Warsaw in 1939 to methodically loot the city's most important museum collections of their archaeological treasures. Unable to fend off other Nazi thieves, he was reassigned in 1940 to a quiet teaching job at an SS officer-training school in Bad Tölz.

Dr. Ernst Schäfer (1910–1992) A headstrong zoologist with a volatile temper, Schäfer led an expedition to Tibet in 1938 to search for proof of an ancient Aryan conquest of the Himalayas. During the war, he lectured in occupied Europe as an exemplar of Nazi science, and in 1942 he accepted command of a military and scientific mission, Special Command K, in the Caucasus. One goal of the mission was to conduct a racial diagnosis of the Mountain Jews to determine whether they should be annihilated.

Wolfram Sievers (1905–1948) A high-school dropout who taught himself European prehistory, Sievers was a gifted administrator who became the Ahnenerbe's managing director. During the war, Sievers headed the Ahnenerbe's Institute for Military Scientific Research and orchestrated its notorious medical experiments on prisoners at Dachau and Natzweiler concentration camps.

Eduard Tratz (1888–1977) One of Salzburg's most respected citizens, Tratz was the founder of the Haus der Natur, a major natural-history museum. An ardent supporter of Nazi policy and doctrine, Tratz joined Paulsen's looting unit in 1939, personally plundering the State Zoological Museum in Warsaw.

Erika Trautmann (1897–1968) The daughter of a wealthy estate owner in East Prussia and a close friend of Hermann Göring, Trautmann was an illustrator and photographer by profession. While working for the world famous Research Institute for Cultural Morphology in Frankfurt, she met the classical historian Franz Altheim, who became her lover. In 1937, the couple conducted archaeological research in Italy and Dalmatia at Himmler's behest, and soon after joined the Ahnenerbe. In 1938, they traveled to Romania and Iraq conducting research for the Ahnenerbe, and gathering intelligence for the SS Security Service.

Karl-Maria Wiligut (also known as Weisthor) (1866–1946) A former psychiatric patient and a self-described expert in runic script, Wiligut traced his pedigree back to the Norse god Thor. He claimed to guard the sacred knowledge of primeval German tribes. He met Himmler in 1933, and in the years that followed, the SS leader gave him an office in RuSHA and consulted him regularly on matters of ancient Germanic traditions.

Dr. Herman Wirth (1885–1981) An eccentric Dutch spendthrift with immense reserves of personal charm, Wirth was a philologist by training and a self-proclaimed expert in script and symbol studies. He became the first president of the Ahnenerbe (1935–1937), and led two expeditions to Scandinavia, where he believed he had found examples of the world's oldest writing system: a lost Aryan script.

Dr. Walther Wüst (1901–1993) A cautious, reserved man, Wüst was an expert on Sanskrit and Old Persian, and a Nazi authority on the ancient Aryans. During his first meeting with Himmler, he read aloud from old Sanskrit scriptures, an experience that enthralled the SS leader. In 1937, Himmler appointed Wüst president of the Ahnenerbe, and later made him the institute's superintendent (1939–1945).

MAJOR SOURCES

INTERVIEWS

Mrs. Ursula Schäfer, Bad Bevensen, Germany	08.11.01
Dr. Dieter Metzler, Münster, Germany	09.11.01
Dr. Klaus Goldmann, Berlin, Germany	11.04.02
Mr. Eike Schmitz, Berlin, Germany	12.04.02
Dr. Dieter Jankuhn, Göttingen, Germany	15.04.02
Prof. Adolf Frisé, Bad Homburg, Germany	16.04.02
Dr. Bernard Caemmerer, Karlsruhe, Germany	16.04.02
Dr. Bruno Beger, Königstein, Germany	17.04.02
Mrs. Gabriele Winckler-Dechend, Constance, Germany	18.04.02
Dr. Lisa Schroeter-Bieler, Freiburg, Germany	19.04.02
Dr. Ulrich Gruber, Munich, Germany	22.04.02
Mr. Paul Rohkst, Kolbermoor, Germany	23.04.02
Dr. Ruth Altheim-Stiehl, Nienberge, Germany	28.04.02
Mr. Gerd Pucka, Bokdorf-Ohze, Germany	29.04.02
Mrs. Leopoldine Jordan, Bad Sassendorf, Germany	30.04.02
Dr. Jan Brøgger, Trondheim, Norway	21.06.02
Dr. Henry M. Hoenigswald, Haverford, PA, United States	24.06.02
Mr. Norm Heinrich-Gales, Mittersill, Austria	25.06.02
Dr. Emmanuel Anati, Capo di Ponti, Italy	15.07.02
Ms. Camilla Olsson, Grebbestad, Sweden	07–08.09.02
Dr. Anders Hagen, Bergen, Norway	10.09.02
Dr. Margit Engel, Michelbach-le-Haut, France	14.09.02

Mr. Christoph Nehring, Essen, Germany — 16.09.02

Dr. Helmut Humbach, Mainz, Germany — 17.09.02

Dr. Harald Jankuhn, Rheinbach-Loch, Germany — 18.09.02

Dr. Helmut Arntz, Bad Honnef, Germany — 19.09.02

Mrs. Elke Pleines, Westerland, Germany — 23.09.02

Mr. Joachim Pleines, Westerland, Germany — 23.09.02

Dr. Raimund Pfister, Munich, Germany — 25.09.02

Dr. Juha Pentikäinen, Helsinki, Finland — 17–18.03.04

Mr. Juhani von Grönhagen, Helsinki, Finland — 19.03.04

Dr. Oebele Vries, Groningen, the Netherlands — 31.03.04

Dr. Tjalling Waterbolk, Groningen, the Netherlands — 15.04.04

ARCHIVAL REPOSITORIES

AUSTRIA

Naturhistorisches Museum, Bildarchiv, Vienna — NM

GERMANY

Bayerisches Hauptstaatsarchiv, Munich — BHA

Bayerische Staatsbibliothek, Mikrofiche, Munich — BStM

Brandenburgisches Landesamt für Denkmalpflege und — BLD
 Archäologisches Landesmuseum, Zossen (Ortsteil Wünsdorf)

Bildarchiv Preußischer Kulturbesitz, Berlin — BPK

Bundesarchiv, Berlin-Lichterfelde — BA

Bundesarchiv, Dahlwitz-Hoppegarten — BA-DH

Bundesarchiv, Koblenz — BAK

Bundesbeauftragte für die Unterlagen des
 Staatssicherheitsdientes der ehemaligen Deutschen
 Demokratischen Republik, Berlin — BstU

Deutsches Archäologisches Institut Archiv, Berlin — DAI

Fritz Bauer Institut Archiv, Frankfurt am Main — FBIA

Frobenius-Institut Archiv, Frankfurt am Main — FIA

Hessisches Hauptstaatsarchiv, Wiesbaden — HHA

Institut für Zeitgeschichte Archiv, Munich — IfZ

Klingspor-Museum Archiv, Offenbach am Main — KMA

Kreismuseum Wewelsburg Archiv, Büren-Wewelsburg — KWA

Lautarchiv, Humboldt-Universität, Berlin — LHUB

Politisches Archiv des Auswärtigen Amt, Berlin — PAAA

Staatsarchiv, Munich — StAMchN

Stadtarchiv, Metzingen — StAMetz

Universitätsarchiv, Johann Wolfgang Goethe-Universität,
    Frankfurt am Main   UAF/M
Universitätsarchiv, Ruprecht-Karls-Universität, Heidelberg   UAH
Universitätsarchiv, Universität der Künste, Berlin   UA KB
Universitätsarchiv, Rostock   UAR

FINLAND
Arkistolaitos, Helsinki   ARK
Suomalaisen Kirjallisuuden Seura, Kansanrunousarkisto, Helsinki   SKS

NORWAY
Kulturhistorisk Museum Archiv, Universitets i Oslo   KHM

POLAND
Panstwowe Muzeum Auschwitz-Birkenau   PM

SWEDEN
Arkiv för svensk kulturminnesvård   ATA
  och antikvarisk forskning,
  Riksantikvarieämbetet, Stockholm

UNITED KINGDOM
Robinson Library, Special Collections, University of Newcastle upon Tyne RLUNT
  Gertrude Bell Collection

UNITED STATES
Hoover Institute Archives, Stanford, CA   HIA
Library of Congress, Washington, DC   LOC
National Archives and Records Administration,
  College Park, MD   NARA
United States Holocaust Memorial Museum,
  Washington, DC   USHMM

SELECT BIBLIOGRAPHY

While researching the history of the Ahnenerbe, I and my research team consulted many hundreds of books on a vast range of subjects. The following list represents only those books, articles, and unpublished papers that were of direct use in the making of this volume. I hope that this bibliography will be of value to other researchers. I should mention that several German books and articles that appear here date to the Nazi era and clearly represent the racist attitudes of their authors. They are by no means authoritative scholarly works: I consulted them only to gain insight into the research of the Ahnenerbe.

Åberg, Nils. "Herman Wirth: En germansk kulturprofet." *Fornvännen* (1933): 247–249.

Ackermann, Josef. *Heinrich Himmler als Ideologe*. Göttingen, Zürich, Frankfurt: Musterschmidt, 1970.

Adam, Peter. "Schenkel der Göttlichen." *Der Spiegel* 34, No. 40. (September 29, 1980).

Das Ahnenerbe. Offenbach am Main: Gebrüder Klingspor, n.d.

Altheim, Franz. *Die Soldatenkaiser*. Frankfurt: Ahnenerbe, Vitorio Klosterman, 1939.

———. "Forschungsbericht zur römischen Geschichte." *Die Welt als Geschichte* 2 (1936): 68–94.

Altheim, Franz, and Erika Trautmann. "Nordische und italische Felsbildkunst." *Die Welt als Geschichte* 3 (1937): 1–82.

Aly, Götz, Peter Chroust, and Christian Pross, eds. *Cleansing the Fatherland: Nazi Medicine and Racial Hygiene*. Translated by B. Cooper. Baltimore: Johns Hopkins University Press, 1994.

Andersch, Alfred. *The Father of a Murderer*. Translated by Leila Vennewitz. New York: New Directions, 1994.

Angrick, Andrej. *Besatzungspolitik und Massenmord. Die Einsatzgruppe D in der südlichen Sowjetunion 1941–1943*. Hamburg: Hamburger Edition, 2003.

"Der Anti Disney," *Der Spiegel*, no. 13 (1959).

Angress, Werner T., and Bradley F. Smith, "Diaries of Heinrich Himmler's Early Years." *Journal of Modern History* 31, no. 3 (September 1959): 206–224.

Arendt, Hannah. *Eichmann in Jerusalem: A Report on the Banality of Evil*. New York: Viking, 1963.

Arnold, Bettina. "The Past as Propaganda: Totalitarian Archaeology in Nazi Germany." *Antiquity* 64 (1990): 464–478.

Aronson, Schlomo. *Reinhard Heydrich und die Frühgeschichte von Gestapo und SD*. Stuttgart: Deutsche Verlags-Anstalt, 1971.

L'Art des Cavernes: Atlas des Grottes Ornées Paléolithiques Françaises. Ministère de la Culture: Imprimerie Nationale, Paris, 1984.

Ascherson, Neil. *Black Sea*. London: Jonathan Cape, 1995.

Asplund, Anneli, and Ulla Lipponen. *The Birth of the Kalevala*. Translated by Susan Sinisalo. Helsinki: Finnish Literature Society, 1985.

Baumann, Eberhard, ed. *Verzeichnis der Schriften, Manuskripte und Vorträge von Herman Felix Wirth Roeper Bosch von 1908–1995*. Toppenstadt: Uwe Berg Verlag, 1995.

Baumgartner, Andreas. *Die vergessenen Frauen von Mauthausen. Die weiblichen*

Häftlinge des Konzentrationslager Mauthausen und ihre Geschichte. Vienna: Verlag Österreich, 1997.

Beevor, Anthony. *The Fall of Berlin 1945.* New York: Viking, 2002.

Beger, Bruno. *Mit der deutschen Tibetexpedition Ernst Schäfer, 1938/39 nach Lhasa.* Wiesbaden: Verlag Dieter Schwarz, 1998.

Berlin and Potsdam. Grieben's Guidebooks, vol. 108. Berlin: A. Goldschmidt, 1931.

Bernadac, Christian. *Le Mystère Otto Rahn. Le Graal et Montségur: du Catharisme au Nazisme.* Paris : Éditions France-Empire, 1978.

Beyerchen, Alan D. *Scientists under Hitler: Politics and the Physics Community in the Third Reich.* New Haven: Yale University Press, 1982.

Bezwinska, Jadwiga, ed. *KL Auschwitz Seen by the SS: Höss, Broad, Kremer.* Translated by Constantine Fitzgibbon and Krystyna Michalik. New York: Howard Fertig, 1984.

Bishop, Chris, and Chris McNab. *Campaigns of World War II Day by Day.* Hauppauge, NY: Barron's, 2003.

Blindheim, Charlotte. "De fem lange år på Universitetets Oldsaksamling." *Viking* 48 (1984). 27–43.

Boas, Franz. *Race, Language and Culture.* New York: Macmillan, 1940.

Bogucki, Peter. "Konrad Jazdzewski (1908–1985) European Prehistorian." *The Polish Review* 31, no. 1. (1986): 73–77.

Bohmers, Assien. "Die Mauerner Höhlen und ihre Bedeutung für die Einteilung der Altsteinzeit." In *Ahnenerbe Jahrestagungen. Bericht über die Kieler Tagung 1939.* Edited by Herbert Jankuhn. Neumünster: Karl Wachholtz, 1944.

Bollmus, Reinhard. *Das Amt Rosenberg und seine Gegner.* Stuttgart: Deutsche Verlags-Anstalt, 1970.

Bolt, Christine. *Victorian Attitudes to Race.* London: Routledge & Kegan Paul, 1971.

Bose, Fritz. "Folk Music Research and the Cultivation of Folk Music." *Journal of the International Folk Music Council* 9 (1957): 20–21.

———. "Law and Freedom in the Interpretation of European Folk Epics." *Journal of the International Folk Music Council* 10 (1958): 27–34.

———. "Typen der Volksmusik in Karelien: Ein Reisebericht." *Archiv für Musikforschung* 3, no. 1 (1938): 96–118.

Bowen, Robert. *Universal Ice. Science and Ideology in the Nazi State.* London: Belhaven Press, 1993.

Bower, Tom. *Blind Eye to Murder: Britain, America, and the Purging of Nazi Germany—a Pledge Betrayed.* London: Little, Brown, 1995.

Bramwell, Anna. *Blood and Soil: Richard Walther Darré and Hitler's Green Party.* Bourne End, Buckinghamshire : The Kensal Press, 1985.

Breitmann, Richard. *The Architect of Genocide: Himmler and the Final Solution.* New York : Alfred A. Knopf, 1991.

Brentjes, Burchard, ed. *Wissenschaft unter dem NS-Regime.* Schöneiche bei Berlin: Lang, 1992.

Breton, Adela. "Proceeding of Americanists' Congress." *Man* (1910).

Breuil, Abbé H. *Four Hundred Centuries of Cave Art.* Translated by Mary E. Boyle. Centre d'Études et Documentation Préhistoriques: Montignac, Dordogne, 1952.

Brissaud, André. *The Nazi Secret Service.* Translated by Milton Waldman. London: Bodley Head, 1974.

Brodrick, Alan Houghton. *Father of Prehistory. The Abbé Henri Breuil: His Life and Times.* New York: Morrow, 1963.

Brøgger, Anton W. "Arkeologien og samfundenes åndelige balanse." *Morgenbladet* 29 (July 1936).

———. Editorial comment. *Viking.* (November 1937): n.p.

Burkhard, Arthur. *The Cracow Altar of Veit Stoss.* Munich: Bruckmann, 1972.

Burleigh, Michael. *Germany Turns Eastwards: A Study of Ostforschung in the Third Reich.* Cambridge: Cambridge University, 1988.

Burleigh, Michael, and Wolfgang Wippermann. *The Racial State: Germany 1933–1945.* Cambridge and New York: Cambridge University Press, 1991.

Cameron, George C. "Darius Carved History on Ageless Rock." *National Geographic* (December 1950): 825–844.

Chamberlain, Houston Stewart. *Foundations of the Nineteenth Century.* New York: John Lane, 1912.

Chandler, Douglas. "Belgium—Europe in Miniature." *National Geographic* (April 1938): 447–449.

———. "Changing Berlin." *National Geographic* (February 1937): 131–177.

Chapman, F. Spencer. *Lhasa: the Holy City.* London: Readers Union Ltd. by arrangement with Chatto & Windus, 1940.

Childe, V.G. *The Aryans.* New York: Knopf, 1926.

"Churchill Sure Himmler Faces a Glowing Future." *The New York Times,* May 17, 1945.

Clapperton, Chalmers M. *Quaternary Geology and Geomorphology of South America.* Amsterdam: Elsevier, 1993.

Clark, Alan. *Barbarossa: The Russian-German Conflict 1941–1945.* London: Weidenfeld & Nicolson, 1995.

Clark, C.M. "West Germany Confronts the Nazi Past: Some Recent Debates on the Early Postwar Era, 1945–1960." *The European Legacy* 4, no. 1 (1999): 113–130.

Cockburn, Aiden, et. al., eds. *Mummies, Disease and Ancient Cultures.* 2nd ed. Cambridge: Cambridge University Press, 1998.

Coles, John, and Lasse Bengtsson. *Images of the Past: A Guide to the Rock Carvings and Other Ancient Monuments of Northern Bohuslän.* Uddevalla: Bohusläns Museum och Bohusläns hembygdsförbund, 1990.

Conot, Robert E. *Justice at Nuremberg. The First Comprehensive Dramatic Account of the Trial of the Nazi Leaders.* New York: Harper & Row, 1983.

Cook, Stephen, and Stuart Flowers. *Heinrich Himmler's Camelot: Wewelsburg—*

Ideological Center of the SS, 1934–1945. Andrews, N.C.: Kressmann-Backmeyer, 1999.

Corten, Adrianus. "A Proposed Mechanism for the Bohuslän Herring Period." *ICES Journal of Marine Science* 56 (1999): 207–220.

Crawford, John Martin, trans. *Kalevala, the epic poem of Finland, into English*. New York: J.B. Alden, 1888.

Czech, Danuta, ed. *Auschwitz Chronicle: 1939–1945*. New York: Henry Holt and Company, 1990.

Darré, Richard Walther. *Das Bauerntum als Lebensquell der Nordischen Rasse*. München: J.F. Lehmann, 1933.

De Bont, Raf. "The Creation of Prehistoric Man: Aimé Rutot and the Eolith Controversy, 1900–1920." *Isis* 94 (2003): 604–630.

De Luca, Anthony R. "'Der Grossmufti' in Berlin: The Politics of Collaboration." *International Journal of Middle East Studies* 10, no. 1 (1979): 125–138.

Dobrynina, Natalia. "Report summarizing Russian Scholarly and Scientific Literature on the History of Four Crimean Cave Cities." Unpublished report.

Dow, James R., and Hannjost Lixfeld, eds. *The Nazification of an Academic Discipline. Folklore in the Third Reich*. Bloomington: Indiana University, 1994.

Dower, John W. *War without Mercy: Race & Power in the Pacific War*. New York: Pantheon Books, 1986.

Eichholtz, Dietrich, and Kurt Pätzold, eds. *Der Weg in den Krieg. Studien zur Geschichte der Vorkriegsjahre 1935/36 bis 1939*. Cologne: Pahl-Rugenstein, 1989.

Eichler, Inge, ed. *Wilhelm Altheim: Bilder aus den Volksleben. Eine Austellung der Frankfurter Sparkasse (Polytechnische Gesellschaft) 1.10-9.11.1979*. Frankfurt: Frankfurter Sparkasse von 1822 (Polytechnische Gesellschaft), 1979.

Elon, Amos. "Death for Sale: Masks, an Attempt about Shoah." *New York Review of Books*, November 20, 1997.

Eppstein, F.T. "War-time Activities of the SS-Ahnenerbe." In *On the Track of Tyranny, Essays Presented by the Wiener Library to Leonard G. Montefiore, O.B.E., on the Occasion of his Seventieth Birthday*. Edited by Max Beloff. London: Valentine, Mitchel, 1960.

Evan, Gershon. *Winds of Life: the Destinies of a Young Viennese Jew 1938–1958*. Riverside, California: Ariadne Press, 2000.

Eyre, Lincoln. "Renascent Germany," *National Geographic* (December 1928): 639–717.

Feliciano, Hector. *The Lost Museum: The Nazi Conspiracy to Steal the World's Greatest Works of Art*. New York: Basic Books, 1997.

Field, Geoffrey G. *Evangelist of Race: The Germanic vision of Houston Stewart Chamberlain*. New York: Columbia University Press, 1981

———. "Nordic Racism." *Journal of the History of Ideas* 38 (1977): 523–540.

Fischer, Eugen. "Sind die alten Kanarier ausgestorben? Eine anthropologische Untersuchung auf den Kanarischen Inseln." *Zeitschrift für Ethnologie* 62 (1930): 258–273.

Fischer, Hans. *Völkerkunde im Nationalsozialismus. Aspekte der Anpassung, Affinität und Behauptung einer wissenschaftlichen Disziplin.* Berlin: D. Reimer, 1990.

Fitzgibbon, Constantine. *Denazification.* London: Joseph, 1969.

Foreign Relations of the United States Diplomatic Papers, The Conference of Berlin (The Potsdam Conference). Vol. II. Washington: United States Government Printing Office, 1960.

Foster, Nigel G. *German Law & Legal Stystem/Deutsches Recht und Deutsches Rechtssystem.* London: Blackstone Press. 1993.

Frauenfeld, Alfred E. *Die Krim. Ein Handbuch.* Aufbaustab für den Generalbezirk Krim, n.d.

Frei, Norbert, ed. *Standort–und Kommandanturbefehle des Konzentrationslagers Auschwitz 1940–1945.* München: K.G. Saur, 2000.

Frischauer, Willi. *Himmler, the Evil Genius of the Third Reich.* London: Odhams Press, 1953.

Fritz, Georg, and Walter Puttkammer. *Berlin, die alte und die Neue Stadt. 80 Zeichnungen von George Fritz.* Berlin: Klinkhardt & Biermann, 1936.

Das Frobenius-Institut an der Johann Wolfgang Goethe-Universität 1898–1998. Frankfurt a. M.: Frobenius Institut, 1998.

Fromm, Bella. *Blood and Banquets: A Berlin Social Diary.* New York: Harper & Brothers, 1942.

Funder, Anna. *Stasiland.* London: Granta Books, 2003.

Funke, Manfred, ed. *Hitler, Deutschland und die Mächte. Materialien zur Aussenpolitik des Dritten Reiches.* Düsseldorf: Droste, 1978.

Fürstenau, Justus. *Entnazifizierung. Ein Kapitel deutscher Nachkriegspolitik.* Neuwied/Berlin: Luchterhand, 1969.

Gamkrelidze, Thomas, and V. V. Ivanov. "The Early History of Indo-European Languages." *Scientific American* 262, no. 3 (March 1990): 110–116.

Gehl, Walther. *Geschichte. 6 Klasse. Gymnasien und Oberschulen in Aufbauform.* Breslau: Ferdinand Hirt, 1940.

Gellhorn, Martha. *The Face of War.* New York: Simon and Schuster, 1959.

Gertsen, Alexandr G., and Yu. M. Mogarytchev. *Krepost'dragotzennostey.* Simferopol: Tavria, 1993.

———. *Peshchernye tzerkvi Mangupa.* Simferopol: Tavria, 1996.

Gilman, Sander. *The Jew's Body.* New York and London: Routledge, 1991.

Godwin, Joscelyn. *Arktos: The Polar Myth in Science, Symbolism, and Nazi Survival.* Grand Rapids, MI: Phanes Press, 1993.

Goldhagen, Daniel J. *Hitler's Willing Executioners: Ordinary Germans and the Holocaust.* New York: Alfred A. Knopf, 1996.

Goodrick-Clarke, Nicholas. *The Occult Roots of Nazism. Secret Aryan Cults and Their Influence on Nazi Ideology: The Ariosophists of Austria and Germany 1890–1935.* New York: New York University Press, 1992.

Gould, Stephen Jay. *The Mismeasure of Man.* New York: W.W. Norton, 1981.

Grant, Madison. *The Passing of the Great Race.* New York: Scribners, 1921.

"A Grave on the Heath." *Time*, June 4, 1954.

Gravlee, Clarence C., et al. "Heredity, Environment and Cranial Form: A Reanalysis of Boas's Immigrant Data." *American Anthropologist* 105, no. 1 (March 2003): 125–138.

Green, Warren. "The fate of the Crimean Jewish Communities: Ashkenazim, Krimchaks, and Karaites." *Jewish Social Studies* 46, no. 2 (spring 1984): 169–176.

Gregor, A. James. "Nordicism Revisited." *Phylon* 22, no. 4 (1961): 351–360.

Greif, Thomas. *Der SS-Standort Waischenfeld 1934–1945. Hilfswerklager und Ahnenerbe.* Erlangen: Palm & Enke, 2000.

Grimsted, Patricia K. "The Fate of Ukrainian Cultural Treasures During World War II. The Plunder of Archives, Libraries and Museums under the Third Reich." *Jahrbücher für Geschichte Osteuropas* 39, no. 1 (1991): 53–80.

Grobba, Fritz. *Männer und Mächte im Orient. 25 Jahre diplomatischer Tätigkeit im Orient.* Zürich, Berlin, Frankfurt a. M.: Masterschmidt, 1967.

Grönhagen, Georg von. "Karelische Zauberbeschwörungen." *Germanien* (1937): 54–57.

Grönhagen, Yrjö von. *Finnische Gespräche.* Berlin: Nordland, 1941.

———. *Himmlerin salaseura.* Helsinki: Kansankirja, 1948.

———. *Karelien, Finnlands Bollwerk gegen den Osten.* Dresden: F. Müller, 1942.

———. "Über der Totenglauben der Finnen." *Archiv für Religionswissenschaft* 37 (1941/1942): 182–195.

Grönhagen, Yrjö, and Herta von. *Das Antlitz Finnlands.* Berlin: Wiking, 1942.

Gruber, Ulrich. "Die Tibetexpeditionen von Dr. Ernst Schäfer." *Tibetforum* 5, no. 1. (1986): 9–12.

Grunberger, Richard. *A Social History of the Third Reich.* London: Weidenfeld & Nicolson, 1971.

Günther, Hans F.K.. *Die Nordische Rasse bei den Indogermanen Asiens.* Munich: J.F. Lehmanns, 1934.

———. *Kleine Rassenkunde Europas.* Munich: J.F. Lehmanns, 1925.

———. *The Racial Elements of European History.* Translated by G.C. Wheeler. Port Washington: Kennikat Press, 1970.

———. *Rassenkunde des deutschen Volkes.* Munich: J. F. Lehmanns, 1924.

Gutman, Israel, ed. *Enzyklopädie des Holocaust. Die Verfolgung und Ermordung der europäischen Juden.* 2nd ed. München: Piper, 1998.

Gutman, Yisrael, and Michael Berenbaum. eds. *Anatomy of the Auschwitz Death Camp.* Published in assocation with the United States Holocaust Memorial Museum. Bloomington: Indiana University Press, 1994.

Guy, P.L.O. "Balloon Photogaphy and Archaeological Excavation." *Antiquity* (June 1932): 148–155.

Habersetzer, Walther. *Ein Münchner Gymnasium in der NS-Zeit. Die verdrängten Jahre des Wittelsbacher Gymnasiums.* München: Verlag Geschichtswerkstatt Neuhausen, 1997.

Hachmeister, Lutz. *Der Gegnerforscher. Die Karriere des SS-Führers Franz Alfred Six.* München: Beck, 1998.

Haffner, Sebastian. *Defying Hitler: A Memoir.* Translated by Oliver Pretzel. New York: Farrar, Straus and Giroux, 2002.

Hagen, Anders. "Arkeologi og politikk." *Viking* 49 (1985/1986): 269–278.

Halbey, Hans Adolf. *Karl Klingspor, Leben und Werk. Ein Beitrag zur Geschichte der deutschen Schriftkunst und des Schriftgiessereiwessens im 20. Jahrhundert.* Offenbach am Main: Vereinigung, "Freunde des Klingspor-Museums" e.V., 1991.

Hale, Christopher. *Himmler's Crusade: the Nazi Expedition to Find the Origins of the Aryan Race.* Hoboken, N.J.: John Wiley & Sons, 2003.

Hammerstein, Notker. *Die Deutsche Forschungsgemeinschaft in der Weimarer Republik und im Dritten Reich. Wissenschaftspolitik in Republik und Diktatur 1920–1945.* München: Beck, 1999.

Hanfstängl, Ernst. *Hitler: the Missing Years.* London: Eyre & Spottiswoode, 1957.

Härke, Heinrich. "Archaeologists and Migrations: A Problem of Attitude?" *Current Anthropology* 39, no. 1 (February 1998): 19–45.

———, ed. *Archaeology, Ideology and Society: The German Experience.* Frankfurt am Main: Peter Lang, 2000.

Hassmann, Henning, and Detlef Jantzen. "'Die Deutsche Vorgeschichte—eine nationale Wissenschaft,' Das Kieler Museum vorgeschichtlicher Altertümer im Dritten Reich." *Offa* 51 (1994): 9–23.

"Haus Atlantis—Die Sammlung Väterkunde." In *Die Böttcherstrasse in Bremen. Eine historische Rundfunkaufnahme von 1932.* Deutsches Rundfunkarchiv/Radio Bremen, 2001.

Heather, Peter J. *Goths and Romans.* Oxford: Oxford University Press, 1991.

Hedin, Sven. *Ohne Auftrag in Berlin.* Kiel: Arndt, 1991.

Heiber, Helmut. *Universität unterm Hakenkreuz.* Munich: K.G. Saur, 1994.

———, ed. *Reichsführer! . . . Briefe an und von Himmler.* Stuttgart: Deutsche Verlags-Anstalt, 1968.

Heim, Susanne. "Research for Autarky: The Contribution of Scientists to Nazi Rule in Germany." In *Geschichte der Kaiser-Wilhelm-Gesellschaft im Nationalsozialismus.* Berlin: Max-Planck-Gesellschaft zur Förderung der Wissenschaften, 2001.

Heinemann, Isabel. *"Rasse, Siedlung, deutsches Blut." Das Rasse- und Siedlungshauptamt der SS und die rassenpolitische Neuordnung Europas.* Göttingen: Wallstein, 2003.

"Herr Hitler's Birthday." *Times of London,* April 20, 1939.

Herwig, Wolfram. *History of the Goths.* Berkeley: University of California Press, 1990.

Heske, Immo. "Von Haithabu nach Kiew und in den Kaukasus—Aspekte des NS-Kunstraubes durch Ur- und Frühgeschichtler." *Nachrichten und Informationen zur Kultur* 1 (1999): 2–6.

Heuß, Anja. *Kunst und Kunstraub: Eine Vergleichende Studie zur Beratungspolitik der Nationalsozialisten in Frankreich und der Sowjetunion.* Heidelberg: Universitätsverlag C. Winter, 2000.

Hielscher, Friedrich. *Fünfzig Jahre unter Deutschen.* Hamburg: Rowohlt, 1954.

"Himmler, Caught, Ends Life by Poison: Arch Criminal Dies." *The New York Times,* May 25, 1945.

Himmler, Heinrich. *Heinrich Himmler Geheimreden 1933–1945 und andere Ansprachen.* Edited by Bradley F. Smith and Agnes F. Peterson. Frankfurt am Main: Propyläen Verlag, 1974.

Hirszowicz, Lukasz. *The Third Reich and the Arab East.* London: Routledge & Kegan Paul, 1966.

Hitler, Adolf. *Hitler's Table Talk 1941–1944: His Private Conversations.* Edited by H.R. Trevor-Roper. London: Phoenix, 2000.

———. *Mein Kampf.* Translated by Ralph Manheim. Boston: Houghton Mifflin, 1971.

Höhne, Heinz. *The Order of the Death's Head; the Story of Hitler's S.S.* Translated by Richard Barry. New York: Coward-McCann, 1970.

Höhne, Rolf. "Die Ausgrabungen der Schutzstaffel." *Germanien* (1938): 224–230.

Holmes, Henrietta Allen. "The Spell of Romania." *National Geographic* (April 1934): 399–450.

Homosexuals: Victims of the Nazi Era 1933–1945. Washington, D.C.: United States Holocaust Memorial Museum, n.d.

Honko, Lauri. ed. *Religion, Myth and Folklore in the World's Epics: the Kalevala and its predecessors.* Berlin, New York: Mouton de Gruyter, 1990.

Hooton, Earnest. *The Ancient Inhabitants of the Canary Islands.* Cambridge, Mass: Peabody Museum of Harvard University, 1925.

Houghton, J.T., et al., eds. *Climate Change 2001: the Scientific Basis.* Cambridge: Cambridge University Press, 2001.

Hufen, Christian. "Beutekunst oder Raubzüge deutscher Archäologen." *Neue Bildende Kunst* 6 (1997): 60–63.

———. "Gotenforschung und Denkmalpflege. Herbert Jankuhn und die Kommandounternehmen des 'Ahnenerbe' der SS." In *'Betr.: Sicherstellung.' NS-Kunstraub in der Sowjetunion,* edited by Wolfgang Eichwede and Ulrike Hartung. Bremen: Edition Temmen, 1998.

Hunger, Ulrich. *Die Runenkunde im Dritten Reich.* Frankfurt am Main: Lang, 1984.

Hüser, Karl. *Die Wewelsburg 1933 bis 1945. Kult- und Terrorstätte der SS.* Paderborn: Bonifatius, 1987.

Huth, Otto. "Die Gesittung der Kanarier als Schlüssel zum Ur-Indogermanentum." *Germanien* (1937): 50–54.

Huxley, Julian S., and A.C. Haddon. *We Europeans: A Survey of 'Racial' Problems.* London: Jonathan Cape, 1935.

Hygen, Anne-Sophie, and Lasse Bengtsson. *Rock Carvings in the Borderlands: Bohuslän and Östland.* Gothenburg: Warne Förlag, 2000.

Jacobs, Jörn. "Peter Paulsen: Ein Wanderer zwischen zwei Welten." In *Prähistorie und Nationalsozialismus,* edited by Achim Leube and Morten Hegewisch. Heidelberg: Synchron, 2002.

Jankuhn, Herbert. *Die Ausgrabungen in Haithabu, 1937–39—ein vorläufiger Ausgrabungsbericht.* Berlin: Ahnenerbe-Stiftung, 1943.

Jazdzewski, Konrad. *Poland.* London: Thames and Hudson, 1965.

Johanson, Donald, and Edgar Blake. *From Lucy to Language.* New York: Simon & Schuster Editions, 1996.

"J.R.R. Tolkien Dead at 81; Wrote 'The Lord of the Rings'." *The New York Times,* September 23, 1973.

Junker, Klaus. *Das Archäologische Institut des Deutschen Reichs zwischen Forschung und Politik: die Jahre 1929–1945.* Mainz: Zabern, 1997.

Kaase, Max. "Demokratische Einstellungen in der Bundesrepublik Deutschland." In *Sozialwissenschaftliches Jahrbuch für Politik.* Vol. 2. Ed. Rudolf Wildenmann. Munich: Olzog, 1971.

Kaienburg, Hermann. *Die Wirtschaft der SS.* Berlin: Metropol, 2003.

Karlsch, Rainer. *Hitlers Bombe.* Munich: Deutsche Verlags-Anstalt, 2005.

Kaschuba, Wolfgang. "Am Ort der Geschichte." In *Prähistorie und Nationalsozialismus,* edited by Achim Leube and Morten Hegewisch. Heidelberg: Synchron, 2002.

Kasten, Frederick H. "Unethical Nazi Medicine in Annexed Alsace-Lorraine: the Strange Case of Nazi Anatomist Professor Dr. August Hirt." In *Historians and Archivists: Essays in Modern German History and Archival Policy,* edited by George O. Kent. Fairfax, Virginia: George Mason University Press, 1991.

Kater, Michael H. *Das "Ahnenerbe" der SS 1935–1945. Ein Beitrag zur Kulturpolitik des Dritten Reiches.* 1st ed. Stuttgart: Deutsche Verlags-Anstalt, 1974.

———. *Das "Ahnenerbe." Die Forschungs- und Lehrgemeinschaft der SS. Organisationsgeschichte von 1935 bis 1945.* Inaugural Dissertation, Heidelberg University, 1966.

———. "Die Artamanen—Völkische Jugend in der Weimarer Republik." *Historische Zeitschrift* 213 (1971): 576–638.

Kaul, Friedrich Karl. "Das 'SS-Ahnenerbe' und die 'jüdische Schädelsammlung' an der ehemaligen Reichsuniversität Straßburg." *Zeitschrift für Geschichtswissenschaft* 16, no. 11 (1968): 1460–1474.

Kemiläinen, Aira. *Finns in the shadow of the "Aryans." Race theories and Racism.* Helsinki: HSH, 1998.

Kerschbaumer, Gert. "Das Deutsche Haus der Natur zu Salzburg." In *Politik der Präsentation*, edited by Herbert Posch and Gottfried Fliedl. Vienna: Turia und Kant, 1996.

Kershaw, Ian. *Hitler: 1889–1936: Hubris*. London: Allen Lane, The Penguin Press, 1999.

———. *Hitler: 1936–1945: Nemesis*. London: Allen Lane, The Penguin Press, 2000.

Kersten, Felix. *The Kersten Memoirs, 1940–1945*. New York: Macmillan, 1957.

———. *The Memoirs of Doctor Felix Kersten*. Edited by Herma Briffault. Garden City, N.Y.: Doubleday, 1947.

Kinder, Elisabeth. "Der Persönliche Stab Reichsführer SS." In *Aus der Arbeit des Bundesarchivs. Beiträge zum Archivwesen, zur Quellenkunde und Zeitgeschichte*, edited by Heinz Boberach and Hans Booms. Boppard am Rhein: Boldt, 1977.

Kirk, George E. *The Middle East in the War*. London: Oxford University Press, 1952.

Kiss, Edmund. "Die Kordillerenkolonien der Atlantiden." *Schlüssel zum Weltgeschehen* 8/9 (1931): 256–265.

———. "Nordische Baukunst in Bolivien?" *Germanien* (1933): 138–144.

———. *Das Sonnentor von Tihuanaku und Hörbigers Welteislehre*. Leipzig: Koehler & Amelang, 1937.

Klee, Ernst. *Auschwitz, die NS-Medizin und ihre Opfer*. Frankfurt am Main: S. Fischer, 1997.

———. *Deutsche Medizin im Dritten Reich. Karrieren vor und nach 1945*. Frankfurt am Main: Fischer, 2001.

Klein, Peter, and Andrej Angrick, eds. *Die Einsatzgruppen in der besetzten Sowjetunion, 1941/42. Die Tätigkeits- und Lageberichte des Chefs der Sicherheitspolizei und des SD*. Berlin: Edition Hentrich, 1997.

Klingspor Karl. *Über Schönheit von Schrift und Druck. Erfahrungen aus fünfzigjähriger Arbeit*. Frankfurt am Main: Georg Kurt Schauer, 1949.

Knopp, Guido. *Hitler's Henchmen*. Translated by Angus McGeogh. Phoenix Mill: Sutton, 2000.

Kogon, Eugen. *The Theory and Practice of Hell: The German Concentration Camps and the System Behind Them*. New York: Berkley Books, 1998.

Kohl, Philip L. and Clare P. Fawcett., eds. *Nationalism, Politics and the Practice of Archaeology*. Cambridge: Cambridge University Press, 1995.

Köhler, Joachim. *Wagner's Hitler: The Prophet and His Disciple*. Translated by Ronald Taylor. Cambridge, U.K.: Polity Press, 2000.

Köhler, M. "Neues vom Nichtwissen. Wissenschaft im Dienst der Nazis: Professor Ernst Schäfer." *Die Zeit*, 20.01.1989.

Königseder, Angelika. "Entnazifizierung". In *Deutschland unter alliierter Besatzung 1945–1949/55. Ein Handbuch*. Edited by Wolfgang Benz. Berlin: Akademie-Verlag, 1999.

Kostrzewski, Jozef. "Osada bagienna w Biskupinie, w pow. zininskim." *Przeglad Archeologiczny* 5 (1938): 121–140.

Kral, Václav. ed. *Die Deutschen in der Tschechoslowakei 1933–1947. Dokumentensammlung.* Prague: Nakladatelstvi Ceskioslovenske Akademie Ved, 1964.

Kroll, Frank-Lothar. *Utopie als Ideologie. Geschichtsdenken und politisches Handeln im Dritten Reich. Hitler-Rosenberg-Darré-Himmler-Goebbels.* Paderborn u.a.: Schöningh, 1998.

Kuhn, Ferdinand, and Delia Kuhn. "Poland Opens Her Doors." *National Geographic* (September 1958): 357–398.

Lachmann, Ernest. "Anatomist of Infamy: August Hirt." *Bulletin of Historical Medicine* 51 (1977): 594–602.

Lang, Hans-Joachim. *Die Namen der Nummern: Wie es gelang die 86 Opfer eines NS-Verbrechens zu identifizieren.* Hamburg: Hoffmann & Campe, 2004.

Lang, Jochen von. *Der Adjutant. Karl Wolff, der Mann zwischen Hitler und Himmler.* Munich: Herbig Verlag, 1985.

Lange, Hans-Jürgen. *Weisthor, Karl-Maria Wiligut. Himmlers Rasputin und seine Erben.* Engarda: Arun Verlag, 1998.

Langsdorff, Alexander, and Hans Schleif. "Die Ausgrabungen der Schutzstaffeln." *Germanien* (January 1938): 6–11.

Large, David Clay. *Berlin.* New York: Basic Books, 2000.

———. *Where Ghosts Walked: Munich's Road to the Third Reich.* New York: W.W. Norton, 1997.

Latour, C.F. "Germany, Italy and South Tyrol, 1938–1945." *The Historical Journal* 8, no. 1 (1965): 95–111.

Lebert, Stephan, and Norbert Lebert. *My Father's Keeper: Children of Nazi Leaders—An Intimate History of Damage and Denial.* Boston: Little, Brown, 2001.

Lehmann-Haupt, Hellmut. *Cultural Looting of the Ahnenerbe: Report Prepared by Monuments, Fine Arts and Archives Section.* OMGUS, March 1, 1948.

Lehrer, Steven. *Hitler Sites: A City-by-City Guidebook (Austria, Germany, France, United States).* Jefferson, North Carolina: McFarland & Company, 2002.

Lengyel, Olga. *Five Chimneys: The Story of Auschwitz.* London: Mayflower, Granada Publishers, 1972.

Lerchenmüller, Joachim. *Die Geschichtswissenschaft in den Planungen des Sicherheitsdienstes der SS.* Bonn: Dietz, 2001.

Lerchenmüller, Joachim, and Gerd Simon. *Im Vorfeld des Massenmords. Germanistik und Nachbarfächer im 2. Weltkrieg.* Tübingen: Verlag der Gesellschaft für Interdisziplinäre Forschung, 1997.

Lester, Robert E. *Art Looting and Nazi Germany: Records of the Fine Arts and Monuments Advisor, Ardelia Hall, 1945–1961.* Part 1, Country Files for Austria, Italy and Germany. Bethesda, Maryland: LexisNexis, 2002.

Leube, Achim, and Morten Hegewisch, eds. *Prähistorie und Nationalsozialismus. Die mittel- und osteuropäische Ur-und Frühgeschichtsforschung in den Jahren 1933–1945.* Heidelberg: Synchron, 2002.

Levenda, Peter. *Unholy Alliance: A History of Nazi Involvement with the Occult*. New York: Arin Books, 1995.

Lifton, Robert Jay. *Nazi Doctors: Medical Killing and the Psychology of Genocide*. London: Macmillan, 1986.

Lixfeld, Gisela. "Das Ahnenerbe Heinrich Himmlers und die ideologisch-politische Funktion seiner Volkskunde." In *Völkische Wissenschaft*, edited by Wolfgang Jacobeit, Hannjost Lixfeld, and Olaf Bockhorn. Vienna: Böhlau, 1994.

Loewenberg, Peter. "The Unsuccessful Adolescence of Heinrich Himmler." *The American Historical Review* 76, no. 3 (June 1971): 612–641.

Losemann, Volker. " 'Die Krise der Alten Welt,' und der Gegenwart. Franz Altheim und Karl Kerényi im Dialog." In *Imperium Romanum. Festschrift für Karl Christ zum 75.Geburtstag*, edited by Peter Kneissl and Volker Losemann. Stuttgart: Franz Steiner, 1998.

Losemann, Volker. *Nationalsozialismus und Antike. Studien zur Entwicklung des Faches Alte Geschichte 1933–1945*. Hamburg: Hoffmann und Campe, 1977.

Lovin, Clifford R. "Blut und Boden: The Ideological Basis of the Nazi Agricultural Program." *Journal of the History of Ideas* 28, no. 2 (April–June 1967): 274–288.

Löw, Luitgard. *Herman Wirth und die Suche nach der germanischen Geistesurgeschichte in Skandinavien*. Unpublished paper.

Lower, Wendy. "A New Ordering of Space and Race: Nazi Colonial Dreams in Zhytomyr, Ukraine, 1941–1944." *German Studies Review* 25, no. 2 (May 2002): 227–254.

Lutzhöft, Hans Jürgen. *Der nordische Gedanke in Deutschland 1920–1940*. Stuttgart: E. Klett, 1971.

Lycett, Andrew. *Ian Fleming*. London: Weidenfeld & Nicholson, 1995.

MacDonogh, Giles. *The Last Kaiser: Wilhelm the Impetuous*. New York: Times Books, 1977.

MacLean, French L. *The Camp Men: The SS Officers Who Ran the Nazi Concentration Camp System*. Atglen, P.A.: Schiffer Military History, 1999.

Macrakis, Kristie. *Surviving the Swastika: Scientific Research in Nazi Germany*. New York: Oxford, 1993.

Maischberger, Martin. "German Archaeology during the Third Reich, 1933–45: A Case Study Based on Archival Evidence." *Antiquity* 76 (2002): 209–218.

Mallory, J.P. "Human Populations and the Indo-European Problem." *The Mankind Quarterly* 33, no.2 (winter 1992): 131–154.

———. *In Search of the Indo-Europeans: Language, Archaeology and Myth*. London: Thames and Hudson, 1989.

Mallory, J.P., and Victor Mair. *The Tarim Mummies: Ancient China and the Mystery of the Earliest Peoples from the West*. London: Thames and Hudson, 2000.

Mandl, Franz. "Das Erbe der Ahnen: Ernst Burgstaller/Herman Wirth und die österreichische Felsbildforschung." *Archäologie und Felsbildforschung. Mitteilungen der ANISA* 19/20, no. 1/2 (1999).

Martens, Jes. "Die Nordische Archäologie und das 'Dritte Reich.'" In *Prähistorie und Nationalsozialismus*, edited by Achim Leube and Morten Hegewisch. Heidelberg: Synchron, 2002, pp. 603–618.

McCann, W.J. "The National Socialist Perversion of Archaeology." *World Archaeology Bulletin* 2 (1988): 51–54.

———. "Volk und Germanentum: The Presentation of the Past in Nazi Germany." In *The Politics of the Past*, edited by Peter Gathercole and David Lowenthal. London: Unwin Hyman, 1994.

Merkel, Eberhard. "Bibliographie Franz Altheim." In *Beiträge zur Alten Geschichte und deren Nachleben*, edited by Ruth Stiehl and Hans Erich Stier. Berlin: Walter de Gruyter & Co. 1970.

Meyer, Harald. *Wewelsburg. SS-Burg, Konzentrationslager, Mahnmal, Prozeß*. Paderborn: Selbstverlag, 1982.

Meyer, Karl E., and Shareen Blair Brysac. *Tournament of Shadows: the Great Game and the Race for Empire in Central Asia*. Washington, D.C.: Counterpoint, 1999.

Mezynski, Andrzej. *Kommando Paulsen. Organisierter Kunstraub in Polen: 1942–45*. Cologne: Dittrich, 2000.

Michael, Robert, and Karin Doerr. *Nazi-Deutsch/Nazi German: An English Lexikon of the Language of the Third Reich*. Westport, Conn.: Greenwood, 2002.

Mitscherlich, Alexander, and Fred Mielke. *Medizin ohne Menschlichkeit. Dokumente des Nürnberger Ärzteprozesses*. Frankfurt am Main: Fischer, 1949.

Mode, Markus. "Altertumswissenschaft und Altertumswissenschaftler unter dem NS-Regime." In *Wissenschaft unter dem NS-Regime*, edited by Burchard Brentjes. Schöneiche bei Berlin: Lang, 1992.

Moltke, Erik. *Runes and Their Origin. Denmark and Elsewhere*. Copenhagen: National Museum of Denmark, 1985.

Montagu, Ashley. *Statement on Race: An Annotated Elaboration and Exposition of the Four Statements on Race Issued by the United Nations Educational, Scientific and Cultural Organization*. New York: Oxford University Press, 1972.

Moser, Doreen et al. "Marine Mammal Skeletal Preparation and Articulation." Poster Presentation at the Biennial Conference on the Biology of Marine Mammals. November 28–December 3, 2002. Vancouver, B.C.

Mosse, George L. *Nazi Culture: Intellectual, Cultural and Social Life in the Third Reich*. New York: Grosset & Dunlop, 1966.

"Mr. & Ms." *Time*, June 16, 1947.

Müller-Hill, Benno. *Murderous Science: Elimination by Scientific Selection of Jews, Gypsies, and Others. Germany, 1933–1945*. Oxford: Oxford University Press, 1988.

Müller-Karpe, Hermann. *Handbuch der Vorgeschichte. Erster Band: Altsteinzeit*. Munich: C.H. Beck'sche Verslagsbuchhandlung, 1966.

Nagel, Brigitte. *Die Welteislehre. Ihre Geschichte und ihre Rolle im "Dritten Reich."* Stuttgart: Verlag für Geschichte der Naturwissenschaften und der Technik, 1991.

Nagy-Talavera, Nicholas M. *The Green Shirts and the Others: A History of Fascism in Hungary and Rumania.* Stanford: Hoover Institution Press, Stanford University, 1970.

Natkiel, Richard. *Atlas of World War II.* North Dighton, MA: World Publications Group Ltd, 2001.

Nicholas, Lynn H. *The Rape of Europa. The Fate of Europe's Treasures in the Third Reich and the Second World War.* New York: Alfred A. Knopf, 1994.

Nicosia, Francis. "Arab Nationalism and National Socialist Germany, 1933–1939: Ideological and Strategic Incompatibility." *International Journal of Middle East Studies* 12 (3) 1980: 351–372

Nohlen, Klaus. *Baupolitik im Reichsland Elsaß-Lothringen 1871–1918. Die repräsentativen Staatsbauten um den ehemaligen Kaiserplatz in Strassburg.* Berlin: Mann, 1982.

Nußsbaumer, Thomas. *Alfred Quellmalz und seine Südtiroler Feldforschung (1940–42). Eine Studie zur musikalischen Volkskunde unter dem Nationalsozialismus.* Innsbruck: Libreria Musicale Italiana, 2001.

Nyiszli, Miklós. *Auschwitz: A Doctor's Eyewitness Account.* Translated by Tibere Kremer and Richard Seaver. Geneva : Ferni Pub. House: Distributed by Pleasant Valley Press, 1979.

Orlow, Dietrich. *The History of the Nazi Party: 1933–1945.* Pittsburgh: University of Pittsburgh Press, 1973.

Osterhuis, Harry. "Medicine, Male Bonding, and Homosexuality in Nazi Germany," *Journal of Contemporary History* 32, no. 2 (1997): 187–205.

Padfield, Peter. *Himmler: Reichsführer-SS.* London: Cassell & Co., 2001.

Pentikäinen, Juha. "The Ancient Religion of the Finns," *Finnish Features.* Helsinki: Ministry of Foreign Affairs, 1985.

The Petroleum Almanac: A Statistical Record of the Petroleum Industry in the United States and Foreign Countries. New York: The Conference Board, 1946.

Petropoulos, Jonathan. *Art as Politics in the Third Reich.* Chapel Hill: University of North Carolina Press, 1996.

Peyrony, Denis. *Les Eyzies: Ses Musées d'art préhistorique.* Paris: Henri Laurens, 1931.

Poliakov, Léon. *The Aryan Myth; A History of Racist and Nationalist Ideas in Europe.* Translated by Edmund Howard. London: Sussex University Press in association with Heinemann Educational Books, 1974.

Poller, Alphons, ed. *Das Pollersche Verfahren zum Abformen an Lebenden und Toten sowie an Gegenständen.* Berlin, Vienna: Urban & Schwarzenberg, 1931.

Posch, Herbert, and Gottfried Fliedl, eds. *Politik der Präsentation. Museum und Ausstellung in Österreich 1918–1945.* Vienna: Turia und Kant, 1996.

Posnansky, Arthur. *Tihuanacu: The Cradle of American Man.* New York: J.J. Augustin, 1945.

Potter, Pamela M. *Most German of the Arts: Musicology and Society from the Weimar Republic to the End of Hitler's Reich.* New Haven & London: Yale University Press, 1998.

Price, David. "Cloak & Trowel." *Archaeology* (September/October 2003): 30–35.

Purin, Bernhard. "Die museale Darstellung jüdischer Geschichte und Kultur in Österreich zwischen Aufklärung und Rassismus." In *Politik der Präsentation: Museum und Ausstellung in Österreich 1918–1945,* edited by Herbert Posch and Gottfried Fliedl. Vienna: Turia and Kant, 1996.

Radke, Gerhard. "Suche nach Antwort: zum Tode des Althistorikers Franz Altheim." *Der Tagespiegel* (October 20, 1976).

Rahn, Otto. *Luzifers Hofgesind. Eine Reise zu Europas guten Geistern.* Leipzig: Schwarzhäupter, 1937.

Reitlinger, Gerald. *The SS: Alibi of a Nation 1922–1945.* Melbourne: Heinemann, 1956.

Reybrouck, David van. "Boule's error: On the Social Context of Scientific Knowledge." *Antiquity* 76 (2002): 158–164.

Rhodes, Richard. *The Making of the Atomic Bomb.* New York: Touchstone, 1988.

———. *Masters of Death: The SS Einsatzgruppen and the Invention of the Holocaust.* New York: Alfred A. Knopf, 2002.

Rössler, Mechtild. *Wissenschaft und Lebensraum. Geographische Ostforschung im Nationalsozialismus.* Berlin: Reimer, 1990.

Rössler, Mechtild and Sabine Schleiermacher, eds. *Der "Generalplan Ost." Hauptlinien der nationalsozialistischen Planungs und Vernichtungspolitik.* Berlin: Akademie-Verlag, 1993.

———. "Himmlers Imperium auf dem 'Dach der Erde': Asien-Expeditionen im Nationalsozialismus." In *Medizingeschichte und Gesellschaftskritik. Festschrift für Gerhard Baader,* edited by Michael Hubenstorf. Husum: Matthiesen Verlag, 1997.

Rössler, Otto. "Kosmopolitische Gelehrsamkeit: zum Tod des Althistorikers Franz Altheim." *Neue Zürcher Zeitung,* October 31/November 1, 1976.

Rosvall, Toivo D. *Finland: Land of Heroes.* New York: E.P. Dutton, 1940.

Roth, John K. and Elizabeth Maxwel, eds. *Remembering for the Future: The Holocaust in an Age of Genocide.* New York: Palgrave, 2000.

Rückerl, Adalbert. *NS-Verbrechen vor Gericht. Versuch einer Vergangenheitsbewältigung.* Heidelberg: C.F. Müller, 1982.

Rude, Erwin. *Deutsche Vorgeschichte im Schulunterricht.* Osterwieck/Harz: A.W. Zickfeldt, 1937.

Rürup, Reinhard, ed. *Topography of Terror: Gestapo, SS and Reichssicherheitshauptamt on the "Prinz-Albrecht-Terrain," A Documentation.* Translated by Werner T. Angress. Berlin: Willmuth Arenhövel, 2001.

Russell, Stuart, and Jost W. Schneider. *Heinrich Himmlers Burg. Das weltanschauliche Zentrum der SS*. Essen: Heitz und Höffkes, 1989.

Sanden, Wijand van der. *Through Nature to Eternity: The Bog Bodies of Northwest Europe*. Amsterdam: Batavian Lion International, 1996.

Schachinger, Barbara. "Tee- und Biergenuß bei den Tibetern. Objekte aus der Sammlung Ernst Schäfer im Staatlichen Museum für Völkerkunde in München." *Münchner Beiträge zur Völkerkunde* 4 (1994): 57–95.

Schäfer, Ernst. *Geheimnis Tibet. Erster Bericht der Deutschen Tibet-Expedition Ernst Schäfer*. Munich: F. Bruckmann, 1943.

———. *Unbekanntes Tibet. Durch die Wildnisse Osttibet zun Dach der Erde*. Berlin: Paul Perey, 1937.

Schleiermacher, Sabine. "Die SS-Stiftung 'Ahnenerbe.' Menschen als Material für 'exakte' Wissenschaft." In *Menschenversuche: Wahnsinn und Wirklichkeit*, edited by Rainer Osnowski. Cologne: Kölner Volksblatt, 1988.

Schleif, Hans. "Ausgrabungen der Schutzstaffeln." *Germanien* (1936) 391–399.

Schmidt, Ina. *Der Herr des Feuers—Friedrich Hielscher und sein Kreis zwischen Heidentum, neuem Nationalismus und Widerstand gegen den Nationalsozialismus*. Dissertation, Hamburger Universität für Wirtschaft und Politik, 2002.

Schmitz-Berning, Cornelia. *Vokabular des Nationalsozialismus*. Berlin: Walther de Gruyter, 1998.

Schnabel, Raimund. *Macht ohne Moral, Eine Dokumentation über die SS*. Frankfurt am Main: Rödenberg Verlag, 1958.

Scholl, Inge. *The White Rose: Munich 1942–1943*. Translated by Arthur R. Schultz. Middletown: Wesleyan University Press, 1983.

Schöttler, Peter. "Einsatzkommando Wissenschaft. Neue Forschungen zum Verhalten deutscher Gelehrter im '3. Reich.'" *Die Zeit*, August 12, 1999.

Schultz, Bruno K. *Taschenbuch der rassenkundlichen Meßtechnik. Anthropologische Meßgeräte und Messungen am Lebenden*. Munich: J.F. Lehmanns, 1937.

Schulze, Winfried, ed. *Deutsche Historiker im Nationalsozialismus*. Frankfurt am Main: Fischer Taschenbuch, 1999.

Schwab, Waltraud. "Die Spuren des Trösters," *Die Tageszeitung*, June 27, 2003.

Schwarz, Gudrun. *Eine Frau an seiner Seite. Ehefrauen in der SS-Sippengemeinschaft*. Hamburg: Hamburger Edition, 1997.

"Search for Himmler Widens." *The New York Times*. May 15, 1945.

Sebald, W.G. "A Natural History of Destruction." *The New Yorker*. November 4, 2002, 66–77.

See, Klaus von. *Deutsche Germanen-Ideologie. Vom Humanismus bis zur Gegenwart*. Frankfurt: Athenäum, 1970.

———. *Barbar, Germane, Arier. Die Suche nach der Identiät der Deutschen*. Heidelberg: Winter, 1994.

Segal, Lilli. *Die Hohepriester der Vernichtung. Anthropologen, Mediziner und Psychiater als Wegbereiter von Selektion und Mord im Dritten Reich.* Berlin: Dietz, 1991.

Seidler, Horst, and Andreas Seidler. *Das Reichssippenamt entscheidet. Rassenbiologie im Nationalsozialismus.* Vienna: Jugend und Volk, 1982.

Seip, Didrik Arup. *Hjemmi og i fiendland 1940–45.* Oslo: Gyldendal, 1946.

Shirer, William. *Berlin Diary 1934–1941.* New York: Knopf, 1941.

———. *20th Century Journey: A Memoir of Life and the Times.* Vol. II, *The Nightmare Years 1930–1940.* Toronto: Bantam Books, 1984.

Silver, Daniel B. *Refuge in Hell: How Berlin's Jewish Hospital Outlasted the Nazis.* Boston: Houghton Mifflin, 2004.

Simpich, Frederick. "Change comes to the Bible Lands." *National Geographic* (December 1938): 695–750.

Smith, Bradley F. *Heinrich Himmler: a Nazi in the Making, 1900–1926.* Stanford, California: Hoover Institute Press, 1971.

Speer, Albert. *Infiltration.* New York: Macmillan Publishing, 1981.

———. *Spandau: The Secret Diaries.* New York: Macmillan, 1976.

Spencer, Frank, ed. *History of Physical Anthropology.* New York: Garland, 1997.

Sroka, Marek. "The University of Cracow Library under Nazi Occupation: 1939–1945." *Libraries & Culture* 34, no. 1. (winter 1999): 1–17.

"SS-Anthropologe Beger zu drei Jahren Freiheitsstrafe verurteilt." *Frankfurter Allgemeine Zeitung,* April 7, 1971.

Steigmann-Gall, Richard. *The Holy Reich: Nazi Conceptions of Christianity, 1919–1945.* Cambridge, U.K.: Cambridge University Press, 2003.

Steuer, Heiko. "Herbert Jankuhn und seine Darstellungen zur Germanen- und Wikingerzeit." In *Eine hevorragend nationale Wissenschaft. Deutsche Prähistoriker Zwischen 1990 und 1995,* edited by Heiko Steuer. Berlin: Walter de Gruyter, 2001.

Stocking, George W., ed. *Bones, Bodies, Behavior: Essays on Biological Anthropology.* Madison, Wis.: University of Wisconsin Press, 1988.

Stoddard, Lothar. *The Rising Tide of Color Against White World Supremacy.* New York: Scribners, 1920.

Strohmeyer, Arn. *Der Gebaute Mythos: das Haus Atlantis in der Böttcherstraße. Ein deutsches Mißverständnis.* Bremen: Donat, 1993.

Strzetelski, Stanislaw. *Where the Storm Broke.* New York: Roy Slavonic Publications, 1942.

Sünner, Rüdiger. *Schwarze Sonne. Entfesselung und Mißbrauch der Mythen in NS und rechter Esoterik.* Freiburg: Herder, 1999.

Tacitus. *The Complete Works of Tacitus.* Translated by Alfred John Church and William Jachson Brodribb. Ed. Moses Hadas. New York: The Modern Library, 1942.

Taschwer, Klaus. "'Anthropologie ins Volk'—Zur Austellungspolitik einer anwendbaren Wissenschaft bis 1945." *In Politik der Präsentation. Museum und Ausstellung in Österreich 1918–1945,* edited by Herbert Posch and Gottfried Fliedl. Vienna: Turia und Kant, 1996.

Taylor, Robert R. *The Word in Stone: The Role of Architecture in the National Socialist Ideology.* Berkeley, CA: Univ. of California Press, 1974.

Thompson, R. Campbell. "The Rock of Bisitun." In *Wonders of the Past,* edited by Sr. J.A. Hammerston. New York: Wise and Co., 1937.

Tieke, Wilhelm. *The Caucasus and the Oil. The German-Soviet War in the Caucasus 1942–1943.* Winnipeg: J.J. Fedorowicz, 1995.

Tighe, Carl. *Gdansk: National Identity in the Polish-German Borderlands.* London: Pluto, 1990.

"Time's Winged Chariot, Heinrich Himmler." *Time,* June 24, 1945.

Todd, Malcolm. *The Early Germans.* Cambridge, MA: Blackwell Publishers, 1992.

Tratz, Eduard. P. *25 Jahre Haus der Natur Salzburg (1924–1949).* Salzburg: Haus der Natur, 1949.

Trautmann, Erika. "Auf den Spuren der Goten in der Dobrudscha." *Germanien* (1939): 145–150.

Trials of War Criminals before the Nuernberg Military Tribunals. Vol. 1, *Case 1: The Medical Case.* Washington, D.C: U.S. Government Printing Office, 1949–1950.

Trials of War Criminals before the Nuernberg Military Tribunals. Vol. 2, *Case 1: The Medical Case.* Washington, D.C: U.S. Government Printing Office, 1949–1950.

Trials of War Criminals before the Nuernberg Military Tribunals. Vol. 2, *Case 2: The Milch Case.* Washington, D.C: U.S. Government Printing Office, 1949–1950.

Trials of War Criminals before the Nuernberg Military Tribunals. Vol. 4, *Case 9: The Einsatzgruppen Case.* Washington, D.C: U.S. Government Printing Office, 1949–1950.

Trigger, Bruce. "The Past as Power: Anthropology and the North American Indian." In *Who Owns the Past,* edited by Isabel McBryde. Oxford: Oxford University Press, 1985.

Trimondi, Victor, and Victoria Trimondi. *Hitler, Buddha, Krishna. Eine unheilige Allianz vom Dritten Reich bis heute.* Vienna: Ueberreuter, 2002.

Trinkhaus, Erik, and Pat Shipman. *The Neandertals.* New York: Alfred A. Knopf, 1993.

Trojan, Rudolf. "Vaterschaftsdiagnose." *Volk und Rasse* 7. München: J.F. Lehmanns, 1942.

Tuchel, Johannes, and Reinhold Schattenfroh. *Zentrale des Terrors. Prinz-Albrecht-Straße 8: Das Hauptquartier der Gestapo.* Berlin: Siedler, 1987.

"Tysk nazistungdom till Halland." *Nytid,* August 19, 1934.

"Tysk professor avgjuter Tanums många hällristningar." *Bohusläningen,* September 23, 1935.

Uliajasczek, Stanley J., and C. Mascie-Taylor, eds. *Anthropometry: The Individual and the Population.* Cambridge: Cambridge University Press, 1994.

United States Fleet Headquarters of the Commander in Chief. "Amphibious Operations: Invasion of Northern France Western Task Force, June 1944." COMINCH Pub 006, October 1944.

Väisänen, A.O. "Song and Instrument Playing in Folk Culture." Translated by Heli Tuomi. *Muskkitieto.* November 1943.

Vasiliev, Alexander Alexandrovich. *The Goths in the Crimea*. Cambridge, MA: The Mediaeval Academy of America, 1936.

Velleius Paterculus. *Compendium of Roman History*. Translated by F.W. Shipley. London: William Heinemann, 1924.

Verbrechen der Wehrmacht: Dimensionen des Vernichtungskrieges 1941–1944. Ausstellungskatalog. Hamburg: Hamburger Institut für Sozialforschung, 2002.

Verschuer, Otmar Freiherr von. "Rassenbiologie der Juden." *Forschungen zur Judenfrage* 3 (1938): 137–151.

Vollnhals, Clemens, and Thomas Schlemmer. *Entnazifierung. Politische Säuberung und Rehabilitierung in den vier Besatzungszonen 1945–1949*. München: Deutscher Taschenbuch Verlag, 1991.

Vrba, Rudolf. *I Cannot Forgive*. Vancouver: Regent College Publishing, 1997.

Vries, Oebele. "Een amateur-archeoloog op zoek naar de prehistorische cultur. Johannes Minnema (1903–1984)." *Archeo-Forum* 4 (1999 / 2000).

Wahle, E. "Geschichte der prähistorsche Forschung." *Anthropos* 45 and 46 (1950–1951).

Waller, John H. *The Devil's Doctor: Felix Kersten and the Secret Plot to Turn Himmler against Hitler*. New York: John Wiley & Sons, 2002.

"Walls Within Walls," *Times of London*, May 18, 1943.

Watt, Robert. *Bibliotheca Britannica: or A General Index to British and Foreign Literature*. Vol. 1. Edinburgh: Printed for Archibald Constable and Company, 1824.

Watts, Larry. *Romanian Cassandra: Ion Antonescu and the struggle for reform, 1916–1941*. Boulder: East European Monographs, 1993.

Webb, James. *The Occult Establishment*. La Salle, IL: Open Court Publishing, 1976.

Wegner, Bernd. *Hitlers politische Soldaten, die Waffen-SS 1933–1945. Studien zu Leitbild, Struktur and Funktion einer nationalsozialistischen Elite*. Paderborn : Schöningh, 1982.

Weinert, Hans. *Entstehung der Menschenrassen*. Stuttgart: F. Enke, 1941.

———. "Unsere letzteiszeitlichen Cro-Magnon Vorfahren und die Frage der Neger-Entstehung." *Germanien* (1937): 326–334.

Weingart, Peter. *Doppel-Leben. Ludwig Ferdinand Clauss. Zwischen Rassenforschung und Widerstand*. Frankfurt am Main: Campus, 1995.

Weinreich, Max. *Hitler's Professors. The Part of Scholarship in Germany's Crimes against the Jewish People*. New York: Yiddish Scientific Institute, YIVO, 1946.

Weisenburger, Elvira. "Der Rassenpapst. Hans Friedrich Karl Günther Professor für Rassenkunde." In *Die Führer der Provinz. NS Biographien aus Baden und Württemburg*. Edited by Michael Kissener. Konstanz: Universitätsverlag Konstanz, 1997.

Wells, Peter S. *The Barbarians Speak: How the Conquered Peoples Shaped Roman Europe*. Princeton, N.J.: Princeton University Press, 1999.

Weltzer, Harald et al. *'Opa war kein Nazi' Nationalsozialismus und Holocaust im Familiengedächtnis*. Frankfurt am Main: Fischer, 2002.

Wendrin, Franz von. *Entdeckung des Paradieses*. Hamburg: Braunschwieg, 1924.

Wendt, Herbert. *It Began in Babel: The Story of the Birth and Development of Races and Peoples*. Translated by J. Kirkup. Boston: Houghton Mifflin, 1962.

Werle, Gerhard, and Thomos Wandres. *Auschwitz vor Gericht. Völkermord und bundesdeutsche Strafjustiz*. München: C.H. Beck, 1995.

Whitehead, Thór. *Íslandsaevintyri Himmlers 1935–1937*. Reykjavik: Vaka-Helgafell, 1998.

Wiegers, Fritz. "Herman Wirth und die Geologie." In *Herman Wirth und die deutsche Wissenschaft*. Edited by Fritz Wiegers. München: J.F. Lehmanns, 1932.

Wiesner, Joseph. "Der Osten als Schicksalsraum Europa und des Indogermanentums." *Germanien* (June 1942).

Williamson, Gordon. *The SS: Hitler's Instrument of Terror*. Osceola, Wis.: Motorbooks International, 1994.

Wippermann, W. *Der "Deutsche Drang nach Osten." Ideologie und Wirklichkeit eines politischen Schlagwortes*. Darmstadt: Wiss. Buchges., 1981.

Wirth, Herman. *Der Aufgang der Menschheit : Untersuchungen zur Geschichte der Religion, Symbolik und Schrift der atlantisch-nordischen Rasse*. Jena: Diederichs, 1928.

———. "Das Geheimnis von Arktis-Atlantis." *Die Woche* 38 (August 29, 1931): 1144.

———. *Die heilige Urschrift der Menschheit*. Leipzig: Koehler & Amelung, 1936.

———. "Die Heiligen Zeichen der Weihenacht unserer Ahnen." *Nationalsozialistische Landpost*, December 21, 1934.

———. *Um die wissenschaftliche Erkenntnis und den nordischen Gedanken*. Berlin: Herman Wirth Society, 1931.

Wistrich, Robert. *Who's Who in Nazi Germany*. London: Weidenfield & Nicolson, 1982.

Witte, Peter et al., eds. *Der Dienstkalender Heinrich Himmlers 1941/1942*. Hamburg: Hans Christians Verlag, 1999.

Wiwjorra, Ingo. "Herman Wirth—Ein gescheiterter Ideologen zwischen 'Ahnenerbe' und Atlantis." In *Historische Rassismusforschung. Ideologen, Täter, Opfer*, edited by Barbara Danckwortt et al. Hamburg: Argument Verlag, 1995.

———. "The Third Reich—executing Germanic continuity?" In *Nationalism and Archaeology in Europe*, Edited by M. Diaz-Andreu and T. Champion. London: UCL Press, 1996.

Wobbe, Theresa, ed. *Nach Osten. Verdeckte Spuren nationalsozialistischer Verbrechen*. Frankfurt am Main: Neue Kritik, 1992.

Wojak, Irmtrude. "Das 'irrende Gewissen' der NS-Verbrecher und die deutsche Rechtsprechung. Die 'jüdische Skelettsammlung' am anatomischen Institut der Reichsuniversität Strassburg." In *"Beseitigung des jüdischen Einflusses . . ." Antisemitische Forschung, Eliten und Karrieren im Nationalsozialismus*, edited by Fritz Bauer Institut. Frankfurt/New York: Campus, 1999.

Wolf, Rotraut. "Nachruf Peter Paulsen 1902–1985," *Fundberichte aus Baden-Württemberg* 10 (1985): 727–728.

Wolfram, Herwig. *History of the Goths*. Translated by Thomas J. Dunlap. Berkeley: University of California Press, 1990.

Wolfram, Sabine, and Ulrike Sommer, eds. "Macht der Vergangenheit—wer macht die Vergangenheit. Archäologie und Politik." *Beiträge zur Ur- und Frühgeschichte Mitteleuropas* 3. Wilkau: Hasslau, 1993.

Wolfslast, Wilhelm. *Die Germanische Völkerwanderung*. Stuttgart: Robert Lutz Nachfolger Inhaber Rudolf Weisert, 1941.

Wüst, Walther. "Die indogermanischen Bestandteile des Rigveda und das Problem der 'urindischen' Religion." In *Zweites Nordisches Thing. Rufer des Thing Ludwig Roselius*. Bremen: Angelsachen Verlag, 1934.

———. *Indogermanisches Bekenntnis—sieben Reden*. Berlin-Dahlem: Ahnenerbe Stiftung, 1943.

Zentner, Christian, and Friedemann Bedürftig, eds. *The Encyclopedia of the Third Reich*. Trans. Amy Hackett. New York: Da Capo Press, 1997.

Zotz, Lothar. "Archaeological News: Germany." *American Journal of Archaeology* 53, no. 2 (1949): 175.

———. "Die Deutsche Vorgeschichte im Film." *Nachrichtenblatt für Deutsche Vorgeschichte* 9 (4) (1933): 50–52.

———. *Erlebte Vorgeschichte: Wie ich Deutschland ausgrub*. Stuttgart: Kosmo Gesellschaft der Naturfreunde, 1934.

INDEX

ACKNOWLEDGMENTS

THE WORD "ACKNOWLEDGMENTS" SEEMS far too slight a term to describe the immense debt of gratitude that I owe to the many people whose generosity, kindness, ingenuity, curiosity, hard work, and passionate commitment made this book a reality. Writing a history on a subject as large and convoluted as that of the Ahnenerbe is a mammoth undertaking by any measure. Its successful completion depends upon so many complicated pursuits—investigating and tracking down reliable sources for interviews, locating important archival materials, searching out rare books from the period, and finding gifted archivists who can, for example, recall at the drop of a hat the exact contents of microfilms that they may have briefly scanned years earlier. In this case of this book, there was an additional challenge. When I first began this project in the summer of 2001, I did not read or speak a word of German.

Will Schwalbe at Hyperion had faith from the beginning that I could overcome all these hurdles. From our very first conversation about Nazi archaeologists, he strongly encouraged me to undertake the project and then arranged for generous financial support from Hyperion. I am eternally grateful to Will for his publishing acumen, his wonderful editorial eye, and his continuing friendship. I am also hugely grateful to my editor, Peternelle van Arsdale, who has been so supportive, patient, constructive, and discerning throughout this past four and a half years, never flinching once from the torrent of e-mails I sent her. She is truly a brick! I would also like to thank Christopher Potter and

Cynthia Good as well as Catherine Heaney at Fourth Estate in London and Diane Turbide at Penguin Canada for their enthusiasm for this book. And I would be greatly remiss, too, if I did not mention all the help, encouragement, and sound advice that I have received from my superb agent Anne McDermid and her associate Jane Warren.

From the outset, I realized that I needed the assistance of a small band of linguistically talented researchers if I was to trace the activities of the Ahnenerbe across Europe, Asia, and South America. While scouting about for suitable people, I contacted Peter Stenberg, the head of the Department of Central, Eastern and Northern European Studies at the University of British Columbia. This was probably the single most important phone call that I made during this project, for it was in this way that I was introduced to the astonishingly gifted Stenberg clan. In the course of this project, every member of the family—polyglots all—jumped in to lend a hand. Peter and Rosa Stenberg brought me new German books hot off the press from their annual summer sojourns in Munich. They loaned me articles and other research materials; assisted with difficult translations; bought me lunches; introduced me to their friend, Auschwitz survivor Rudolf Vrba; invited me to Christmas Eve parties; and valiantly served as readers of the final manuscript, critiquing its contents and picking out many of my errors. I can't thank them enough. Their youngest son, Josh, took time out from his studies of Mandarin and Russian at Harvard to dig out and photocopy rare German scientific books in the university's library. Their oldest son, Erik, offered expert advice on matters to do with the war and the German military. Rachel and Anja greatly brightened my days by their visits to the office.

But the family's greatest contribution to this book has come from Charlotte Stenberg. A wonderfully resourceful researcher blessed with an abundance of linguistic talent, Charlotte has labored full-time on this project since its inception. She immersed herself in Nazi German, a lexicon in its own right; translated thousands of pages of archival and court documents from German, Swedish, French, and Norwegian into English; doggedly tracked down photographs from archives across Europe and the United States; meticulously checked and double-checked the book's notes for accuracy; presided over a bulging mini-archive of copied documents in our office; and rescued me more than once from the crashes of my recalcitrant iMac. It's very difficult for me to imagine how I could have completed this book without her sunny temperament and the vast energy that she brought to her work.

It would also be difficult to adequately thank another superb researcher, Sabine Schmitt. A Berliner with a doctoral degree in modern German history, Sabine was of invaluable assistance with the German archival research. In addition, she helped us contact family members and friends of the Ahnenerbe researchers and accompanied me to nearly all the interviews that I did in Germany and Austria, sometimes at great personal and emotional cost. Despite these difficulties, she proved to be a wonderful travel companion and an astute guide to modern German life, kindly assisting me in finding apartments for my stays in Berlin, introducing me to everything from the complexities of German train schedules to the nuances of Turkish cuisine, and giving me impromptu primers on such diverse subjects as the spread of neo-Nazism in modern Germany.

In addition, I'd like to extend my gratitude to three other members of the research team. For more than a year, archaeologist Sibylle Günther combed the libraries of Berlin, searching out forgotten books written by Ahnenerbe researchers and poring over their texts, trying to make sense of them. It was from Sibylle that I first learned of the "poison closet," a section in German libraries where hate-mongering literature is stowed for the use of serious researchers only. In the libraries of Reykjavík, a young Icelandic student, Haflidi Saevarsson, located some very valuable research for us on Himmler's interest in Iceland, while in St. Petersburg, an old friend, Natasha Dobrynina, found and translated important Soviet sources on the Goths of the Crimea. Thanks are also due to Canadian cartographer Signy Fridriksson-Fick, who labored diligently to create the four original maps that grace this book, and to John Masters, Birgitt Bischof, and Peter Bennett for their patient labors on digital scans and other photographic matters.

A small army of archivists and librarians went out of their way to help us ferret out documents, microfilms, photographs, films, sound recordings, and obscure books from their collections. I would particularly like to thank James Kelling and Niels Cordes at the National Archives and Records Administration in College Park, Michael Hollmann and Simone Langner at the Bundesarchiv in Berlin, Gregor Pickro at the Bundesarchiv in Koblenz, Christoph Bachmann at the Staatsarchiv Munich, Editha Platte and Peter Steigerwald at the Frobenius-Institut Archiv in Frankfurt, Ulrike Talay and Dr. Klaus A. Lankheit at the Institute für Zeitgeschichte in Munich, Ms. Krug at the Deutsches Ärchaeologisches Institut in Berlin, Michael Maaser at the Universität Archiv in Frankfurt, Ms. Galonska at Bundesbeauftragte für Unterlagen des Staatssicherheitsdientes der ehemaligen Deutschen Demokratischen Republik, Gerhard Keiper at the Politisches Archiv des Auswärtigen Amt in Berlin, Jürgen Mahrenholz at Humboldt University in Berlin, and David Smith at the New York Public Library. Mette Hide, the archivist at the Kulturhistorik Museum Archiv in Oslo, was especially warm and cordial, inviting me into her home to sample traditional Norwegian cuisine and dashing off frequent letters and postcards of encouragement while I was writing the book.

Many other people freely gave of their time and their expertise, making this a much better book than it would have been otherwise. Achim Leube at Humboldt University in Berlin was an unfailing source of leads and a wonderful guiding light, always finding time from his own numerous historical projects to answer my questions. Rudolf Vrba in Vancouver generously shared his insights on the operations at Auschwitz, while Hans-Joachim Lang in Tübingen kindly helped me piece together some of the more troubling details of the Ahnenerbe's skeleton collection. And I will never forget the fabulous weekend I spent with Luitgard Löw and Camilla Olsson in Sweden, touring the famous rock art of Bohuslän and drinking German wine and talking late into the night in the little house by the sea at Grebbestad.

I could never adequately thank all the others who patiently answered so many questions and e-mails, correcting misconceptions and supplying much valuable new information. I am sure many groaned every time they saw yet another e-mail or fax or telephone message from our team, so I'd like to take this opportunity to extend my deepest appreciation to the following individuals: Andrej Angrick, Michael Balter, Achim Barsch, Margit Berner, David Braund, Anton Brøgger, Jan Brøgger, Frank M. Clover, John Peter Collett, Dee R. Eberhart, Martijn Eickhoff, Isrun Engelhardt, Frode Færøy, Dietmar Feldhaus, Jens Fleischer, Espen Grøgaard, Walther Habersetzer, Anders Hagen, Elmar Hammerschmidt, Brian Hayden, Henry M. Hoenigswald, Niklas Holzberg, Mary Ison, Irma-Riitta Järvinen, Olav Sverre Johansen, Frederick H. Kasten, Hans Ewald Keßler, Stefan Klein, Serge Lebel, Sylvette Lemagnen, Freddy Litten, Katharina Lommel, Volker Losemann, Wendy Lower, Bob Martinson, Michael Meyer, Waldemar Nehring, Erle Nelson, Karl-Heinz Nickel, Javier Nuñez de Arco, Linda Owen, Taina Partanen, Werner Renz, Perry Rolfsen, Wijnand van der Sanden, Ina Schmidt, Eike Schmitz, Winfried Schultze, William E. Seidelman, Gerd Simon, Kurt Singer, Jiri Svoboda, Sid Tafler, Maria Teschler-Nicola, Claudia Theune-Vogt, Annette Timm, Bruce Trigger, Gretchen Vogel, Hector Williams, Ingo Wiwjorra, and Ingrid Zwerenz.

Moreover, I can't thank enough those scholars who took time out from their busy research and teaching schedules to read parts of the manuscript: Victor H. Mair at the University of Pennsylvania in Philadelphia; Juha Pentikäinen at the University of Helsinki; Alexander Gertsen at Taurian National University in Simferopol; Hannu E. Sinisalo from the University of Tampere in Finland; Guido Lombardo from Universidad Peruana Coyetano Heredio in Lima; Tjalling Waterbolk from the University of Groningen; Juhani von Grönhagen in Helsinki; Luitgard Löw in Gothenburg; Camilla Olsson in Trondheim; and Sabine Schmitt in Berlin. They saved me from the scourge of many an error: those mistakes still embedded in the text are entirely my own.

Last, but certainly not least, I would like to express my gratitude to my father for his personal recollections of the final days of the war and the bombing campaigns of the Royal Air Force, Kathleen Hodgson for her amiable concern during the grueling months I spent writing the darkest chapters of this book, and my brother Alex and sister-in-law Sheila, and their wonderful clan— Thomas, Anna, and Sarah—for their unflagging encouragement and curiosity about this book. I'm also indebted to John Masters and Andrew Nikiforuk for their research suggestions, unstinting friendship, and moral support.

Most of all, however, I want to thank my husband Geoff, who thought from the very beginning that a book on the Ahnenerbe was a great idea. He had no inkling that he would spend the next four and a half years listening to bizarre blow-by-blow accounts of the research, sitting through numerous German documentaries and films on the Nazi regime, and browsing through nearly all the books I brought home on Himmler, the SS, and the Second World War. But even if he had suspected all that he was in for, I feel certain he would have been game. Geoff is at heart a historian, and his immense fascination for the past has been a continuing source of delight and inspiration to me. I wrote this book with him in mind. I can think of no one else that I would rather share my life with.